The Grand Canyon's remote Toroweap Overlook rises 3,000 feet over the Colorado River.

SECOND EDITION

THE 10 BEST OF EVERYTHING

NATIONAL PARKS

800 TOP PICKS FROM PARKS COAST TO COAST

NATIONAL GEOGRAPHIC

WASHINGTON, D.C.

CONTENTS

Head to the Yellowstone River, near Gardiner, Montana, for family-friendly white-water rapids.

INTRODUCTION

The first edition of *The 10 Best of Everything National Parks* was published in 2011. Since then, more than 30 park units have been added to the National Park System, for a total of more than 400 units spanning an astounding 85 million acres. This revised edition highlights many of these new properties, while also introducing updated experiences throughout the park system, unique areas of interest, fascinating cultural highlights, and deep dives into the history of our parks.

Whether you're searching for a breathtaking hike worth the climb, a historic site that'll stir your conscience, or a sunny spot in which to laze away an afternoon, you'll find the 10 best opportunities for all of this—and more—in this one-of-a-kind guide.

Explore sites where kids are bound to be wowed (page 92), like Great Sand Dunes National Park and Preserve, where they can sandboard safely across the unique landscape, or Mammoth Cave National Park, with a kid-friendly cave tour that delves deep into spectacular underground sights. Garner inspiration at sites that honor trailblazing women (page 341), from First Ladies National Historic Site in Ohio to the Mary McLeod Bethune Council House, a national historic site dedicated to the civil rights champion and founder of a school for African American girls in Washington, D.C.

History lovers will be in their element with a list of sites as varied as the nation, from ancestral Puebloan culture at Chaco Culture National Historical Park (page 302) to early settlements at De Soto National Memorial (page 311) and iconic Gettysburg National Military Park (page 314). Pay homage to barrier breakers (page 347), or take a literary pilgrimage to the homes, writing rooms, and libraries of some of our finest wordsmiths (page 295).

Plus, find the latest improvements and opportunities throughout our park system. Our list of 10 best accessible parks (page 115) highlights areas with elevated paths, interpretive signage, and even cave tours made reachable by elevators and ramps. Discover engaging ranger programs (page 282), high-thrill rafting (page 165), lodging that's worth a spot on your bucket list (page 388), and the plethora of mountains, mammals, and monuments that make the parks famous.

While you're planning your next park getaway, follow our pointers offered throughout these pages, including booking reservations when needed. With more than 312 million visitors per year—a number that soared following the COVID-19 pandemic—many of these parks are seeing record numbers. In order to preserve these exceptional experiences (including the solitude many people seek in these special places) and limit the impact on natural resources, some park managers have implemented systems for admittance, camping, parking, and scenic drives. You'll find expert tips to plan ahead—an effort that pays off and will help ensure a smooth experience when it's time for your trip.

As we noted in the first edition, to sum up the vastness and diversity of the National Park System into lists of "10 best" is nearly impossible. Thus, this book is merely a starting point, rather than a checklist, for first-time explorers and regular adventurers alike to plan their next great quest to explore the U.S. story.

THE NATIONAL PARK SYSTEM

North Cascades N.P.

Olympic N.P.

WASHINGTON

Mt. Rainier N.P.

Glacier N.P.

MONTANA

OREGON

Crater Lake N.P.

IDAHO

Yellowstone N.P.

Redwood National and State Parks

Grand Teton N.P.

Lassen Volcanic N.P.

CALIFORNIA

NEVADA

Great Basin N.P.

UTAH

Arches N.P.

Yosemite N.P.

Kings Canyon N.P.

Canyonlands N.P.

Capitol Reef N.P.

PINNACLES

Pinnacles N.P.

Sequoia N.P.

Bryce Canyon N.P.

Zion N.P.

Death Valley N.P.

Mes Verd N.

PACIFIC OCEAN

Channel Islands N.P.

Grand Canyon N.P.

Petrified Forest N.P.

ARIZONA

Joshua Tree N.P.

MEXICO

Saguaro N.P.

HAWAI'I

PACIFIC OCEAN

50 mi
50 km

Haleakalā N.P.

Hawai'i Volcanoes N.P.

PACIFIC OCEAN

National Park of American Samoa

ARCTIC OCEAN

Kobuk Valley N.P.

Gates of the Arctic N.P. & Preserve

CANADA

Yukon

ALASKA

Bering Sea

Denali N.P. & Pres.

200 mi
200 km

Lake Clark N.P. & Pres.

Wrangell-St. Elias N.P. & Preserve

Kenai Fjords N.P.

Katmai N.P. & Pres.

Glacier Bay N.P. & Pres.

PACIFIC OCEAN

CANADA

NORTH DAKOTA

Theodore Roosevelt N.P.

MINNESOTA

Voyageurs N.P.

Isle Royale N.P.

MICHIGAN

CANADA

MAINE

Acadia N.P.

VT.

N.H.

WYOMING

SOUTH DAKOTA

Wind Cave N.P.

Badlands N.P.

WISCONSIN

MASS.

RHODE ISLAND

CONNECTICUT

NEW YORK

Rocky Mountain N.P.

Black Canyon of the Gunnison N.P.

COLORADO

NEBRASKA

IOWA

ILLINOIS

INDIANA

OHIO

PENNSYLVANIA

NEW JERSEY

DELAWARE

MD.

D.C.

Indiana Dunes N.P.

Cuyahoga Valley N.P.

Shenandoah N.P.

W. VA.

VIRGINIA

Great Sand Dunes N.P. & Pres.

KANSAS

MISSOURI

Gateway Arch N.P.

New River Gorge N.P. & Preserve

KENTUCKY

NORTH CAROLINA

Mammoth Cave N.P.

TENNESSEE

NEW MEXICO

OKLAHOMA

ARKANSAS

Hot Springs N.P.

Great Smoky Mts. N.P.

SOUTH CAROLINA

Congaree N.P.

White Sands N.P.

Carlsbad Caverns N.P.

Guadalupe Mts. N.P.

MISSISSIPPI

ALABAMA

GEORGIA

TEXAS

LOUISIANA

FLORIDA

ATLANTIC OCEAN

Big Bend N.P.

MEXICO

Albers Conic Equal-Area Projection

200 mi

200 km

Gulf of Mexico

Everglades N.P.

Biscayne N.P.

Dry Tortugas N.P.

ATLANTIC OCEAN

100 mi

100 km

PUERTO RICO

U.S. Virgin Islands N.P.

U.S. VIRGIN ISLANDS

Caribbean Sea

National park

National lakeshore or seashore

National monument

National preserve or reserve

National recreation area

National river or national wild and scenic river and riverways

National parkway

National scenic trail

National battlefield, battlefield park, battlefield site, or military park

National historical park

National historic site

National memorial

Other National Park Service property

Roaring Fork Stream tumbles over moss-lined rocks in Great Smoky Mountains National Park.

EXPLANATION OF DESIGNATIONS

The diversity of sites within the National Park System is reflected in the variety of titles given to them. In its official index, the National Park Service provides the following explanations for the various designations used to identify the more than 420 properties under Park Service protection:

Areas added to the National Park System for their natural values are expanses or features of land or water of great scenic and scientific quality and are usually designated as national parks, monuments, preserves, seashores, lakeshores, or riverways. Such areas contain one or more distinctive attributes like forest, grassland, tundra, desert, estuary, or river systems; they may contain windows into the past and a view of geological history;

At 200 to 330 feet in diameter, Grand Prismatic Spring is Yellowstone's largest hot spring.

Rafters meander through the Grand Canyon on the Colorado River.

they may contain imposing landforms like mountains, mesas, thermal areas, and caverns; and they may be habitats of abundant or rare wildlife and plant life.

Generally, a **national park** contains a wide variety of natural resources and encompasses large land or water areas to help provide adequate protection of those resources. At times, the parks also include significant historical assets.

A **national monument** is intended to preserve at least one nationally significant resource. It is usually smaller than a national park and lacks its diversity of attractions. Presidents can declare landmarks, structures, and other historic or scientific objects situated on government owned or controlled lands as national monuments.

In 1974, Big Cypress and Big Thicket were authorized as the first **national preserves**. National preserves have similar characteristics to national parks in that they protect at least one significant natural resource; however, unlike in national parks, activities like hunting and fishing or the extraction of minerals and fuels may be permitted. **National reserves** are similar to the preserves. Management may be transferred to local or state authorities. The first national reserve, City of Rocks, was established in 1988.

Preserving shoreline areas and offshore islands, **national lakeshores** and **national seashores** focus on the preservation of both developed and primitive areas and their natural resources, while at the same time providing water-oriented recreation. Although national lakeshores can be established on any natural freshwater lake, the existing ones are all located on the Great Lakes. The national seashores are on the Atlantic, Gulf, and Pacific coasts. **National rivers** cover a wide range of categories, including **wild and scenic riverways**, **national river and recreation areas,** and **national scenic rivers.** These designations preserve free-flowing streams and their immediate environment with at least one outstandingly remarkable natural, cultural, or recreational value. They must flow naturally without major alteration of the waterway by dams, diversion, or otherwise. Besides protecting and enhancing rivers, these areas

Citizen scientist programs help national park visitors connect with nature and assist with data collection for scientists.

provide opportunities for outdoor activities like hiking, canoeing, and hunting.

National scenic trails are generally long-distance footpaths winding through areas of natural beauty. **National historic trails** recognize original trails or routes of travel of national historical significance. There are more than 55,000 miles of preserved trails in the National Park System, authorized under the National Trails System Act of 1968.

Although best known for its great scenic parks, more than half the areas in the National Park System preserve places and commemorate persons, events, and activities important in the history of the United States. These range from archaeological sites associated with prehistoric Indigenous civilizations to sites related to the lives of modern Americans. Historical areas are customarily preserved or restored to reflect their appearance during the period of their greatest historical significance.

Most places designated a **national historic site** have been given the authorization by acts of Congress, though they can also be established by secretaries of the Interior. This title means a place contains a single historical feature of significance. A wide variety of titles—**national military park, national battlefield park, national battlefield site,** and **national battlefield**—has been used for areas associated with American military history. But other areas like **national monuments** and **national historical parks** may include features associated with military history. National historical parks are commonly areas of greater physical extent and complexity than national historic sites. The lone **international historic site,** Saint Croix Island, is relevant to both U.S. and Canadian history.

The title **national memorial** is most often used for areas that are commemorative. They need not be sites or structures historically associated with their subjects. For example, the home of Abraham Lincoln in Springfield, Illinois, is a national historic site, but the Lincoln Memorial in the District of Columbia is a national memorial.

Several areas whose titles do not include the words "national memorial" are nevertheless classified as memorials. These are Franklin Delano Roosevelt Memorial, Korean War Veterans Memorial, Lincoln Memorial, Lyndon Baines Johnson Memorial Grove on the Potomac, Theodore Roosevelt Island, Thomas Jefferson Memorial, Vietnam Veterans Memorial, Washington Monument, and World War II Memorial in the District of Columbia; Perry's Victory and International Peace Memorial in Ohio; and Arlington House in Virginia.

Originally, **national recreation areas** were units surrounding reservoirs impounded by dams built by other federal agencies. The National Park Service manages many of these areas under cooperative agreements. The concept of recreational areas has grown to encompass other lands and waters set aside for recreational purposes by acts of Congress and now includes major areas in urban centers. There are also national recreation areas outside the National Park System that are administered by the U.S. Forest Service and U.S. Department of Agriculture.

National parkways encompass ribbons of land flanking roadways and offer an opportunity for driving through areas of scenic interest. They are not designed for high-speed travel.

In addition to the National Park System, five area designations—**Authorized Areas, Affiliated Areas, National Heritage Areas,** the **National Wild and Scenic Rivers System,** and the **National Trails System**—are linked in importance and purpose to areas managed by the National Park Service. These areas are not all units of the National Park System, yet they preserve important pieces of the nation's heritage.

A quiet moment at the Vietnam Veterans Memorial in Washington, D.C.

NATURAL WONDERS

Sunrise paints the landscape at Oxbow Bend in Grand Teton National Park.

LANDMARKS

Landmarks have long served as beacons for park visitors, providing markers on great journeys and standing as symbols of meaningful places. Modern maps tell us how to get somewhere, but landmarks tell us where we are, from the striated rise of Devils Tower to the towering peaks of Denali. They also reveal unique histories, geography, and geology throughout our national parks.

1 Grand Canyon
NATIONAL PARK
Arizona

Angels don't need windows, but if they ever wanted to frame a great view, they might choose the North Rim's **Cape Royal** and its noble companion parapet, **Angels Window.** Thrust far above the immense luminous space of the canyon, this natural arch overlooks the big bend where the canyon turns west, carving ever deeper into the heart of the Kaibab Plateau. No viewpoint offers a better perspective on the contrast between the dizzying verticality of the gorge and the horizontal rock layers through which it was carved. Red-and-yellow cliffs march across bays and escarpments for mile after astounding mile. Wotans Throne is in the foreground. That distant green ribbon is the Colorado River. The southern horizon is the South Rim, about nine miles away and almost 1,000 feet lower.

Adding to its appeal, the North Rim is forested, wildflower strewn, and pleasantly cooler in summer than the South Rim. Cape Royal is a prime spot to watch cloud formations sail across the void, but beware of thunderstorms. Angels may be a matter for faith, but lightning strikes are high-voltage reality at this most exposed geological extremity.

2 Devils Tower
NATIONAL MONUMENT
Wyoming

As a landform, **Devils Tower** seems almost impossible. From the relatively flat surrounding land, the tree stump–like tower's sides form smooth upward arcs, drawing thoughts to the sky. The summit, hovering 867 feet above its base, is flat, not visible from below, and therefore mysterious. Plains tribes— Lakota, Shoshone, Crow, Blackfeet, Kiowa, Arapaho, and others— consider the tower a sacred object and call it by evocative names like Bear Lodge, Grey Horn Butte, and Tree Rock. Legends tell of heroes, creation, and redemption. The tower's ongoing importance is reflected by ceremonies and rituals conducted annually by regional tribes. The geological story, not fully understood, credits an intrusion of molten igneous rock that took shape beneath overlying sedimentary layers, where it hardened, only to be eventually exposed by erosion. In the process of cooling, the rock formed vertical hexagonal columns that, parallel but separate, give the tower its distinctive striated appearance. Rock climbers find the columns irresistible. Most are happy to gaze upward from the base where, in 1906, President Theodore Roosevelt proclaimed the first national monument.

Considered a sacred site by many Native American peoples, Devils Tower is also America's first national monument.

Hiking the Narrows in Zion National Park requires sure footing and a willingness to get wet.

3 | Zion
NATIONAL PARK
Utah

Arguably one of the most extraordinary hikes in Utah, the **Narrows** at Zion National Park is a journey through the narrowest section of Zion Canyon. Humbling at best and distressing at worst, the Narrows is a gorge carved by water over millions of years, with walls towering some 2,000 feet and, in places, only 20 to 30 feet apart. The bold Virgin River is the undauntable trail, and hikers—guaranteed to get wet—can choose from a variety of lengths. Along the way, patches of lush hanging gardens, a haven of biodiversity in the desert, speckle the canyon walls. Archaeological evidence suggests that people inhabited the canyon as early as the Archaic period. The Paiute settled this place in 1250 and called it Mukuntuweap, or "straight canyon." Indeed. For a truly immersive—albeit strenuous—experience, well-prepared and permitted hikers can complete the 16-mile trek from the top of the Narrows down, or opt for a two-day journey with an overnight stay deep in the canyon's blissfully eerie quiet. Rangers implore all hikers to heed weather warnings: Flash floods are common and water levels can rise rapidly.

4 | Chimney Rock
NATIONAL HISTORIC SITE
Nebraska

Days could get long for 19th-century migrants headed to Utah, Oregon, and California. Starting at Independence, Missouri, where wagon trains formed up so people could travel together, trundling toward the sunset at the pace of a walking ox, settlers entered a world more open than most could imagine: no trees, little water, and grass that grew thinner as the miles went by. What Francis Parkman described in 1846 as the "same wild endless expanse" stretched through tomorrow into forever. On a route with few notable mileposts, **Chimney Rock** certainly stood out. Most diarists commented on the sight of it. Quite a few people climbed the slope at its base to scratch their names in the soft sandstone. Needle-shaped, 326 feet high (by today's measurements), and a short walk from their camps on the North Platte River, the rock told travelers that they were nearing the end of the prairies and would soon be in the mountains. Hooray, you have made it this far. Carry on bravely!

5 | Yosemite
NATIONAL PARK
California

Like all reliable landmarks, **Half Dome** is an eye magnet. It towers over the other grand monoliths of Yosemite Valley and demands attention. The others in the pantheon, including El Capitan, Sentinel Rock, and Cathedral Spires, are no less illustrious; however, there's something special about Half Dome's stage presence. View it from the valley floor, beside the winding Tuolumne River. Or drive up to Glacier Point to watch it glow in the sunset as night falls. Best of all, see it from its own bald top. The trail, which takes in the glories of Vernal and Nevada Falls along the way, ends on a cable-protected pathway nailed to smooth granite slabs. Eight miles and 4,800 vertical feet each way, it promises a long day for those with permits and preparation—but the path is entirely worth the effort.

What's in a Name?

For more than a century a naming debate plagued today's Denali National Park and Preserve. When Mount McKinley National Park was established in 1917, its moniker paid homage to William McKinley, who had never set foot in Alaska and would go on to become the 25th U.S. president. A New Hampshire–born gold prospector had named the peak in 1896, but Indigenous groups had for generations denoted the significance of this spot with their own names. Despite the federal designation, Alaskans continued to call the mountain Denali, the Athabaskan word for "high one." Local legislative efforts to petition for a formal change began in the mid-1970s, and in 1980, Mount McKinley National Park was incorporated into the much larger Denali National Park and Preserve, with its iconic mountain retaining the name of McKinley. It took another three decades of work, but in 2016, marking the 100th anniversary of the National Park Service, the name of the continent's highest peak finally received its official title of Denali.

6 | Grand Teton
NATIONAL PARK
Wyoming

Grand Teton, the central crag of the Teton Range, scrapes the clouds nearly 7,000 feet above the Wyoming valley floor. Together with the other mighty **Central Peaks** crags surrounding the 13,775-foot summit, they compose a formidable alpine stronghold of snow, rock, and ice, a seemingly untouchable and remote world. But looks can be deceiving. This is a place of legend. Once a spot for mysterious Native American vision quest sites found high on "The Grand," these peaks are now the place for modern feats of endurance and skill, such as hiking the major summits in the Grand Traverse. Coupling this history with some time on even the lower trails makes it clear: That far summit is a human place after all.

7 | Mount Rainier
NATIONAL PARK
Washington

Now you see it, now you don't. True to its name, **Mount Rainier** disappears behind cloud banks, stays hidden for days and weeks at a time, and reappears in most dramatic fashion. Sometimes, the active volcano floats above the clouds, visible only to mountaineers on its glacier-decked slopes and to thrilled passengers aboard flights speeding south from Seattle. When

▶ *Climbing the "high one"— aka Denali—is a feat. In fact, fewer than a thousand people reach its summit each year. A task for experienced mountaineers only, the journey takes two weeks or more, often with time added for inclement weather.*

weather permits, 14,410-foot-high Rainier is visible from most of western Washington and far out to sea. It looms above the skyline of downtown Seattle as if its glaciers were invading suburban neighborhoods. Of course, the best encounters are from park roads and trails, notably on the south side in the popular area called Paradise, known for its wildflower meadows, views of the mountain, prodigious snowfall, and occasional rainstorm.

8 | Yellowstone
NATIONAL PARK
Wyoming

For the Kiowa, **Grand Prismatic Spring** was the "place of hot water"; for the Crow, the "land of the burning ground." Indigenous tribes inhabited this abundant region more than 11,000 years ago—living,

hunting, and traveling through what would become the world's premier national park. European American explorers frequented the area in the early 1800s, sending home tales of "boiling mud holes and exploding geysers," and in 1871 the Hayden Expedition began mapping the scientifically significant Midway Geyser Basin. Yellowstone has more than 10,000 hydrothermal features, with hot springs being its most common and Grand Prismatic Spring the largest. Located in the northwestern corner of Wyoming, this rainbow of thermal brilliance—from vibrant red, orange, and yellow to deep green and a penetrating blue—measures 200 to 330 feet in diameter and more than 121 feet deep, with a center hot enough to boil flesh from bone in seconds. Its colors come from bacteria, while its rings are the result of fluctuating temperatures within the hot spring. A surrounding boardwalk gets visitors close to the kaleidoscope, or, for a more aerial view that most certainly adds to the awe, join others in this colorful quest on Fairy Falls Trail to a lookout point.

9 | Cumberland Gap
NATIONAL HISTORICAL PARK
Kentucky, Tennessee & Virginia

This landmark may not hold high prominence now, but to American immigrants in the late 1700s, the **Gateway to the West** was an extremely important geographic feature. Settlers eager to push into the bluegrass region of Kentucky met resistance from Native American tribes, who prized the area as a hunting territory, in addition to the challenge the physical barrier of the Cumberland mountains presented. Eventually, war and politics ended the claims of Indigenous people, and a flood of settlers poured through the Cumberland Gap. The route was originally a Native American footpath called the Warriors' Path. In 1775, Daniel Boone hacked out a wider track that became famous as the Wilderness Road. By 1820, despite sporadic warfare and the inherent challenge of life on the frontier, some 300,000 settlers had passed through the gap on their way west. Today the highway runs underground, leaving the gap as peaceful as ever. Modern travelers get a fine view of it and the surrounding mountains from the Pinnacle Overlook, a four-mile drive along Skyland Road from the park visitor center.

10 | Denali
NATIONAL PARK & PRESERVE
Alaska

Denali sprawls across the Alaska tundra like half a planet, gleaming white and broad shouldered. How big is it really? It's hard to tell by looking. And one can read the facts, and accept them, and still not know the measure of the place. Indigenous Athabaskans expressed their awe with a single word, Denali, which means "high one." The summit towers 20,320 feet above sea level, more than 18,000 feet above the base. This gives the mountain an all-in-one-view vertical rise more than a mile greater than Mount Everest, which begins its grand ascent at an already lofty elevation of about 17,000 feet. But comparisons are good only for discussion. A true understanding of the mountain and its relationship to those gazing at it in wonderment lies somewhere in the experience of being near it. Climbers, hikers, and travelers of all types have tried to understand it. It's safe to say, as with Everest, that no one has fully succeeded.

Catch the reflection of Mount Brooks in Denali's Wonder Lake at dawn.

TREES

The parks contain some of the biggest, tallest, girthiest, and oldest trees in the world, from the moist hardwood forests of the Smokies to austere stands of saguaros in Arizona. But for all the wonders of the grand trees, one fact is clear: Trees don't stand alone—their health signals the health of the environment.

1 | Redwood
NATIONAL & STATE PARKS
California

When John Steinbeck stood in the presence of the tallest living thing on the planet, he described the sensation as a "cathedral hush." The writer was describing a reverential awe more than any sound-damping effect. Among the **coast redwoods** preserved in Redwood National Park, itself a quilt of national and state park parcels, are the oldest, largest, and tallest of coast redwoods—"ambassadors from another time," Steinbeck called them, because their relatives stood in the Jurassic era. Only about 4 percent of their historic two-million-acre range remains, mostly in the park. All of the trees are magnificent, and it's almost impossible to gauge and compare their height without special instruments. The tallest tree, the 379-foot Hyperion tree, was only "discovered" in the backcountry in 2006, but in an effort to protect it, the national park encourages visitors not to seek it out. Dozens of easy and moderate trails offer stunning redwood giants, all of which elicit that cathedral hush.

2 | Big Cypress
NATIONAL PRESERVE
Florida

About a third of Big Cypress is covered with **bald cypress trees,** but not many of them still qualify as "big." Most of South Florida's giant bald cypress trees, 900-year-old, 150-foot behemoths, fell to loggers' axes in the first half of the 20th century. The trees were coveted for their water resistance and used for pickle barrels, decking, stadium seats, and even early World War II PT boats. Though now most of the preserve's cypresses are a dwarf variety, some giants do remain. The best place to see them is on the Gator Hook Trail off the Loop Road south of Monroe Station, but this is a muddy, water-to-the-knees walk even in the winter dry season, particularly since cypresses grow from water-filled depressions. Slosh on in to see the buttressed, moss-draped giants, or opt for the preserve's two scenic drives, where plenty big groupings of the dwarf variety impress.

3 | Joshua Tree
NATIONAL PARK
California

They are the "most repulsive tree in the vegetable kingdom," said Capt. John C. Frémont in 1844, but when you see **Joshua trees** silhouetted in the soft light of a desert sunset, or their creamy white-green springtime flowers, or a Scott's oriole nesting in the

Otherworldly Joshua trees stand silhouetted against a night sky at the national park named after them.

crook of a tree's branches, you begin to gain affection for these hardy desert survivors. Legend has it that they got their name from Mormon settlers in the mid-19th century who saw in their outstretched limbs the biblical figure Joshua guiding them westward. The trees are a species of yucca (*Yucca brevifolia*) and grow only in the high desert, mainly in the Mojave, between 3,000 and 4,000 feet. They grow in sparse stands loosely called forests throughout the north part of this southeastern California park, but not in the lower elevations of the south. The densest grouping of Joshua trees can be found in Upper Covington Flats—as can the largest.

4 | Saguaro
NATIONAL PARK
Arizona

Saguaros are technically cacti, not trees, but their stoic majesty, virtually the symbol of the Old West, earns them a rightful place in this chapter. And to the envy of many a tree, they stand as tall as 50 feet and live up to 200 years. The biggest can sprout up to 40 arms. Their range, however, is small—only in the Sonoran Desert in Arizona and only below 4,000 feet. And many fell victim to cattle grazing in years past—it takes 10 years for one to grow to the size of a fist. Saguaros are basically water storage tanks; they'll soak up 200 gallons after a brief, intense rain. The park's two units near Tucson contain two million saguaros, so you can't miss them, but the greatest concentration is along the Bajada Loop Drive in Saguaro West, with the densest stand right behind that unit's visitor center. The Valley View Overlook Trail gives a great overhead view of the giant stiff-armed cactus, too.

5 | Petrified Forest
NATIONAL PARK
Arizona

Dead trees typically decompose quickly. But if they become buried in the right sort of sediment, one fine-grained and rich in silica, decomposition is arrested. Water soaks into the wood, carrying minerals that gradually replace the organic matter, literally turning it to stone. The woody structure remains, but the original colors have been replaced by a glittering crystal rainbow of reds, blues, and yellows. At Petrified Forest National Park visitors can walk among the scattered remains of **fossilized trees** that grew 225 million years ago, seeing and (carefully) touching these significant windows into the past. Although you can tour the park by car, for the full experience, set out by foot on the well-maintained trails or into the backcountry. Near the park's southern entrance, the Rainbow

Up In Smoke

More so than earthquakes, floods, or volcanoes, wildland fires dramatically change park landscapes. Whether the result of lightning strikes, arson, or negligence, unplanned and unregulated blazes in the past decade have burned significant portions of the National Park System, sometimes even forcing properties to temporarily close.

While melting glaciers are a telltale sign of climate change, the increase in frequency and intensity of these wildfires is especially revealing. Stressed by rising temperatures, drier landscapes act like kindling. One of the most destructive wildfires in recent park system history was the 2018 Carr fire, which burned 97 percent of Whiskeytown National Recreation Area. In 2020, the Caldwell fire burned 70 percent of Lava Beds National Monument, and in 2021, the Dixie fire burned 69 percent of Lassen Volcanic National Park. Sequoia National Park and Sequoia National Forest lost an estimated 10,000 iconic trees in the 2020 Castle fire, leading park staff to use aluminum wraps in the fall of 2021 as more wildfires encroached.

And the destruction isn't just happening in the West: In 2016, a wildfire near Great Smoky Mountains National Park burned 11,000 acres, coming less than two miles from the historic LeConte Lodge.

The Indigenous practice of prescribed burns is once again becoming a popular tool for fire management, sometimes in combination with mechanical thinning. Scheduled blazes at parks like Great Smoky Mountains, North Cascades, and Yosemite prevent overgrowth and repair the critical natural fire cycle, restoring native species and providing habitat for wildlife.

Forest Museum gives a primer to petrified wood, plus a launching off point to trail loops like the aptly named Giant Logs and Long Logs. A few miles from the museum, the Crystal Forest Trail has thousands of petrified logs worth seeing.

6 | Great Basin
NATIONAL PARK
Nevada

The world's longest living trees appear nothing like mighty sequoias or giant oaks. **Bristlecone pines** are twisted, gnarled, forlorn looking, and stand only 15 to 30 feet high. But these heroic survivors date from the time of the pyramids. They grow on just a few high, dry mountain slopes in the West. Some of the oldest are in Great Basin National Park. Because they grow on exposed slopes that get hammered by wind and snow, many appear to be dead. Some have only a single thin strip of bark and a single living branch. But they're alive, and actually live longest where conditions are the harshest. That includes high on 13,063-foot Wheeler Peak. One tree removed from the Wheeler Peak grove in 1964 was more than 4,900 years old. In 2012, it was out-aged by one that proved to be a whopping 5,065. The best places to see the trees are along the gentle Sky Islands Forest Trail and the longer 2.8-mile Bristlecone Trail.

Take in the scale of the world's largest trees—some as tall as a 26-story building— at Sequoia National Park.

7 | Sequoia
NATIONAL PARK
California

The world's largest living thing merits the short walk it takes to pay respects. The **General Sherman Tree** is the largest representative of the largest species of tree in the world as measured by volume: the giant sequoia. (Its cousin, the coast redwood, grows taller, but its trunk is much more slender.) The Sherman Tree is 274.9 feet tall, 102 feet in circumference, 36.5 feet in diameter, 52,000 cubic feet in volume, and estimated at 2,200 years old. Why is such a giant protected by a low fence? Because sequoia roots are quite shallow; over the years, the high volume of trampling could damage it. But the Sherman Tree doesn't stand alone: Other giant sequoias, which are rarely found in such abundance, grow all around in

Sequoia National Park's **Giant Forest.** Giant sequoias grow only on the western slopes of the Sierra Nevada from 4,000 to 8,000 feet, in a narrow belt that extends about 250 miles.

8 | Great Smoky Mountains
NATIONAL PARK
North Carolina & Tennessee

While some other parks feature a superstar tree species, Great Smoky Mountains is all about **abundance** and **diversity.** The conditions are perfect: elevation ranging from 875 feet to 6,643 feet and higher; sunny south-facing ridges as well as moist north-facing slopes; terrain that was never scoured by glaciers; and abundant rainfall of up to 85 inches annually. The park is 95 percent forested, of which 20 percent is old-growth and contains 15 national champions and eastern hemlocks taller than 170 feet. The trees, in fact, are responsible for the blue haze for which the park was named; it's the result of the natural transpiration of vapors from trees. Albright Grove is one of the best places to see old-growth stands, including eastern hemlock trees that are threatened by blight elsewhere. The grove also contains champion maples up to 140 feet high. In the Cove Hardwood Forest, trails wind among some of the country's largest tulip, yellow buckeye, Fraser magnolia, red maple, and black cherry trees.

Sunrise over the saw grass wetlands of Big Cypress National Preserve (p. 24)

9 | Everglades
NATIONAL PARK
Florida

The emblematic tree of the tropics, certainly of tropical swamps, is the **mangrove,** and nowhere else in the Western Hemisphere does the tree grow in such abundance as in the Everglades. Mangroves can be an acquired taste. Their bizarre growth habits and sheer density make them appear foreign, almost sinister. The spidery taproots of red mangroves look like tiny legs about to stage a march. Black mangrove roots are even stranger—they sink into the muck, then send pneumatophores back above the surface to obtain oxygen. But the mangroves foster an amazing habitat that is home to the American crocodile; to sport fish like snook, redfish, and tarpon; and to exotic birds like roseate spoonbills and mangrove cuckoos. It's hard to miss these trees since they dominate the Everglades coastlines. They're easily seen afoot on the West Lake Trail, on canoe trails such as Noble Hammock and Hells Bay, and throughout the extensive Wilderness Waterway labyrinth.

10 | Congaree
NATIONAL PARK
South Carolina

Some of the tallest trees in the eastern United States are in Congaree National Park, home to 11,000 acres of extremely dense old-growth bottomland (meaning subject to river flooding) hardwood forest. In fact, Congaree contains one of the tallest **temperate deciduous forest canopies** in the world. The park has 90 species of trees that thrive on rich nutrients provided by the flooding of the Congaree River, which happens an average of 10 times a year. Among its more than two dozen national champions are a swamp tupelo, a loblolly pine, and a water hickory. Walk any of the park's 20 miles of trails, including the wonderful 2.6-mile Boardwalk Loop Trail, and you'll encounter giant trees. Even in summer, their shade and a breeze blowing across the water combine to provide refreshing natural air-conditioning.

WATERFALLS

From thundering cataracts to fine veils of mist, falling water has the power to rest our eyes and hearts. We stop. We gaze. We pose in front of them, arms linked with friends and family, and take pictures for posterity. The great ones strike us with awe. The small, misty ones speak to us of peace and contentment.

1 Yosemite
NATIONAL PARK
California

It's no coincidence that great scenery and great waterfalls go together. The water needs something to fall from, which usually means cliffs and mountains and canyon walls. The Yosemite Valley in California claims them all and produces spectacular waterfalls at every turn. They fall from dizzying heights. **Yosemite Falls** drops an astonishing 2,425 feet over two vertical drops (Upper Fall and Lower Fall) separated by a cascade. **Bridalveil,** near the entrance to the valley, is 620 feet high. A torrent in spring, it becomes a diaphanous, shifting veil as the summer wears on. **Vernal** and **Nevada** come as a pair. For close views, hike the Mist Trail to the top of Vernal, then onward to the strenuous top of Nevada: 5.4 miles round-trip with a 2,000-foot elevation gain. Other falls visible without hiking include well-named **Ribbon Fall,** 1,612 feet high, and **Horsetail Fall,** 1,000 feet high and known for its fiery glow during winter sunsets.

2 Haleakalā
NATIONAL PARK
Hawaii

Waimoku Falls could be a natural haiku, minimal and eloquent but at the same time powerful. It slips rather than falls, sliding 400 feet down a near-vertical lava wall painted with bright green vegetation. Getting there is a delight. Waimoku is located in the Kīpahulu area of the park, on the southeast coast of the island of Maui, reached by driving the Hana Road, which itself is a sort of waterfall alley. Practically every watercourse along the coast offers one or more falls and cascades. When you come to Oheo Gulch, also known as Seven Sacred Pools, set off hiking up the Pīpīwai Trail, renowned as one of the best short hikes on the island and very popular. Be especially aware that heavy rain can cause flash floods. The trail climbs 650 feet through green forest, including several groves of bamboo—non-native, but visually striking—along a chain of rock-rimmed pools, waterfalls, and cascades. Along the way, the trail passes 185-foot-tall Makahiku Falls, a fine sight in its own right.

3 Cuyahoga Valley
NATIONAL PARK
Ohio

When water levels are low in Cuyahoga Valley National Park, 60-foot **Brandywine Falls** makes the best use of it. Hardly more than a trickle is needed in the smooth sandstone bed of Brandywine Creek above the falls to become

Framed by autumn's colors, Cuyahoga Valley's Brandywine Falls tumbles 65 feet in a bridal veil pattern.

a lovely sheet of water. Below the sandstone, a series of steplike shale ledges at the top spread the water, then set it loose to whisper down the sloping ramp to a lucid pool at its base. In times of high flow, the falls becomes a thundering spectacle. Yet there is something eminently satisfying in the elegant beauty of the low-water flow. A wooden walkway (partially accessible) leads an eighth of a mile from the parking area under the forested rim of the gorge to a viewing platform perfectly situated halfway down the falls; it's part of a 1.5-mile loop trail that circles the gorge. The popular Brandywine is one of nearly 70 falls that can be found in Cuyahoga Valley. Parking is quite limited during peak times, so park staff suggests taking the five-mile round-trip hike from the Boston Mill Visitor Center.

4 Zion
NATIONAL PARK
Utah

Speaking of ephemeral flows, a wondrous magic happens in the Southwest desert during heavy rainstorms. Dry streambeds suddenly burst into life. Waterfalls appear where hours ago there was nothing. Where cliffs are high, the result is unforgettable, as in the **Temple of Sinawava** at the upper end of Zion Canyon in Utah. The red-rock natural amphitheater is a stunning place any time. High walls glow with reflected light, giving a particular luminescence to leaves of trees and flowers, be they green in the summer or gold in the fall. But when it rains—and summer thunderstorms can be torrential—an elegant and very high bridal veil falls tumbles from far overhead, falling into the upper reaches of the North Fork of the Virgin River. Appreciate the opportunity it presents; in a few hours it will be gone.

5 Yellowstone
NATIONAL PARK
Wyoming

Waterfalls often mark the point in a river valley where hard rock gives way to softer rock. The river erodes the soft rock more quickly while the hard rock resists erosion. It's possible that nowhere is this more dramatically illustrated than in the **Grand Canyon of the Yellowstone** in Wyoming, where the Lower Falls drops 308 feet to the bottom of a bright yellow canyon. The surrounding stone is volcanic rhyolite, the hard, dark rock that covers most of the park. In this case, an ancient geyser basin altered the rhyolite through heat and chemical action. In effect, the rock was cooked until soft and yellow, and easily eroded. Viewing the falls from Grandview Point, you can see that the rock at the brink of the falls is unaltered dark rhyolite, while everything downstream—so fortunate for artists and photographers—is a palette of yellows and earth tones. You can see this up close, from a dizzying perspective, by hiking the paved, yet strenuous, Brink of the Falls Trail, or from various stops on the South Rim Trail. While Lower Falls is the tallest waterfall in the park, other options are also breathtaking, including the canyon's Upper Falls and neighboring Crystal Falls.

Birds of the Mist

If you look closely at a waterfall—not so much at the falling water but rather at the things around it—you may see a small gray bird: a dipper, or water ouzel. A fairly common sight in Yellowstone National Park streams, the dipper is short-legged with a stubby tail. It perches on mist-drenched rocks, making little bouncing motions. It doesn't look like a waterbird, but it rivals any duck for riparian affinity. Watch closely; now you see it, now you don't. In a twinkle, the bird vanishes beneath the water, where it walks on the gravel bottom snapping up aquatic bugs. Then *poink!* up it comes, wings whirring, to perform its best avian trick. Flying straight at the falling curtain of water, it disappears, gone behind the weighty sheet of thundering water as if through a puff of smoke, to its nest on an unseen ledge. Its hatchlings grow up in a zone of security where no predator can follow.

Brown bears time their next meal just right at Brooks Falls in Katmai National Park and Preserve.

Glacier
NATIONAL PARK
Montana

With its many high mountains, Glacier National Park in northwestern Montana is a wonderland of waterfalls. They include **McDonald Falls,** a roadside cascade thundering through beautifully carved rock walls. Other gems in the Many Glacier area are **Redrock Falls,** an easy hike, and the multistage **Apikuni Falls,** which is often less crowded. **Bird Woman Falls** can be seen along the Going-to-the-Sun Road. A high ribbon of white that falls nearly 1,000 feet in two main drops, Bird Woman is typical of the park's numerous hanging valley waterfalls. Hanging valleys result when a large glacier cuts deeper than a tributary glacier. The tributary, in effect, is left hanging high above the main valley, a perfect recipe for dramatic waterfalls. Bird Woman could be a top choice, but **Running Eagle Falls** in the Two Medicine area edges it by dint of a peculiarity: It was once called Trick Falls because it doubles its height depending on water level. Early in the summer, when runoff is abundant, the water comes down in a single frothing curtain. Later, when flows diminish, most or all of the water follows a secondary channel that bypasses the high rim and emerges from a cave in the middle of the rock face. At high water, the upper stream completely obscures the lower; at low water, there is no upper falls at all. A good trick.

7 | Katmai
NATIONAL PARK & PRESERVE
Alaska

Brooks Falls might warrant no special attention were it not for the brown bear show. Together with leaping salmon and the Alaska landscape, the scene ranks among the world's top wildlife-viewing experiences. The falls is a simple ledge a few feet high that spans the Brooks River. In summer, great numbers of sockeye salmon return from the ocean to spawn. Making their way upriver, they encounter the falls—and a gang of hungry bears waiting for them, sometimes numbering a dozen or more. The bears plunge into the pool below the falls where the fish mass, waiting for a chance to jump the falls; or they stand on top and catch the airborne fish as they literally fly into their mouths. A viewing platform allows you to survey the action at close range. There is no road access to the falls. You normally fly from Anchorage to the settlement of King Salmon, then transfer to a floatplane or boat for the last leg.

8 | Grand Canyon
NATIONAL PARK
Arizona

In the enormous sharp-edged gorge of the Grand Canyon, there exist hidden grottoes rounded by water, decorated with tender moisture-loving vegetation, and frequented by creatures like tree frogs and warblers. The epic landscape hides sweet delicacies. Places like Elves Chasm, Thunder River, and Havasu Canyon stand out. In fact, Havasu Falls, so beautifully set among red cliffs and travertine terraced pools, would probably take top honors if it were not just outside the park boundary on the Havasupai Reservation. But that doesn't take anything away from the beauty of **Deer Creek Falls.** Pouring down from the North Rim, it falls over cliffs and waters the roots of cottonwood groves until just before it hits the Colorado River, where it cuts a smooth slot canyon in the limestone. Its last act is a fitting climax. Seeming to burst from the cliff face, it falls in one clean sweep 100 feet to a beautiful pool at river level. Most who visit the falls are on wilderness river trips. It's also possible to hike down from the rim, a multiday trek.

▶ *Katmai National Park and Preserve's male brown bears are some of the largest in the world. They can weigh more than 1,000 pounds and catch and consume up to 30 fish per day.*

9 | Great Smoky Mountains
NATIONAL PARK
North Carolina & Tennessee

Mountains everywhere are rain catchers, and the higher the better. On average, the upper elevations of the Great Smoky Mountains in Tennessee and North Carolina pull down 85 inches of rain each year. The runoff feeds more than 2,100 miles of streams, which makes for a lot of waterfalls. Seeing the best of them requires some hiking. It's a pleasant feeling, on a hot summer day, to arrive in a zone of cool, forest-fragrant mist. The delightful **Grotto Falls** drops 25 feet from a ledge that overhangs enough for the Trillium Gap Trail to go behind the water curtain; it's a three-mile round-trip. A shorter walk on the Deep Creek Trail (1.6 miles round-trip) takes hikers to 25-foot-high **Indian Creek Falls.** Along the way, the trail passes **Toms Branch Falls,** a side stream that slides gracefully down a series of ledges. **Laurel Falls** rewards hikers on one of the park's most wildly popular trails, a paved 2.6-mile round-trip path named after the falls. And popular **Rainbow Falls,** also on an eponymously named trail, drops 80 feet, turning the waters of Le Conte Creek into a broad veil that hikers can walk behind. In winter, the veil becomes a freestanding ice pillar.

10 | Little River Canyon
NATIONAL PRESERVE
Alabama

They call Little River a mountaintop river because it flows for most of its length on Lookout Mountain in northeast Alabama. But it's also a canyon river, running beneath sandstone walls hundreds of feet high. **Little River Falls** stands in the middle. Upstream, the country is gentler, the river relatively placid. Downstream, it becomes a significant white-water river with boulder-bed rapids. The 45-foot-high falls pours over a 100-foot-wide ledge that spans the river. In autumn, when water levels are low, the falls is a peaceful white curtain surrounded by bright yellow and red foliage. Spring floods transform it into a roaring maelstrom. That would also be the time to visit **Grace's High Falls,** a 133-foot seasonal free-falling plume of water that can be seen from an overlook.

Shoot Like a Pro

To capture *the* shot at a national park, you have to think like a photographer.
Here are a few tips from the pros who spend their days on the other side of the lens.

It's summer at the Grand Canyon. You arrive midday, road weary but eager to see the legendary wonder. The air is hot. The family is cranky after hours in the car. You hike out to an overlook and there it lies, the awesome gulf in its full glory. Well, not quite full. Although the Grand Canyon is an impressive sight at any time, you might feel a bit let down, squinting in the glare of noon light. Where are the bright colors, the luminous voids, the rainbows and billowing clouds? It sure doesn't look like the pictures in travel magazines.

But the canyon often *does* look that way. The pictures are not exaggerations. Rather, they hold out a promise that applies to all the national parks: If you think like the people who made those pictures—that is, like a photographer—you'll not only get memorable images, but you'll also experience the parks at their inspiring best.

RAIN OR SHINE

Jeff Foott knows this as well as anyone. As a filmmaker, photographer, and naturalist, he has spent his life shooting for the likes of *National Geographic,* BBC, Discovery, and the International League of Conservation Photographers. "For drama," Foott says, "weather is key. Sometimes when it looks like it's going to be worst, it's the best time to go out. There might be lightning, cloud formations, other interesting atmospheric things happening."

The last hour before sunset and the first hour after sunrise are particularly sought-after for their light. "Golden hour is real and—if the weather agrees—it's particularly magical in parks," says Katie Dance, a photo editor at National Geographic Books. Dance prefers to shoot on an overcast day when colors can pop. Whether you're shooting with a DSLR, a smartphone, or both, Dance advises that getting the shot involves planning, positioning, and packing the right gear.

If the skies are bright, all's not lost. "Just avoid taking photos in direct sunlight at high noon—you'll get heavy shadows because the sun is straight over you," Dance explains. "If you can't avoid it, find shade so that your camera lens or phone is not in direct sunlight." And if your goal is to embrace the sun and capture those nice rays, lower your exposure for a more contrasting shot.

TRICKS OF THE TRADE

If you really want to get technical, there are tools to help you do it. Among Foott's go-tos are a compass and astronomical charts. Suppose you want to see Delicate Arch, the iconic symbol of Utah's red-rock country, bathed in warm sunlight against a deep blue sky framing a snowy view of the La Sal Mountains. "To get it right, you need to know that the arch faces northwest, and where the sun will be," says Foott. Online solar calculators provide the needed details, even to the extent of indicating when the full moon will rise perfectly positioned behind the arch.

Also consider the framing. "Don't be afraid to lie down on the ground for an alternative perspective,"

Safely photographing the Grand Canyon from dramatic Toroweap Overlook requires careful planning.

says Dance. For waterfalls, for example, she suggests shooting upward from the base of the falls to enhance the impact of the grand cascades. "Try to get several different perspectives—think wide, medium, and detail shots."

Travel is the time to rely on solid gear and tech, says Dance, who suggests the following: For phones, pop-out knobs and mobile tripods can both help with stabilization. Ensure you'll have enough power with a solar phone charger and DSLR battery banks. Dry packs can save you lots of problems, but be mindful of your lens and dry it with a microfiber towel. Lastly, save yourself from losing time and memories by backing up your SD cards throughout your trip. "Taking the time to preserve your pictures properly will go a long way in helping you share and recall your epic adventures."

THE GOLDEN RULE

The most important guideline of all? No matter when you go and how much you plan, it's imperative that you follow park safety regulations at all times. Also, human impact can be detrimental to ecosystems. Remember to zoom with your lens, not your feet. Park guidelines dictate people stay 100 yards from predators like bears and wolves, and at least 25 yards from everything else. Safety ultimately calls the shot.

10 BEST PARKS FOR
VOLCANIC MARVELS

Some of the most enthralling and alien landscapes in the national parks are volcanic. Many of them defy conventional notions of beauty, but they have a particular allure of their own. Fortunately, the volcanic parks do an outstanding job of conveying their geology, and a compelling story of ongoing creation begins to emerge.

1 Lassen Volcanic
NATIONAL PARK
California

Located in northeastern California, this national park gladdens the hearts of volcano lovers because among its 60 identifiable volcanoes are four major types: **plug dome, shield, cinder cone,** and **composite.** Add to that the thrill of numerous **thermal areas,** making it a highly concentrated showcase of volcanism. The crown jewel of the park is Lassen Peak, among the world's largest plug dome volcanoes. Its last major activity occurred between 1914 and 1917, peaking in 1915; it has been quiet since 1921. A five-mile round-trip hike climbs nearly 2,000 feet to the 10,457-foot summit. The Bumpass Hell thermal area is the park's most popular attraction, full of colorful boiling pools and the noisy Big Boiler fumarole. The three-mile round-trip trail takes about two hours at a relaxed pace. Because of

massive snowpack, the best time to plan a visit here is July to October. Devil's Kitchen is equally compelling but draws fewer people. Count on a strong rotten-egg smell at both places. If all this activity suggests, well, volcanic activity, it's true; scientists study Lassen and other Cascade Range volcanoes closely to predict future eruptions.

2 Chiricahua
NATIONAL MONUMENT
Arizona

Chiricahua is one of Arizona's "sky islands"—discrete, isolated volcanic mountain ranges that rise 9,000 feet or so above the surrounding desert grasslands. "Isolated" is an operative word. This is a quiet park of 12,000 acres. Volcanically speaking, the park's most remarkable feature is its whimsical land of **rhyolitic rock pinnacles**—balanced rocks and stacks hundreds of feet tall that look like giant stone totem poles. The Apache called this the "land of standing-up rocks." The formations are composed of rhyolite—superheated and solidified volcanic ash. The park's eight-mile scenic drive features overlooks of the pinnacles at Echo Canyon and Massai Point. For a closer look, hike the strenuous 7.3-mile Heart of Rocks Trail. From September to May, take the

Known as the Wonderland of Rocks, Chiricahua National Monument is home to hundreds of rhyolitic pinnacles.

free shuttle to Echo Canyon and hike down to the visitor center. Rangers drive the van, so you'll get a crash course in volcanism, then a great look at the rocks—some weigh thousands of pounds yet are perched on bases just inches around.

3 | Crater Lake
NATIONAL PARK
Oregon

Crater Lake is such a dazzling sight that it's easy to forget its origins. The lake, which is one of the 10 deepest in the world, comprises snowmelt and rainwater that filled a 1,932-foot-deep **crater** left by the eruption of Mount Mazama about 7,700 years ago. Drive the 33-mile Rim Drive (or take the Trolley Tour)—it traces the caldera of the volcano with that dimension in mind and you get a sense of the mountain's immense scope. The top of the

The steaming terraces of Mammoth Hot Springs in Yellowstone National Park

mountain would have been 12,000 feet high; the surface of the lake today is 6,173 feet wide. Mazama's eruptions had 42 times the force that Mount St. Helens had in 1980. Later eruptions inside the crater created Wizard Island and the submerged Merriam Cone. If you can't quite picture a volcano here, drive to Pumice Desert north of the lake. The swath of pumice and ash, 50 feet deep, covers 5.5 square miles.

4 | Yellowstone
NATIONAL PARK
Wyoming

Yellowstone country in northwest Wyoming is a gigantic volcanic playground, comprising **three**

calderas produced by three super eruptions between 3,000,000 and 600,000 years ago. The northwest quadrant of the park is the nexus of an immense hot spot, a growing underground bulge that spreads hundreds of miles, and which is effectively one of the largest volcanoes in the world. Because its magma chamber is just a couple of miles below the surface, a constant source of heat is continuously meeting circulating water that migrates down through cracks, and presto: **geyser.** Also here is the world's most astounding concentration of **thermal features,** 10,000 of them, including hot springs, mud pots (a blend of sulfuric acid, silica, and clay), and fumaroles. See them on

vivid display at the park's two main geyser basins, Upper and Norris, and the Mammoth Hot Springs area. Norris contains the world's tallest active geyser, Steamboat; Upper is the home of the world-famous Old Faithful geyser and many others; Mammoth's flowing waters have formed beautiful travertine limestone terraces.

5 | Sunset Crater Volcano
NATIONAL MONUMENT
Arizona

Sunset Crater, a classically shaped **cinder cone,** erupted circa 1040–1100, making it the most recent volcanic event in north-central Arizona's San Francisco Volcanic Field. When explorer John Wesley Powell first saw the cone in 1885, he remarked that it "seems to glow with a light of its own"—the result of the cinders being oxidized to reddish, sunset-like colors. Hence the cone's name, which Powell is credited with bestowing. Only sparse vegetation has grown around the cone's lower slopes, and the top is so pristine and so little eroded that it is off-limits—a trail closed in the 1970s still scars its slopes. But you can hike to nearby Lenox Crater—a steep half-mile path leads to the top—and across the **Bonito Lava Flow** at the base of Sunset. The visitor center shows a video reenactment of the crater's eruption as it would have been witnessed by Indigenous people nearly a thousand years ago.

6 | Craters of the Moon
NATIONAL MONUMENT & PRESERVE
Idaho

"Where's the volcano?" is high atop the list of FAQs at Craters of the Moon. Instead of a single peak, look for **spatter cones** and **cinder cones** aligned along what is known as the **Great Rift**—a 50-mile fissure that permitted lava to well up and flow across half a million acres of landscape. Spatter cones are the monument's

signature feature, created by globs of lava that shot up through the rift and adhered to one another. Trails lead to several of them, including one called Snow Cone that can hold ice all summer. A walk on Caves Trail leads to four different lava tubes—underground chambers through which rivers of lava once ran—including massive Indian Tunnel, 800 feet long with a 30-foot ceiling. Adventurous backpackers with permits can follow the Wilderness Trail to Echo Crater, where they can savor the extraordinary experience of sleeping inside a volcanic crater. Be sure to check the park's website as some sections will close due to seismic activity.

7 | Haleakalā
NATIONAL PARK
Hawaii

Haleakalā is more than the most popular attraction on the island of Maui. It's also something of a volcano's greatest hits: a massive (10,023 feet) volcanic mountain with a gaping crater that displays all manner of lava types and formations. A road winds to the top, where the big picture is revealed: This enormous mountain—which actually rises 30,000 feet from the ocean's floor—is in fact a **cinder cone vent** that once spewed volcanic particles like popcorn from a popper. Traversing a series of cinder

Know Your Volcanoes

Among the fascinations of visiting volcanic parks is the opportunity to play amateur geologist, which entails a detective hunt for clues about the origins of the rocks and formations you encounter. For example, Lassen Volcanic National Park in California is famous for having all four major volcano types—formations that also occur throughout the parks. So how do you identify them, and what do they mean?

A *cinder cone* is an accumulation of volcanic fragments spewed out through a vent. They can be small, like the cones lined up along the Great Rift in Craters of the Moon National Monument and Preserve, or large, like in Sunset Crater Volcano National Monument.

A *shield volcano* is a broadly sloping mountain built up by the eruption of fluid lava. Prospect Peak in Lassen

is one. Kīlauea in Hawai'i Volcanoes National Park is an example of one that remains under construction.

A *plug dome* is a dome-shaped volcano formed by viscous magma. Domes often erupt from the flanks of composite volcanoes, which was the case with Lassen Peak, a plug dome that formed on its ancestor, a mountain called Tehama that was 1,000 feet taller than Lassen.

A *composite*, or *stratovolcano*, is created by the eruption of viscous lava flows, tephra (fragments), and pyroclastic flows. It is typically built up over thousands of years and consists of a number of vents that may spew separate cones and domes on the composite's flank. Broke-off Mountain in Lassen is a composite volcano, as is Mount St. Helens.

Fiery lava erupts from Halemaʻumaʻu in Hawaiʻi Volcanoes National Park.

vents, the Keoneheʻeheʻe (Sliding Sands) Trail leads past sharp aa lava, ropy pahoehoe, and globby lava bombs (lava formations as large as a small car) spewed out by eruptions peppered across the landscape. The bombs are now old enough to be peeling away, revealing their inner structure.

8 | Hawaiʻi Volcanoes
NATIONAL PARK
Hawaii

It's not every day one gets to see land being born, which is why a walk near flowing lava in Hawaiʻi Volcanoes National Park is a thrilling experience. Start by taking Crater Rim Drive to the top of **Kīlauea's** often billowing caldera—which in 2018 had one of the largest eruptions in centuries, followed by more in 2020, 2021, 2022, and 2023. There you can gaze into Halemaʻumaʻu, the 400-foot crater pit and traditional home of Pele, goddess of the volcano, an experience all the more stunning after dark. Everything about walking here is subject to the caprices of nature. Lava seekers might walk a mile from a vehicle to the flowing lava, or several dicey miles across craggy hardened lava, or never get close at all. But to get close is a total sensory experience. Don't miss taking the Chain of Craters Road, a meandering 18-mile venture past lava fields, as it takes you to the sparkling ocean.

▶ *From 1908, when it was named a national monument, to 2013, when President Barack Obama designated it a national park, Pinnacles was expanded from 2,500 acres to 26,606 acres.*

Although it hasn't erupted in more than 30 years, the park's **Mauna Loa** is still one of the world's most active volcanoes, and the Mauna Loa Road carries visitors to its lookout site. A reminder: The park posts daily updates and cautions, and the sight of active lava is never guaranteed.

9 | Lava Beds
NATIONAL MONUMENT
California

The name may be Lava Beds, but **lava tube "caves"**—underground conduits through which lava once flowed—are the star of the show here. The monument is home to more than 800 caves, of which two dozen are accessible. That doesn't mean every cave is a cakewalk. The caves are rated, and a down-and-dirty experience is guaranteed in a challenging cave—duckwalking, crawling, ceilings as low as 12 inches. Permits, helmets (the park sells "bump hats"), kneepads, and

flashlights are de rigueur. But the features are similar in all the caves, so the less adventurous can opt for Mushpot Cave or Skull Cave, with its chilly bottom floor coated permanently with ice. These caves played a part in history, serving as hiding places for the Modoc people during the Modoc War of 1872–73.

10 | Pinnacles
NATIONAL PARK
California

Close to 23 million years ago, volcanic eruptions rolled out along the San Andreas Fault, 195 miles south of where Pinnacles National Park is today. Tectonic plates moved these **volcano remains,** shifting the ancient rocks about an inch per year, currently placing the geological wonderland among the Gabilan Range. Weather and animals have also made their mark on the rust-colored spires and pinnacles. Today, hikers, climbers, and even majestic California condors find a haven in the park's spires, caves, and canyons. More than 30 miles of trails traverse the park. Take the Condor Gulch Trail for spectacular views of the High Peaks. At just 1.7 miles one way, the trail also features the Condor Gulch Overlook, which separates the moderate section of hiking from a strenuous stretch. The challenging Juniper Canyon Loop rewards hikers with a switchback into the heart of the park's magnificent focal point, High Peaks.

10 BEST PARKS FOR
ANCIENT ARTIFACTS

Imprints from people, creatures, and plant life continue to give us clues into the past, even millions of years after they lived on these lands. Fossil remains of dinosaurs, birds, fish, ferns, insects, and even algae combine with painted rock art to piece together a fuller picture of ancient America.

1 Florissant Fossil Beds
NATIONAL MONUMENT
Colorado

The mayfly practically defines its order name, Ephemeroptera, from the Greek *ephemeros,* or short-lived. Yet here it is, 34 million years later, the lightest of gossamer preserved between thin layers of shale. Florissant Fossil Beds is extraordinarily rich in **Eocene epoch fossil species;** about 1,700 have been identified, including the leaves of ferns, cypress, maple, mahogany, laurel, cocoa, and many others. Among the insects are dragonflies, cicadas, lacewings, beetles, caddisflies, and even cockroaches. Fish are abundant, birds and mammals less so. Not all the fossils are delicate. Several huge, petrified redwood stumps still stand upright behind the visitor center, which is loaded with fossil displays. (A note: Don't expect dinosaurs here—the volcanic activity that deposited rich ash and mud happened 30 million years after their disappearance.) The park has a roster of hands-on paleontology activities; check the website or call ahead for schedules and availability.

2 Canyonlands
NATIONAL PARK
Utah

Some of North America's oldest and most impressive rock art is found in one of the park system's most secluded segments: Horseshoe Canyon in central Utah. Rendered in ocher and white, the ethereal pictographs of the **Great Gallery**—20 humanoid images spread along 200 feet of rust-colored rock—are masterpieces of the Barrier Canyon Style. Devoid of arms and legs, the life-size tapered figures are thought to date from the Late Archaic period (2000 B.C.–A.D. 500), when Paleo-Indians occupied the canyon. A separate unit from the rest of Canyonlands, Horseshoe Canyon is accessed via an unpaved road between the town of Green River and Utah Highway 24 (near Capitol Reef National Park). Drive time is roughly 90 minutes from Green River. From the west rim parking area, a footpath and equestrian trail leads down to the Great Gallery. The strenuous seven-mile round-trip hike averages five hours and can be dangerously hot in summer. You can join ranger-guided walks to the Great Gallery in spring and autumn.

3 Dinosaur
NATIONAL MONUMENT
Colorado & Utah

Dinosaurs existed in a huge variety of sizes and shapes; however, it's the big ones that stir our imagination. This monument is best known for its spectacular **quarry,**

Canyonlands' False Kiva is a human-made stone circle at a remote archaeological site, only viewable from a distance.

a steeply angled rock wall that displays a dense concentration—we're talking 1,500—of partially excavated fossil bones. The bones are in the Morrison formation, a mix of sandstone and conglomerate deposited during the Jurassic period 150 million years ago. The concentration here is thought to have been a sandbar in an ancient river where carcasses came to rest during floods. Among the bony roll call: *Stegosaurus,* with its protruding spinal plates and spiked tail; *Apatosaurus,* also called *Brontosaurus,* whose long neck and longer tail gave it a total length up to 75 feet; and *Allosaurus,* the nightmarish relative of *Tyrannosaurus rex.* Quarry Exhibit Hall, located on the Utah side of the park, is open year-round. Start at the Quarry Visitor Center for an engaging film and a seasonal shuttle.

The McKee Springs site in Dinosaur National Monument (p. 42) boasts large anthropomorphic petroglyphs.

4 | Petroglyph
NATIONAL MONUMENT
New Mexico

With more than 25,000 images, this desert park on the western outskirts of Albuquerque safeguards one of the largest and most significant rock art anthologies in North America. Created by both ancestral Puebloans and Spanish settlers between 700 and 400 years ago, the **petroglyphs** reflect differences between the groups. The ancestral Puebloans used rock art to preserve and pass along information about their culture, history, and spiritual beliefs. Their favorite subjects were human and animal figures and geometric designs. Some of the ancient motifs—like the flute-playing Kokopelli—are still popular in local culture. The Spaniards, who came later, were more likely to render Christian crosses or livestock brands in the dark basalt faces. The park is divided into three petroglyph-viewing units along a stretch of the volcanic West Mesa Escarpment. Rinconada Canyon's 2.2-mile loop yields more than 300 petroglyphs and geological features. The Boca Negra section offers three short, paved trails and a hundred images. Farther north, the Piedras Marcadas Canyon boasts 400 more. Guided walks and Native American cultural events are park staples.

5 | Hagerman Fossil Beds
NATIONAL MONUMENT
Idaho

Across the Snake River from Hagerman, Idaho, this monument boasts one of the world's richest deposits of **horse fossils** from the late Pliocene epoch, some four to three million years ago. Chief among them is the Hagerman horse (*Equus simplicidens*), a small and strongly built horse with a thick neck and muscular forequarters, quite like today's zebra. The horse became extinct 10,000 years ago at the end of the last ice age, along with many other large species. Twenty complete skeletons and many partial ones have been excavated since a local rancher discovered the bones in 1928. Besides horses, more than 220 other species (105 vertebrates) have been located in these strata, representing a variety of ancient habitats, including grassy plains and wetlands. There were mastodons, saber-toothed cats, ground sloths, giant marmots, camels, bears, otters, muskrats, and peccaries. Of two dog species, one was the ancestor of modern coyotes. The other, *Borophagus,* has no living descendants. It sported a powerful, shortened jaw like that of a hyena, which explains its unofficial name: bone-crushing dog. The Thousand Springs Visitor Center is the only place to spot fossils at the monument.

6 | Petrified Forest
NATIONAL PARK
Arizona

The **trees** of Petrified Forest grew during the late Triassic period about 225 million years ago, early in the age of dinosaurs. Growing conditions were good. Giant conifers rose 200 feet and higher above thickets of lush vegetation. Their fossil remains are scattered among the colorful clay hills of the Chinle formation. Trunks of the genus *Araucarioxylon* stand out for their size (up to 190 feet long) and their awesome petrified beauty. Other fossils help fill out the primal scene: early dinosaurs,

crocodile-like reptiles 40 feet long, half-ton amphibians, horseshoe crabs, clams, and freshwater sharks. Mammals were far in the future, so you won't see many ancient signs of them here—though modern ones may be spotted.

7 | Guadalupe Mountains
NATIONAL PARK
Texas

Far back in the shadows of the Permian era, 265 million years ago, western Texas was near the Equator. Much of it was covered

by the Permian Basin, an inland sea. Around its margins formed a reef—roughly circular, 400 miles long, built up by tiny lime-secreting organisms. Today, this part of Texas is a desert and the reef is an imposing limestone mountain range that ranks as one of the world's best examples of a **fossilized marine reef.** Its fine-grained stone contains fossils of sea urchins, crinoids, trilobites, nautiloids, and much more. They are important, in part, because they lived before the Great Dying, a catastrophic series of events that drove some 96 percent of Earth's

Art on the Rocks

Hard to date—and often even harder to interpret—ancient rock art, like chipped or carved petroglyphs and painted pictographs, can provide extraordinarily valuable reflections of a park's Indigenous populations. At Chaco Culture National Historical Park, don't miss the etched astronomical calendar that marked lunar and solar cycles, the quarter-mile Petroglyph Trail, and Una Vida, an eclectic petroglyph gallery of human figures, geometric designs, and handprints on a sandstone panel. In Capitol Reef National Park, the Fremont culture (600–1300) left extensive and elaborate markings deep in Capitol Gorge, while late 19th-century settlers etched their names and dates on what became known as Pioneer Register. And at Lava Beds National Monument, petroglyphs were rendered on lakeside cliffs and pictographs on the large lava tubes. The art at Lava Beds is especially unique for its astronomical motifs.

marine species into extinction. Hikers can get a geological tour of the results via the Permian Reef Trail (8.4 strenuous miles round-trip).

8 | John Day Fossil Beds
NATIONAL MONUMENT
Oregon

Scattered among the park's three units are more than **750 known fossil sites,** revealing the development of flowering plants and mammals spanning 40 million years. They include 14 genera of ancient horse; also prehistoric turtles, alligators, tapirs, rhinos, peccaries, cougars, bears, hippos, and more. Among plant fossils and imprints are primitive horsetails, ferns, and ginkgoes, but also flowering species like dogwood, magnolia, beech, and laurel. Don't miss the Thomas Condon Visitor Center at the Sheep Rock Unit, where you can watch scientists work with fossils found at the monument.

9 | El Morro
NATIONAL MONUMENT
New Mexico

While some may not consider them to have the artistic flair of today's taggers, those who rendered the **thousands of inscriptions** on El Morro some 300 years ago

▶ *Sixteen National Park System units have been established specifically for the protection of important fossils, but the geological history of plants and animals is preserved in more than 280 units.*

collectively created the nation's most important graffiti gallery. Drawn by the fresh water of an oasis-like pool at the base of the 250-foot headland, generations of Spanish explorers and American soldiers and settlers paused here to drink and presumably reflect on what they were doing in the middle of the Southwest desert. They carved their names, dates, and messages in the soft sandstone, creating a record of who passed this way and to some extent what they were thinking at the time. They may have been inspired by the petroglyphs that already existed at El Morro, the artwork of the ancestral Puebloans who once lived in mud-brick dwellings atop the headland. Exhibits at the visitor center trace 700 years of human history at El Morro. The half-mile Inscription Trail leads to the main gallery, which has more than 2,000

individual "tags." Another trail leads to the crest of the bluff and the ruins of Atsinna ("place of writing on rock") Pueblo, where the area's earliest artists lived.

10 | Waco Mammoth
NATIONAL MONUMENT
Texas

In 1978, Paul Barron and Eddie Bufkin found what appeared to be a massive leg bone while digging for arrowheads and fossils near the Bosque River north of Waco. They took the bone to Baylor University, where it was identified as a fossilized femur of the Columbian mammoth. Soaring 14 feet at the shoulder, these extinct creatures were unarguably enormous. But what made the discovery so significant was that it wasn't singular—future archaeological digs at this important Ice Age site unearthed an **intact mammoth nursing herd** of more than 20 females and juveniles that perished together, likely in floodwater, approximately 72,000 to 65,000 years ago. Other finds included a camel, large male mammoth, dwarf antelope, American alligator, giant tortoise, and saber-toothed cat. Officially recognized as a national park unit in 2015, Waco Mammoth offers tours that include a visit to the Dig Shelter, where fossils have been left in situ—in their original position in the bone bed.

Take the Headland Trail to see the 13th-century Atsinna Pueblo in El Morro, home to the ancestral Puebloans.

ROCKS & ARCHES

Arches, hoodoos, natural bridges, domes, and other gravity-defying shapes belie the notion that rock is an inanimate material lying hard beneath our feet. The most remarkable natural sculptures are mere remnants, the last bits of large formations that have been carved away by erosion, able to disappear in a geological flash.

1 Capitol Reef
NATIONAL PARK
Utah

All rocks have stories, some more interesting than others. The rock of Capitol Reef in south-central Utah ranks fairly high among geological narratives, but not because of any complex twists of plot. Rather, it's the elegant finish that makes the tale satisfying. This one begins with a rock warp. Ten thousand feet of sedimentary rock—sandstone and shale—were folded when a deep fault moved beneath them to create a monocline, a warp or dip in a region of horizontal rock layers. Fifty to 70 million years of erosion later, the **Waterpocket Fold,** as the monocline is called, is a steeply angled line of cliffs 100 miles long. From one end to the other, this is elite rock work, spangled with arches, narrow canyons, soaring rock fins, and—explaining the other half of the park's name—gleaming high, white **domes** crafted out of beautiful Navajo sandstone. Geologists and poets call the stone aeolian, for its origin as dunes. Finely sifted and shaped by the wind in a Sahara-like desert, this stone has an unusual purity and a tendency to form graceful, rounded shapes, of which Capitol Dome is an ideal example. The park's 7.9-mile Scenic Drive roughly parallels a portion of the fold and provides a good introduction to the wondrous formations.

2 Devils Postpile
NATIONAL MONUMENT
California

Whoever gave Devils Postpile its name might have been picturing the pillars of Hades—black in color, born in fire, massed heavily underfoot. The posts, however, are quite beautiful and better described by their geological names, **hexagonal basalt columns.** They formed when a lava flow pooled behind a glacial moraine to a depth of several hundred feet. Slow and uniform cooling created the necessary conditions for columnar jointing. Exposed by glacial erosion, the columns appear to have been bundled neatly together. In fact, it's simply the way the rock cracked while cooling. A short trail from the parking area provides views of the exposed cliff of vertical posts, which are up to 60 feet long. You can also walk on top of the pile, where the posts appear like well-fitted tiles.

3 Arches
NATIONAL PARK
Utah

No place on the planet rivals the red-rock country of the American Southwest for the number and beauty of its **unusual rock formations.** The explanation lies in the

Dramatic sunsets illuminate the natural stone formations in Arches National Park.

geological history of the Colorado Plateau, which covers southern Utah and parts of adjacent states. Uplifted and rumpled in places, the plateau is a hotbed of erosion. Its soft and easily carved horizontal layers of rock lie exposed to the weather, with underlying softer layers—mudstones and shale—readily undercutting harder layers that collapse along vertical lines, creating the imposing cliffs and layer-cake structures characteristic of the Southwest deserts. At the center of this geological expanse is Arches, the best natural sculpture garden in America. Its signature piece, **Delicate Arch,** is most famous (pictured on Utah license plates), but others of its **more than 2,000 arches** are equally marvelous. Half the pleasure is getting to them. Park roads lead to highlights like Double Arch, Courthouse Towers, and Parade of Elephants. For trails, try Devils Garden where the long, slender strand of Landscape Arch challenges belief; or hike to Delicate Arch for a sunset that will never fade from memory.

4 | Bryce Canyon
NATIONAL PARK
Utah

The rock of Bryce Canyon is limestone of the Claron formation and although it does hold some arches, it is most famous for its tightly clustered spires, or **hoodoos.** Seeing them from a distance, high against the rim of the forested Paunsaugunt Plateau, it's clear that Bryce is not a canyon but rather a series of **eroded amphitheaters.** The plateau is being demolished by erosion in a most spectacular fashion. Peering into the hoodoos from the rim, that impression is intensified, and so is the desire to get down amid the fiery colors to experience them close at hand. Numerous trails provide access. The Queens Garden Trail is among the least demanding and most popular; it leads to a formation that reminds some people of Queen Victoria. Combining it with the Navajo Loop adds a little distance and takes in other landmarks, including Wall Street (a narrow passage through the rock), Thor's Hammer, and Two Bridges. A more strenuous route, the four-mile round-trip Hat Shop Trail, drops off Bryce Point to a cluster of balanced rocks perched like white hats on red pedestals.

Two hikers venture into Bryce Canyon's fantastical world of fiery colored hoodoos.

The Rim Trail offers any number of wonderful views without the cost of having to climb back up.

5 | Canyonlands
NATIONAL PARK
Utah

A parade of dramatic rocks marches through remote Canyonlands, where you could view the entire park as one dramatic sculpture. Yet if choices must be made, head for the Needles District. The paved UT 211 ends amid a cluster of reddish and tan sandstone pinnacles: the **Needles.** From here, trails enter a three-dimensional playground of smooth rock, narrow canyons, wildflower gardens, shallow streams, and hidden meadows. Distances are moderate, and every mile is a delight. Highlights include the narrow subterranean cracks of the Joints Trail and the full-day excursion to massive **Druid Arch.**

6 | Rainbow Bridge
NATIONAL MONUMENT
Utah

Graceful **Rainbow Bridge** in southern Utah seems as remote as the legendary pot of gold. Before the creation of Lake Powell initially made it possible to get here by motorboat on a day trip, reaching Rainbow Bridge involved a rugged overland foot trip through the Navajo Nation, or a float down the

Colorado River and then a hike up Bridge Canyon. Either way took at least a week and usually more. In recent years Lake Powell has had dramatic fluctuations in water level, so check ahead regarding dock access and concession-operated boat tours to the Rainbow Bridge Trail. For those who are able, it's worth the effort. In fact, the Navajo and other tribes consider this a significant sacred site. Made of beautiful Navajo sandstone, it stands in near-perfect symmetry—290 feet high and 275 feet across at the base. One of the world's largest known

natural rock bridges, Rainbow Bridge could be called an arch, except for the manner in which it was born. Unlike arches, bridges are formed by flowing water. True to their names, they span existing or former watercourses.

7 | Zion
NATIONAL PARK
Utah

Zion National Park is blessed with a stunning array of rock shapes. Most visitors head straight for Zion Canyon itself, where they find a scenic feast for their eyes and camera.

Fallen Arches

Arches seem to defy gravity: With nothing but hollow space beneath them, how do they stay standing? Answer: They don't. Arches seem durable only because the human scale of time is so short. Some arches last longer than others, yet in geological terms, they come and go in a flash. Wall Arch, one of the better-known spans at Arches National Park, collapsed in 2008. Ten years later, Rainbow Arch, at the same park, also collapsed. At Hawai'i Volcanoes National Park, a 90-foot lava rock formation called Hōlei Sea Arch lost a key section of its base in July 2022. Landscape Arch—an impossibly slender, nearly horizontal strand in Arches—spans 290 feet, with a thickness at one point of just six feet. Large chunks of rock have fallen from it, with one incident captured on video. More than any other major arch, it is hard to comprehend what keeps this one in place. But some day the twin forces of gravity and erosion will bring it down.

32 USA
Utah 1896

Some make the roundabout trip to the park's remote northwest corner, the **Kolob Canyons Unit.** From the road, it's a full day's hike of 14 miles round-trip to **Kolob Arch,** a flat-topped alcove arch similar to Landscape Arch in Arches but considerably more massive and just a few feet shorter (287 feet) in horizontal span. For a quicker option nearby, the Taylor Creek Trail offers a five-mile round-trip to **Double Arch Alcove,** a huge overhanging wall beautifully painted with seeping springs.

8 | Badlands
NATIONAL PARK
South Dakota

Seen from their base at sunrise, the ragged peaks of the South Dakota badlands rise up, seemingly enormous, and glow with an unearthly light. But they are not tall; they only have the proportions of grander summits. It's a lesson in erosion: Large or small, given the right material you get similar forms. As evidenced in Badlands National Park, wind and water have sculpted the rock into **buttes, spires,** and **pinnacles** that are a colorful torte of volcanic ash, river sands, floodplain mud, and ocean-deposited shale, with layers in shades of yellow, black, brown, and gray. Dating from 75 to 28 million years ago, they are also loaded with early mammal fossils from the late Eocene and

▶ *The hoodoos of Bryce Canyon National Park—the greatest concentration of the geological feature on Earth—range in size from five to 150 feet.*

Oligocene epochs. The moderate 0.75-mile Door Trail offers a good introduction to the park's geological formation, entering through an opening in the rock—the **Door**—and meandering through rugged, eroded badlands. For a longer exploration, hikers will find solitude and some steep formations on the 10-mile Castle Trail, accessed from the parking area for the Door.

9 | Natural Bridges
NATIONAL MONUMENT
Utah

Three easily visited **natural bridges** stand in two canyons in southeast Utah. A nine-mile loop road meanders along the rims of White and Armstrong Canyons, and short trails drop in for closer views. The canyons are beautifully convoluted, curving tightly around narrow **rock fins** that make it easy to visualize the creation of a natural bridge. It starts with a bend in the stream. As the water cuts deeper, the bend becomes a hairpin loop. The rock

inside the loop becomes thinner as erosion eats away from both sides until the stream punches through and a bridge is born. By twisting itself in serpentine curves, water finds the straight path, the shortest distance. Hydrologists provide explanations for this; yet the bridges, in their airy improbability, encourage thoughts of natural mysteries. These three—Owachomo, Kachina, and Sipapu—are prime subjects.

10 | Chiricahua
NATIONAL MONUMENT
Arizona

It boggles the mind how these **rock pinnacles**—sometimes rising hundreds of feet into the air—can balance on such insignificant bases. Called rhyolites, the fine-grained volcanic rocks pile in seemingly impossible pinnacles that number in the hundreds throughout Chiricahua National Monument's Heart of Rocks area. Other wonders include the organ pipe–like monument pillars created from a massive volcanic eruption 27 million years ago, when ash and debris fissured and eroded. Choose the Heart of Rocks Loop, Sarah Deming Trail, or Big Balanced Rock Trail to see the famous Big Balanced Rock, a tremendous boulder standing on a five-foot surface, or Pinnacle Balanced Rock. These hikes can be strenuous, but the payoff is profound.

10 BEST PARKS FOR
CAVES

Caves lure us into utter blackness and, with the aid of lanterns and flashlights, reveal exquisite wonders: great halls, cavernous rooms, and long corridors. You'll also find rivers thundering over high cascades, lakes, hills, and deep yawning pits. Meanwhile, creatures that prefer the darkness mingle amid gleaming crystals and flowstone formations.

1 Carlsbad Caverns
NATIONAL PARK
New Mexico

Like other solution caves, New Mexico's Carlsbad was carved from limestone bedrock—but this one got a power boost. Hydrogen sulfide gas, rising from deep petroleum deposits, combined with underground water and subterranean microbes to form a potent dissolving bath of sulfuric acid. When the water table fell, the chambers were left high and relatively dry. About a million years ago, fresh water trickling down from the surface began decorating the maze of passageways. Carlsbad is rich in **speleothems** (the technical word for cave formations), including staple items like stalactites, columns, draperies, and ribbons; also helictites, soda straws, popcorn, and lily pads. For most, the Big Room, also called Hall of the Giants, is the central

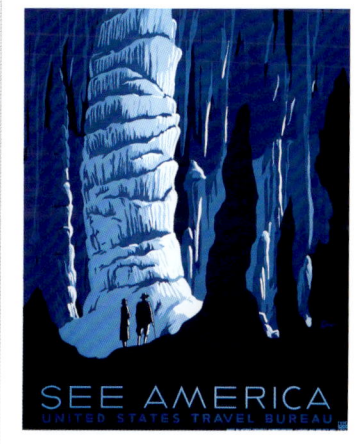

SEE AMERICA
UNITED STATES TRAVEL BUREAU

attraction. You can take an elevator below, or hike down on your own through the Natural Entrance, a descent of more than 750 feet. Carlsbad is also quite famous for its flights of **Brazilian free-tailed bats.** After sunset from spring through October, hundreds of thousands of the half-ounce creatures leave their roosting sites and pour out of the cave in spiraling clouds, headed for an evening of foraging. Visitors gather at an amphitheater nightly for the surreal show.

2 Jewel Cave
NATIONAL MONUMENT
South Dakota

High in the Black Hills of South Dakota stretches the world's fourth longest cave, where more than 215 miles have been surveyed. Jewel gets its name from the blunt **calcite spar crystals** that cover most of the cave walls, giving them a glittery sheen. It contains other **calcite speleothems**—stalactites, flowstone, and such—but its greater treasures include unusual gypsum flowers, needles, spiders; also rare hydromagnesite balloons and aragonite frostwork so delicate it seems that a baby's breath could destroy it. The popular Scenic Tour, which begins with an elevator descent

With more than 400 miles of passageways, Mammoth Cave in Kentucky is the world's longest known cave system.

and follows a half-mile loop on a paved trail, is a comprehensive look at the cave's chambers, passages, and crystals. During summer months, rangers in historical uniforms hand out lanterns for a step back into 1930s cave exploration. For adventure, join the Wild Caving Tour, a strenuous hard-hat experience that includes crawling in tight spaces.

3 | Mammoth Cave
NATIONAL PARK
Kentucky

Mammoth is just that—the **world's longest known cave system,** measuring 420 miles of known passageways. No end has been found to this vast multilayered labyrinth. The cave also superbly exemplifies the formation of limestone solution caverns. Limestone deposited in

the sea is pushed above sea level, causing the seeping fresh water to hollow out spaces, decorating them with baroque extravagance. Mammoth has some huge chambers with ceilings that disappear into darkness overhead. The space called Mammoth Dome is 192 feet high; you climb down it on stairways. There are tight squeezes, too: The walls at Fat Man's Misery have been polished by thousands of bodies squeezing through. Depending on the season, more than a dozen cave tours are offered, lasting from one to several hours. The Frozen Niagara Tour (0.25 mile) offers an easy introduction, passing a large waterfall-like formation of flowstone. On the moderate-intensity Historic Tour (two miles), adventure buffs traverse tunnels humans have navigated for thousands of years. Caves are perfect places for ghost stories, too. What strange things lie in wait beyond the reach of a light? Plenty, according to the stories told about Mammoth, including unexplained sounds, vague lights in the blackness, and the occasional nudge from behind when no one is around.

4 Wind Cave
NATIONAL PARK
South Dakota

Caves breathe. Wind Cave, comprising more than 150 miles of passageways under the rolling hills of South Dakota, was named for its barometric puffing, the result of changing atmospheric pressure on the surface. One of the largest **barometric wind caves** in the United States, its outstanding feature is **boxwork**, a sort of irregular honeycomb structure with delicate calcite fins separating the cells. Other wonders include Calcite Lake, a shallow pond where thin sheets of

▶ Some national park caves may close in summer to protect maternal bat colonies and their vulnerable pups, or in winter to shield hibernating bats.

calcite float like lily pads. Gleaming white clusters of frostwork seem to grow best where air flows strongly. Gypsum crystals sprout like alien cotton candy. Helictites are found in such density that cavers call them helictite bushes. Among the many ranger-led cave tours is a candlelight tour and an adventurous four-hour wild cave tour that gives enthusiasts ample opportunity for crawling.

5 Lava Beds
NATIONAL MONUMENT
California

Limestone caves are created by water—first hollowing, then filling—over long silent underground years. Lava tubes are at the opposite extreme; they are the products of short-lived violence. Molten lava, flowing in rivers from a volcano, hardens on top but continues to move beneath the surface crust. When the flows ebb, the lava drains away, leaving tunnels that can be miles long. Located in Northern

California, Lava Beds National Monument features the highest concentration of lava tubes in North America; more than 800 have been counted. They were created over the past half a million years by periodic eruptions of the Medicine Lake volcano, a huge shield volcano. The **lava tube caves** range from crawling size (claustrophobes beware) up to 60 feet in diameter. Some have beautiful ice formations, notably a large frozen waterfall in Crystal Ice Cave. Others, in sections where collapsed roofs let in light, protect gardens of ferns and other plants that would not survive on the surface. Pictographs, animal bones, and in one case two human skeletons tell of past human habitation. Most caves in the park are open to the public. More than two dozen have been developed with trails to make visiting easier, but bring a flashlight (or two), and a hard hat is recommended. Ranger-guided tours are offered in the summer.

6 Pinnacles
NATIONAL PARK
California

Talus caves occur when boulders and blocks of rock accumulate in chasms, sometimes forming a ceiling as, over thousands of years, water slowly erodes deep tunnels beneath. There are few known such caves within the park system, making Pinnacles' two prime sets—Balconies in its west and Bear Gulch in its east—even more fascinating. No roads connect the two sides of the park, but both cave collections can be explored via the trails and stairways built in the 1930s by the Civilian Conservation Corps. Although no spelunking skills are necessary, preparation is still key for cave explorers. Rangers recommend both a headlamp and boots due to darkness and the potential for standing or flowing water and sometimes uneven surfaces as you navigate narrow paths, tight crevices, and, naturally, giant boulders. Like much of nature, caves are vulnerable, active, and sometimes subject to closures. In the case of Pinnacles, this can be due to bat sensitivity, flooding, vandalism, or disease transmission. Bear Gulch is home to a large colony of big-eared bats, which limits its opening to seasonal service. Park staff recommend checking online ahead of travel.

7 | Great Basin
NATIONAL PARK
Nevada

Hidden in Nevada's Great Basin National Park, **Lehman Caves** is a relative boutique among caves: Its entire known length is about two miles. It makes up for its small size with an unusual richness of **speleothems,** however, and a few special treasures. Chief among its rarities are cave shields, large

Guided tours are required to enter Lehman Caves in Great Basin National Park.

disk-shaped structures resembling flat clamshells, their two matching halves separated by a hairline crack. Standing out from the walls at various angles, they become draped with flowstone of their own making. Lehman has more than 300 of them scattered among a superabundance of dripstone, flowstone, columns, draperies, popcorn, helictites, ribbons, and more. Despite the crowded appearance, the jumble of speleothems is remarkably harmonious. The park offers only ranger-guided tours, a highlight of which includes the Grand Palace and its famous shield called the Parachute. The cave itself once played Mars in the comically bad sci-fi movie *Horrors of the Red Planet.*

8 | Oregon Caves
NATIONAL MONUMENT & PRESERVE
Oregon

Hidden beneath the lush old-growth rainforest of southwestern Oregon, the Oregon Caves are unusual for being set in **marble.** The rock began as limestone but was metamorphosed by heat and pressure to become the smooth-grained, white material of classic sculpture. The 1.5-hour tour covers only half a mile, but it is rated moderately strenuous because the route includes some 500 stairs, many of them steep and uneven. Highlights include Angel Falls, a drapery formation that glows under UV light; Paradise Lost, the cave's most decorated room; and a spiral staircase created by an ancient waterfall. In the summertime, a strenuous three-hour caving adventure to undeveloped areas serves as an introduction to caving—how to move through a cave without damaging either yourself or the underground environment. It can be messy work: Think wet mud, loose rocky slopes, and water dripping from the ceiling. Lots of fun for the right people.

9 | Timpanogos Cave
NATIONAL MONUMENT
Utah

Interestingly, before entering Timpanogos, visitors must climb up. The 1.5-mile access trail to

Timpanogos ascends more than 1,000 feet above American Fork Canyon in Utah's Wasatch Mountains, affording a grand vista most of the way. Then, into the intricate dark it goes. Three main chambers are accessible only by ranger-led tours: **Hansen Cave, Middle Cave,** and **Timpanogos.** The most prominent feature is a tapered flowstone speleothem hanging from the ceiling called the Great Heart of Timpanogos. The cave is also known for its uncommon abundance of helictites (calcite formations that branch, curve, or spiral, seemingly in defiance of gravity) and its rare natural coloration—greens and yellows caused by nickel in the calcite. Because snow covers the trail for much of the year, the cave is open only in warmer months. The park usually recommends that you set aside three hours for the round-trip hike and cave tour.

10 | Cumberland Gap
NATIONAL HISTORICAL PARK
Kentucky, Tennessee & Virginia

Located near where Kentucky, Tennessee, and Virginia meet, **Gap Cave** offers a historical element in addition to the geological. A classic solution cave, its features include the 65-foot Pillar of Hercules and the pearly white Frozen Niagara. The ranger-led tour begins with a mile-long walk on the old Wilderness Trail, pioneered by Daniel Boone along the older Warrior's Path, a Native American route to the hunting grounds of Kentucky. After the Revolutionary War, water from the cave powered the mills of the new town of Cumberland Gap. Inside the cave, a rock face bears signatures of Civil War soldiers from both sides who sheltered here. It was once called Cudjo's Cave, after a character in an anti-slavery novel of the same name (Cudjo had escaped enslavement but was killed in the cave by Confederates). Today, tours are led by lantern light to enhance the historical atmosphere.

Cave Dwellers

Often feared, but mostly just misunderstood, nearly four dozen different species of bats make their homes in our national parks. The only mammals that fly, bats have carved out an integral role by eating insects, pollinating plants, and spreading seeds, making them a hardworking, albeit behind-the-scenes, part of the parks' ecosystems. That's not to mention the allure of their bucket list–worthy flights at sites like Carlsbad Caverns National Park. Because female bats typically have only one pup per year, a relatively slow rate of reproduction, bats are especially vulnerable to the impact of disease. White-nose syndrome—a fatal infection in cave-dwelling bats that has already reached more than half of the United States—is caused by a fungus that can be spread by humans. Since learning of the disease and its ability to quickly wipe out populations, the Park Service has partnered with conservation groups and government agencies to research ways to slow its impact. Guidelines vary from park to park but may include cave closures and gear decontamination efforts—all worth the effort to save these winged wonders.

WETLANDS & SWAMPS

More water than land, swamps are beguiling territory. The South claims the best, where dense canopies of cypress and tupelo shelter watery mysteries. Shy creatures disappear behind curtains of Spanish moss. Chorusing birds, frogs, and insects make a happy racket. And then there are marshes, estuaries, mires, bogs ...

1 Big Thicket
NATIONAL PRESERVE
Texas

The word "thicket" barely hints at the biological diversity in the scattered units of this **biosphere preserve.** Forests of pine, cypress, oak, and other hardwoods mingle with meadows, swamps, bogs, lazy rivers, and even some Southwest desert. It's an East-meets-West sort of place, where prickly pear cacti encounter blackwater swamps and midwestern plains find themselves unexpectedly close to the Deep South. Scientists credit cool conditions during the last ice age with compressing different habitats closer together. This fusion leads to unusual biological wealth exemplified by the preserve's status as an international biosphere reserve and a globally important bird area. Bird-watchers have a lot to look at with numerous species in the area. The preserve includes three Texas

Paddling Trails and more than 40 miles of hiking trails. Walkers will like the Turkey Creek Unit, particularly the short Pitcher Plant Trail, named for its carnivorous botany, which blooms yellow in springtime. The curious might also want to probe a baygall: a highly acidic type of bog named for the commonly found sweet-bay magnolia and gall-berry holly and said to be the most dense, impenetrable habitat in the Big Thicket.

2 Everglades
NATIONAL PARK
Florida

Top wetland honors belong to the Everglades, at the southern tip of Florida. Here America dips its toe into the exuberant life of the subtropics. There's no topography to speak of—the park's highest point is just a few feet above sea level—yet the Everglades presents a stunning diversity of plants and animals. Foremost are the marshy rivers called freshwater sloughs, which make up the famous **River of Grass** that channels water to the sea through fields of saw grass and other marsh vegetation. Sloughs are punctuated by forested islands called hammocks. Cypress trees spread their graceful, buttressed roots in shallow water. Mangrove forests thrive in the tidal zone where fresh water mingles with seawater. A tangle of channels, bays, estuaries, and small islands complicates the seashore, providing home to a richness of critters. Unfortunately the wetlands have become severely threatened, and in recent years the park has embarked on extensive restoration efforts. Everglades features some remarkable (and accessible) trails and boardwalks that are ideal during the dry season: Anhinga, Gumbo Limbo, Pahayokee, Mahogany

Melting glaciers formed Gull Pond, one of the numerous kettle ponds located within Cape Cod National Seashore.

Hammock, and others. Better yet, explore by boat. Excellent water trails allow seasoned canoers and kayakers to move as smoothly and quietly as alligators, and to stay overnight thanks to the park's system of elevated tent platforms called chikees.

3 | Cumberland Island
NATIONAL SEASHORE
Georgia

Accessible by ferry, private boat, or kayak, this barrier island, the largest along Georgia's coastline, is known for its **historic sites** and miles of open sand beach. The

wetlands tucked behind the dunes and maritime forest offer significant rewards, including a chance to see river otters, raccoons, mink, manatees, and even bottlenose dolphins in addition to the full catalog of **coastal wetland birds,** notably pelicans, herons, shorebirds, gulls, and terns. One of the best areas for wildlife-viewing opportunities lies between the old cemetery near Dungeness (ruins of a Carnegie family estate) and the beach, where a boardwalk skirts the salt marsh. Most of Cumberland is a quiet place. The heart of the island is a restricted wilderness area, so vehicles are limited to park staff and

the few private residents. Instead, hike the 50-mile-long trail system or bicycle along designated roads through forest draped in Spanish moss. The island has both wilderness and developed campsites for people wishing to overnight.

4 | Cape Cod
NATIONAL SEASHORE
Massachusetts

The sea constantly works at the land. In 1605, the French explorer Samuel de Champlain could sail his ship in what is now the **Nauset Marsh** in Massachusetts. Since then, the Atlantic Ocean has built an intervening barrier of sand, the Nauset Spit. Seawater still pours in and out with the tides, but the entrance has grown smaller. The Fort Hill Trail offers good access points for viewing the marsh, as does the Salt Pond Trail, which skirts one of Cape Cod's many **kettle ponds:** depressions created by huge blocks of ice left by retreating Ice Age glaciers. Once a body of fresh water, Salt Pond is now fed by the same tidal flows that support the teeming life of Nauset Marsh.

5 | Big Cypress
NATIONAL PRESERVE
Florida

Big Cypress shares many of the natural features of its more famous neighbor and partner, Everglades

National Park. Water originating from Lake Okeechobee, augmented by thunderstorms, hurricanes, and storm surges along the coast, flows through the parks. The preserve, measuring 729,000 acres, is a big place but its name honors the fragile and threatened nature of these unique **wetlands.** Most of the big **bald cypresses**—meaning the old and tall ones—were felled more than half a century ago, long before the preserve was established in 1974. But there are several large stands of bald cypresses—hence the "big" in the park's name—and the forest covers a lot of ground. Given time, its relatively young trees will achieve distinction. Meanwhile, the symphony of the **swamps** plays in full richness here. Anhingas spread their wings to dry as alligators, inscrutably patient, wait for opportunity. Swamp rabbits, raccoons, bobcats, deer, river otters, and even critically endangered Florida panthers go about the mysterious business of being furred creatures in a scaly world. Birds are everywhere, some flaunting their presence and others moving secretively through the brush. For visitors, two driving tours on the south side of the preserve provide a good sampling. For a full immersion into the watery wonderland, a number of paddling trails lead quietly to the island-studded coast.

6 Timucuan
ECOLOGICAL & HISTORIC PRESERVE
Florida

Three Florida rivers—the St. Johns, Nassau, and South Amelia—converge on a tangle of creeks between Jacksonville and the Atlantic Ocean. The result is 46,000 acres of **forest, grassland,** and **coastal wetlands** tucked behind a line of barrier islands. A handful of parks protect the islands and adjacent land. In addition to its watery wilds, Timucuan offers a 16th-century French fort, a 19th-century plantation, and a 20th-century farmstead turned natural area. Despite the surrounding urban development, the preserve is a remarkably good place to spot wildlife, with many unique habitats. Brown pelicans sail overhead in graceful phalanxes. Black-crowned night herons haunt the marsh plants along with egrets, ibises, and other wonderfully named avian species: semipalmated plovers, greater yellowlegs, roseate spoonbills, painted buntings, pileated woodpeckers, anhingas, whimbrels, godwits, and willets. The park offers numerous coastal saltwater paddling routes for canoes and kayaks, including portions of the

Swamp Gas

Swamps are strange places. Not quite water, not quite land, they lie in a twilight zone of bizarre possibility. Legends tell of unexplained and spooky things prowling the murky damp—things swallowed up only to be revealed again in altered form. Enter the phenomenon called swamp gas—or, more lyrically, fox fire, will-o'-the-wisp, corpse candle, jack-o'-lantern, and ignis fatuus. The names refer to dimly visible, ghostly lights described for centuries but never well explained. We know that swamps, like those found in Georgia's Okefenokee National Wildlife Refuge, produce methane gas from decaying organic matter. Poke the muddy bottom and a nose-wrinkling whiff of it, mixed with the odors of decay, will escape. Methane is flammable. It burns easily with a yellow or blue color. This would explain the lights except for the igniting part. Is there a natural force capable of torching methane on calm, steamy nights in the bayous? Or is it something else?

Bald cypresses thrive in the seasonally flooded bottomlands of Congaree National Park.

Florida Sea Islands Paddling Trail. Nonboaters can enjoy the natural scene from trails, viewpoints, and bird observation platforms.

7 | Congaree
NATIONAL PARK
South Carolina

Technically a **floodplain,** not a swamp, this park occupies a bottomland near the junction of the Congaree and Wateree Rivers, on the coastal plain of South Carolina. With an elevation around 100 feet, and as many miles from the ocean, seasonal water tends to pool here under the big trees that rely on it for good growing conditions. The Congaree River floods the relatively flat landscape 10 times a year, on average, bringing a wealth of nutrient-rich silt that remains once the waters recede. Floodplain forests were once common in the Southeast but today they are scarce. Congaree happily demonstrates the value of a flooded forest to those who paddle through the old-growth bald cypress and tupelo trees when the waters are high. A 2.6-mile elevated boardwalk, standing six feet above the ground, provides access in any weather, and even when the park is not flooded, there are creeks, oxbow lakes, and two rivers to explore. The trees are tall, commonly around 100 feet, with six national champions, including the tallest loblolly pine (169 feet), water hickory, and swamp tupelo. Birds are abundant as seasonal migrants flood through like the water. Deer, bobcats, river otters, opossums, and raccoons are of interest to all. Snakes, biting insects, and spiders demand their fair share of attention here, too.

8 | Indiana Dunes
NATIONAL PARK
Indiana

If remoteness from human impact gives a wild area extra significance, is the opposite extreme also true? A natural system functioning in the shadow of a great city as if it existed in some distant roadless region is arguably a treasure. It stands as an example and inspiration not only to hardy adventurers but to the millions who can experience the place with a Sunday picnic. The wetlands of Indiana Dunes, which once earned the name **Great Marsh,** have faced depletion, but restoration is a primary goal of the park, which abounds in biodiversity. Trails at Cowles Bog—so diverse that it has been designated a national natural landmark—and Pinhook Bog help you to appreciate what exists and to see the rebirth of the marsh's fens, sedge meadows, and wet prairies. Tag along on a ranger-led hike to witness this especially fragile habitat with a knowledgeable guide.

▶ *Jean Lafitte National Historical Park and Preserve is part of the Atchafalaya National Heritage Area, which stretches across 14 parishes in southern Louisiana and includes America's largest freshwater swamp.*

9 | Jean Lafitte
NATIONAL HISTORICAL PARK & PRESERVE
Louisiana

Barataria is a 26,000-acre bayou-riddled tract south of New Orleans on the shores of Lakes Cataouatche and Salvador. A Choctaw word, "bayou" describes a very slow-moving body of water, usually in a former river channel or lake, and commonly surrounded by swamp or marshland. Practically speaking, a bayou is a way to get deeper into a swamp and, one would hope, out of it as well. A good start at exploring Barataria is to walk the half-mile, accessible Bayou Coquille Trail It starts on dry ground, leads gently through stands of live oak and palmetto to a bald cypress swamp, and ends at a floating prairie. Hurricane Ida damaged parts of the Barataria in 2021, but most of the trails have now reopened. The intrepid can paddle a pirogue (which elsewhere might be called a canoe; the Louisiana variety has a flat bottom) on one of several water trails meandering through bayous, canals, and trenasses. Warning from the park: At twilight, it can be hard to converse over the din of celebrating frogs.

10 | Glacier
NATIONAL PARK
Montana

Glacier National Park in northwestern Montana has the odd **fen**—odd not only because fens are not common in the northern Rockies, but also because they are unusual among wetlands. Fens are essentially soggy sponges of peat that offer very low nutrient levels to the plants that live in them—cotton grass, buckler fern, blueberries, stunted shrubs, and sphagnum moss. And if you're a small insect, beware the murderous bladderwort. Those little tendrils with their floating balloon structures are the plant's roots; it has no others, and they are voracious. Brushing the trigger hairs will cause them to inflate suddenly, sucking little bugs to their doom. In Glacier, the drama unfolds at McGee Meadow on the Camas Creek Road, sometimes attracting moose.

10 BEST PARKS FOR

GLACIERS

Not many things move mountains. Glaciers do. Ponderous and powerful, glaciers can reduce them to rubble. Everything about a glacier is spectacular. The ice groans and cracks, revealing its deep blue interior. Great slabs calve with mighty roars. Clear rivulets cross its surface while dark, silt-laden subglacial rivers pour from its snout.

1 | Glacier Bay
NATIONAL PARK & PRESERVE
Alaska

We begin with an obvious choice, Glacier Bay in southeast Alaska, where **more than 1,000 glaciers** grind their way to sea level, giving passengers of cruise ships and tour vessels thrilling close-up views of towering ice walls and gleaming slabs crashing into the bay. In such a dynamic landscape, climate change and its effects on environment are a natural subject of conversation. In 1794, Capt. George Vancouver described no bay at all, just an unremarkable ice-bordered recess five miles deep. When John Muir visited in 1879, the ice had retreated northward, leaving 48 miles of open water. The glaciers have continued to shrink ever since. To see the park, the great majority travel by water, in everything from giant cruise ships to sea kayaks. They come for the glaciers,

but also the extravagant mountain scenery and wildlife. Orcas, humpback whales, minke whales, harbor seals, Steller sea lions, and sea otters are all present. Kayakers and hikers interested in camping can arrange with the Bartlett Cove day-tour boat to be dropped off at selected locations.

2 | Wrangell–St. Elias
NATIONAL PARK & PRESERVE
Alaska

Alaskans like to point out that **Malaspina Glacier** would cover the state of Rhode Island. Malaspina is truly huge, about 28 by 40 miles and in places almost 2,000 feet thick. Its size is symbolic of the vast rock-and-ice wilderness of which it is a part. Wrangell–St. Elias and Glacier Bay National Parks together with Canadian neighbors Kluane National Park and Tatshenshini-Alsek Provincial Park compose a

24-million-acre contiguous protected World Heritage site. Malaspina is a piedmont glacier, the largest of its type in North America. "Piedmont" refers to its forming in an open area unrestrained by mountain walls. It can't be seen in its entirety from closer than orbit. From satellite photos, it is possible to see how it flows through the St. Elias Range and spreads out on the coastal plain in a vast sloping disk. In fact, space photos are the only way most of us will ever see it. Access in this exceptionally remote part of the world is difficult in the extreme.

3 | Great Basin
NATIONAL PARK
Nevada

You won't find much ice in **Wheeler Peak Glacier**—in fact, it's the lack of ice that makes it noteworthy. Nor is the size outstanding; it measures

A boat is dwarfed by Margerie, a tidewater glacier in Glacier Bay National Park and Preserve.

about two acres. The glacier occupies a cirque at about 11,500 feet elevation on the shaded northeast side of Wheeler Peak. High quartzite walls rise above a steep talus slope ending in a jumble of sharp-edged boulders. There it is: a rock glacier. Below the rocks, in the spaces between them, there is ice. It freezes, thaws, expands, and contracts depending on the weather. It pushes and lubricates the rocks and slowly, like a more conventional glacier, the whole mass moves downslope. Getting there involves taking a first-rate scenic road, the Wheeler Peak Scenic Drive, which climbs nearly 5,000 feet from the valley bottom to a cool alpine basin. From there, a pleasant two-mile trail leads to the glacier and an added bonus: Just shy of the glacier stands a magnificent grove of bristlecone pines, many of them more than 5,000 years old. They have weathered the millennia, recorded the centuries in their gnarled shapes, and witnessed the transient nature of glaciers.

4 | Mount Rainier
NATIONAL PARK
Washington

Put high mountains together with wet climates and glaciers will happen. Mount Rainier gets a lot of precipitation. The annual average snowfall, measured at Paradise, on the mountain's south side, is

Exit Glacier in Kenai Fjords National Park flows out of the Harding Icefield.

640 inches. (The record is 1,122 inches, more than 93 feet, received during the winter of 1971–72.) If you consider that Paradise is 9,000 feet below the summit, and higher elevations usually attract more snow, the numbers get more impressive. Rainier supports **25 major glaciers,** the greatest single-peak glacial system in the lower 48 states. Of these, **Nisqually** is the most easily accessed and, being close to the park's main visitor center in the popular Paradise area, the best known. It has been monitored since the mid-1850s. During that time, the ice has advanced and retreated several times. Currently, like glaciers all over the world, its growth is negative, meaning that more ice melts in the summer than the glacier

receives in winter—even a Rainier winter. From the Paradise Henry M. Jackson Memorial Visitor Center, Nisqually Vista is an easy 1.2-mile round-trip hike. Glacier Vista, along the Skyline Trail, offers a closer view among lush wildflowers and a broad mountain panorama.

5 | Grand Teton
NATIONAL PARK
Wyoming

Skillet Glacier is not large. Few visitors to the Tetons ever get to appreciate it from close range, and it looks more like a snowfield than a glacier. But there are good reasons to include it among the best, all related to your vantage point. Skillet clings to the steep east face of Mount Moran, a kingly sight from miles around. The glacier's name refers to its long-handled frying pan shape, although to some modern eyes it looks more like an electric guitar. The prime viewing location is the Oxbow Bend Turnout on U.S. 89/191/287, a spot along the Snake River just below the Jackson Lake Dam. Here, a slow meander in the river provides an entrancing foreground—sometimes mirror smooth, often busy with wildlife. On cool autumn mornings, the mountains float on an ethereal blanket of mist. The Willow Flats Overlook, a short distance up the same road to the north, is a worthy second choice. Instead of water, the foreground is an expanse of shrub willows that turn warm gold in the fall. Moose, elk, and bears commonly wander into the scene as swans glide past. Few glaciers can claim such a theatrical setting.

6 | Kenai Fjords
NATIONAL PARK
Alaska

Looking at glaciers from a distance is good; standing on one is better. **Exit Glacier** offers an up-close-and-personal encounter, easily accessible by road from the town of Seward. A short, paved walk from the parking area leads through a stand of cottonwood trees to a panoramic overlook of a living Alaska glacier that is falling in spectacular slow motion from the Harding Icefield. The ice is cracked by its rapid descent, rather like a river tumbling down a cascade. Crevasses glow deep blue. Melting creates fantastic shapes. Dark, silt-laden water rushes from the glacier terminus. A moderately strenuous mile-long side trail leads to the edge of the glacier for a side-on look at towering blue ice. If the outflow stream is not too high, it's also possible to cross the gravel outwash plain to the actual toe of the glacier. In addition to staying clear of overhanging ice, obey all posted signs as ice fall hazards are often identified. Those wishing to visit the glacier's birthplace can take the 8.2-mile round-trip Harding Icefield Trail (gaining approximately 3,500 feet on the ascent) for expansive views of the vast icy plain.

7 | Glacier
NATIONAL PARK
Montana

You expect glaciers to be a prominent feature at **Many Glacier** in Glacier National Park. But there is so much more. Three spectacular valleys branch upward toward the Continental Divide. Each boasts a string of icy lakes fed by cascading streams. The wildflower display in July is unsurpassed. The high peaks—Gould, Siyeh, Apikuni, and others—are the very definition of mountain gravitas. And then there are the glaciers, especially **Grinnell,** which is reached by one of the finest hiking trails the park system has to offer. The trail begins in a deep coniferous forest, skirts the clear green waters of Grinnell Lake, and climbs high up the Garden Wall to where the glacier sits beside a small lake in the cirque of its own making. The glacier is far smaller than it was just a few decades ago. By some estimates, Glacier National

Weighing Glaciers

How much ice is in a glacier? The answer is important to scientists investigating climate change, but it's not as simple as checking to see if a particular glacier is getting longer or shorter, or even thinner. Glaciers behave in complicated ways that defy simple, short-term measurements. The most useful number for climate study is what researchers call the mass balance, a comparison of annual snow accumulation (which adds mass) with annual melting and evaporation, or ablation (which reduces mass). Scientists derive accurate figures from periodic measurements of snowfall, snow depth, runoff, and rate of ablation throughout the year. But they can get an idea of a glacier's condition just by looking at it in late summer. If 60 to 70 percent of its surface area is not covered by snow—that is, if too much snow has melted from the ice during the summer—then the glacier is in negative balance and is losing mass.

Park is warming at twice the global average. Accounts say just 25 active glaciers remain, down from 150 in 1850. Naturalists make the point that Glacier was named more for the effects of ice—the hanging valleys, cirques, moraines, and other features—than for the glaciers themselves. The ice may disappear for a while, but the park won't have to change its name.

8 | Rocky Mountain
NATIONAL PARK
Colorado

If the value of something rises with scarcity, then **Tyndall Glacier** has become exceedingly treasured. Once a large alpine glacier, Tyndall survives now as a small cirque glacier in the shelter of Hallett Peak's steep north face. Hikers can get close to it by trekking to Emerald Lake from the Bear Lake Trailhead. Recommendation? Start very early. (And consider the free park shuttle during the height of tourist season.) This trail is one of the park's most popular hikes, for good reason. Emerald is the highest in a chain of small alpine lakes, strung together along Tyndall Creek beneath the sheer walls of surrounding mountain ridges. Among the scenic highlights of Rocky Mountain, the Bear Lake area ranks near the top, and people flock to it. Not all of the glacier is visible from the end of the trail. The canyon

▶ *While there are more than 660 officially named glaciers in Alaska,* **The Alaska Almanac** *estimates that the state has upwards of 100,000 total glaciers.*

steepens just beyond the lake, and mountaineering skills are needed to climb into the cirque itself. But the beauty of Emerald Lake is ample consolation, a blue jewel of a tarn ringed by stands of conifers. For a less crowded but much more strenuous hike, follow the signed trail for **Andrews Glacier** from the Glacier Gorge Junction, revealing not only one of the park's few remaining glaciers, but an emerald green pool at its base.

9 | Olympic
NATIONAL PARK
Washington

Olympic National Park boasts **200 glaciers,** most of them draped over the central peak, 7,980-foot Mount Olympus. Taken together, they form one of the largest glacier systems in the lower 48, a mass of ice explained in part by the Olympic Peninsula being the wettest place in the continental United States. It's not easy to get within arm's reach of a glacier here—hiking and climbing are required, as 95 percent of the

park is officially designated wilderness—but the park offers a fine opportunity to see both ends of the glacial process. First, check out the glittering panorama of Mount Olympus from Hurricane Ridge. Then drive to Hoh Rain Forest, where the Hoh River carries icy meltwater from the glaciers through giant trees and a rain-soaked wonderland of lush vegetation. The 17.3-mile Hoh River Trail climbs through temperate rainforest (some of the last remaining in the United States), montane forest, and subalpine meadows to Glacier Meadows on the shoulder of Mount Olympus. Above, **Blue Glacier** reaches down from the mountain's heights. A short trail (that gets longer each year the glacier retreats) leads to the glacier's toe, and the starting point for many ascents of Mount Olympus.

10 | North Cascades
NATIONAL PARK
Washington

The glacier-topped mountains in the North Cascades are volcanic, well watered, and steep, with names as dramatic as the setting: Fury, Terror, Despair, and Triumph. There's an old saying among climbers and hikers that in the Cascades, a step forward is a step up or down. Nothing is on the level. To see some of the complex's **more than 300 glaciers,** this means a lot of walking through big-tree forests

Glacier-draped Mount Shuksan rises above Picture Lake in North Cascades National Park.

on the valley bottoms, followed by hard climbs into alpine meadows above tree line. Views like the one from Whatcom Pass across the valley at **Challenger Glacier,** which covers much of the north side of Mount Challenger, are a happy reward for the long miles. But the

view of **Mount Shuksan** is at least as good, and it can be had from the Mount Baker Highway. Studded with sharp crags, dripping with glaciers, and reflected in the mirror-smooth waters of Picture Lake, Shuksan has become an iconic image of the

Pacific Northwest. Compared to the near-perfect volcanic cones in the area, Shuksan has a wilder look, more craggy, more mysterious. Several glaciers are visible on the western face. The most prominent is Upper Curtis and, below it, White Salmon.

BY LAND

The lush forests of the Appalachian Trail offer peaceful seclusion.

SHORT BACKPACKING TRIPS

In our national parks a two- to four-day backpacking trip into backcountry can deliver all the sensations of a full wilderness experience. The following hikes are more illustration than prescription. There's seldom a single magical "right" route. Use these as touchstones, not step-by-step blueprints, and be sure to check ahead for required reservations and permits.

1 | Yosemite
NATIONAL PARK
California

If Yosemite has a gentle side, it's **Wawona** in the southern part of the park. Compared to the High Sierra, these elevations are lower, crowds fewer, and the hiking season longer; it can stretch well into October. This is still majestic country, featuring granite-ringed lakes and tumbling streams, but the setting is more rolling forests of firs and Jeffrey pines than barren alpine of the higher mountains. To forge a 22-mile counterclockwise loop, start hiking from Wawona to Buena Vista Pass (9,300 feet) by way of Chilnualna Falls—a stunning series of foamy tumbles. Once you drop down from the pass you can savor serenity at a choice of lakeside campsites. Buena Vista Lake is a classic alpine cirque at the foot of Buena Vista Peak (9,709 feet). Royal Arch Lake is beautifully set against slabs of exfoliating granite. On the way back, either Johnson or Crescent Lake makes a great camp within a day's hike of the trailhead.

2 | Olympic
NATIONAL PARK
Washington

Long coastal hikes are rare in the United States, which is why the 23 wild miles of coastline between **Cape Alava** and **Rialto Beach** are so compelling, even in a park known for mighty forests and mountains. It's no stroll on the beach, however. Scrambling over headlands is hard, slippery work, while walking around them might mean waiting on a tide change and slogging through deep, coarse sand. But the superb scenery of the wild "Shipwreck Coast," with its mighty sea stacks, scads of seabirds, and colonies of sea lions is worth the effort. From the Ozette Ranger Station, start for the coast on a three-mile boardwalk through a boggy forest of Sitka spruce and western redcedars. At Cape Alava, scan the Pacific for migrating gray whales in the spring and fall. Farther south, look for middens—piles of shells several feet thick—and the so-called Wedding Rocks, covered in petroglyphs. After Yellow Banks, there's a four-mile run of slippery boulders and ankle-deep gravel— hiking poles are a big help. Permitted camping areas are scattered along the route, some right on the beach; check with rangers beforehand to see which ones allow driftwood fires. The end of the hike is near the point where Hole in the Wall, a massive arch cut into a headland that can only be passed through at low tide, comes into view. If the tide's high, hikers have no choice but to wait patiently and soak up the wild salt air.

3 | Rocky Mountain
NATIONAL PARK
Colorado

Gasping for oxygen is a way of life in this Colorado park where the *low* elevation is 8,000 feet. Still, with some acclimation, fit hikers can handle the 26-mile **North Inlet/Tonahutu Creek Trail Loop** that starts at 10,000 feet and tops out at 13,000—one of the country's greatest high-altitude backpacking trips. The hike circles monolithic 12,274-foot Snowdrift Peak in the park's remote southwest corner. On the first night, follow the North Inlet Trail from Grand Lake to July Campsite. Get an early start, because the trail stays in exposed tundra all day, and thunderstorms are always possible. The next day, switchback through subalpine forest and alpine tundra in the shadow of Flattop Mountain (12,324 feet), then drop down to a Tonahutu Area campsite. Complete the loop on the equally dazzling Tonahutu Creek Trail, where views

Grand Teton National Park gifts backcountry hikers incredible views at dusk.

take in Ptarmigan Mountain to the south, Cascade Mountain to the west, and shimmering glacier tarn lakes in between.

4 | Grand Teton
NATIONAL PARK
Wyoming

At just over 20 rigorous miles, the **Paintbrush Canyon–Cascade Canyon Loop** is epic in both size and scope. It hits many of the revered highlights of Grand Teton National Park, including iconic Jenny Lake, Lake Solitude, and Cascade and Paintbrush Canyons. Along the way, ever changing views present the gamut of backcountry scenery: dense forests, roaring waterfalls, sparkling glaciers, as well as panoramas of the unparalleled Tetons. Given its 4,100 feet of elevation gain, the trail is quite strenuous, and many choose to do it as a two- or even three-day hike. Check in first at the Jenny Lake Ranger Station for the latest weather updates. Parts of the trail can maintain snowfields through July, so cleats and an ice pick may be prudent for trips before late summer. The String Lake Trailhead is your start (and end) point, and while the direction you take will be based on the campsite you reserve, most hikers prefer to take the loop counterclockwise. Keep in mind that this region is bear country; bear-resistant food canisters can be checked out from stations issuing permits.

5 Grand Canyon
NATIONAL PARK
Arizona

It has been said that you're not a Grand Canyon hiker until you've plunged from rim to river and back. Experienced backpackers can do it minus the mobs via the North Rim's **Nankoweap Trail,** which drops a dramatic 6,000 feet in 14 miles. Allow one (long) day for the plummet and two for the return ascent, and be sure to spend a day at the Colorado River. After passing through cool groves of aspen and ponderosa pine, the trail enters the canyon proper, corkscrewing down steep faces of Supai sandstone and redwall limestone. Look for cairns when the trail is less than obvious. Be sure to pause to enjoy the amazing canyon views of buttes, hoodoos, and red-rock temples. At 10.6 miles the trail meets perennial Nankoweap Creek. Follow it to the Colorado, where camping is allowed on a sandy beach near Nankoweap Rapids. On your layover day, explore the thousand-year-old granaries built into the cliffs 750 feet above the river by the ancestral Puebloan people—you'll see the

Packs for the Short Haul

Long-weekend backpacking trips are popular not only because of time constraints but also because they allow trail lovers to hike fast and lean without a monster pack. The ideal load hauler for such a hike is an internal-frame pack with a capacity of 3,000 to 4,000 cubic inches, which in itself saves weight versus an expedition-style pack.

Fit is crucial to the happiness and well-being of any backpacker, so it's a good idea to buy a pack from a mountaineering store; they'll find the perfect gear to fit a person's size and frame, then fine-tune it to suit them to a tee. A well-fitting pack rests comfortably on the hips, with the shoulders taking less of the weight, while a sternum strap keeps the load snugged to your center of gravity. Even a three-day weekend calls for some weight, so be sure the pack has a feature that helps deliver most of it to the hips—either built-in stays or a stiff plastic frame sheet. Otherwise it will feel like a potato sack is hanging from your shoulders.

Most backpacks are top loading and can be expanded vertically to accommodate bigger loads. A separate sleeping-bag compartment aids organization, as do zippered top and front pockets for quick-access snacks. A built-in pouch for a hydration bladder is a handy bonus. Most packs do not have side pockets because the packmaker assumes the hiker wants their arms free to swing—or to wield poles if ski touring. Don't sweat the actual pack material too much. Most packs are made of extremely strong nylon. Nylon, however, is not waterproof, so be sure to buy a waterproof shell or liner if one is not built into the backpack itself.

ruins from camp, just downstream. On your way out of the canyon, plan to camp at Nankoweap Creek, and be sure to top up your water bottles for the last 10.6 dry miles back to the rim—not only is the Nankoweap one of the most waterless trails at Grand Canyon, but it's also unarguably the most difficult of the park's named routes. Wise planning is not only worthwhile, for the Nankoweap it's imperative.

6 Great Smoky Mountains
NATIONAL PARK
North Carolina & Tennessee

You'll need four days for this one-way trip along the **Big Creek and Hemphill Bald Trails,** a route as diverse as the park itself. Many hikers opt for a late start on the Big Creek Trail so they can work up enough sweat to relish a dip in the creek's chilly Midnight Hole, a paradisiacal plunge just 1.4 miles from the trailhead. The next four miles to Campsite 37 are easy. On day two, hikers pick up the Gunter Fork Trail, ford Big Creek, and ogle Gunter Fork Falls as they hike six steep miles to Laurel Gap Shelter. Day three involves hiking eight miles to Campsite 39 by way of the Balsam Mountain and Palmer Creek Trails. Day four requires an early start for the nine-mile hike out through Cataloochee Valley, which affords surprising sights like elk grazing beside late

▶ *Isle Royale National Park's remoteness is a window back in time to the Great Lakes landscape before the impact of development. Before you zip your tent for the night, check the sky—Isle Royale is one of the few national parks where the northern lights may be visible.*

19th- and early 20th-century cabins and buildings, including Beech Grove School and Palmer Chapel. The hike concludes by following the Big Fork Ridge Trail to the Hemphill Bald Trail, finally ending at The Swag, a luxury lodge whose comforts are a balm to weary hikers.

7 Isle Royale
NATIONAL PARK
Michigan

A sense of adventure always adds an extra charge to a backpacking trip. In Isle Royale, backpackers aim to hike the length of the isolated island that gives the park its name by way of the 40-mile **Greenstone Ridge Trail** that traces its spine. For the most rewarding approach, start from the southwest corner and work

northeast. This way the hike begins in deep woods—birch, aspen, maples—and climbs gradually into mixed conifer forests. The payoff comes when the trail breaks out of the woods for amazing views of Isle Royale, Lake Superior, and sundry barrier islands, most notably from 1,394-foot Mount Desor. Moose are likely to be seen along the way— more than 500 of them hang out on the island, feeding on aquatic plants beside swamps and inland lakes, or simply ambling along the trail—and the howl of wolves is often heard at night. Campsites are scattered along the trail beside water sources, allowing the hike to be broken into three or four segments. For a bit of cold refreshment, many side trails lead down to Lake Superior. The Greenstone Ridge Trail terminates in a dead end, so hikers not wishing to backtrack drop down at the Three Mile Trail junction, which leads to Rock Harbor and boat transportation back to the mainland.

8 Petrified Forest
NATIONAL PARK
Arizona

What Petrified Forest's backcountry lacks in marked trails it makes up for in solitude, scenery, and stars. Because there are no designated campgrounds in this Arizona park, the only option for overnight stays is backpacking into Petrified

Forest National Wilderness Area. A semiarid landscape with elevations topping 5,500 feet, the park's backcountry has steep hills, narrow canyons, and plenty of preserved logs. Grab a free permit from a visitor center and drive to the iconic Painted Desert Inn at Kachina Point for parking and a trailhead. From here a dirt trail zigzags less than a mile before leaving you to forge your own path through the north unit of the wilderness area, where you'll meander around mesas and stand in awe of the striking orange and red **Painted Desert.** A few additional permits are given weekly for those who'd like to explore the whimsical formations of Devil's Playground—worth the wait if your schedule can be nimble. The Park Service requires hikers to camp at least one mile from roads, but staying mindful of the distance only adds to the sense of adventure.

9 | Glacier
NATIONAL PARK
Montana

A three- or four-day, approximately 20-mile hike in the southeast part of Glacier, starting at the **Cut Bank trailhead** and ending at **Two Medicine,** delves spectacularly into wild Continental Divide country with the option of summiting a couple alpine peaks. From Cut Bank, hike four miles to Atlantic Creek Campground. Be

The alpine backcountry on the Sahale Arm Trail in North Cascades National Park

sure to allow time to toss flies at the cutthroat trout in Medicine Grizzly Lake. Ambitious hikers can take the next morning (or, better, a layover day) to day-hike the spine of the continent at Triple Divide Pass. It's the only place in the United States where water drains not only east to the Atlantic or west to the Pacific but also north to Hudson Bay. Watch for bighorn sheep along the way. The next leg of the hike roughly parallels the North Fork Cut Bank Creek and heads south to Morning Star Campground, where more cutthroat lurk in nearby Katoya Lake. The third segment of the hike takes trekkers over Pitamakan and Dawson Passes, the latter on the Continental Divide; in fact, the trail parallels the divide for a stretch between the passes. If

time permits, stow your backpack at Dawson to climb Flinsch Peak (9,225 feet) by its south face, and then drop down to camp at No Name Lake. The hike out traces the shore of Two Medicine Lake, where many hikers snag a lift on a tour boat for the final few miles.

10 | North Cascades
NATIONAL PARK
Washington

Climbers, glacier aficionados, and weekend backpackers alike love to hike the **Sahale Arm Trail** to the alpine backcountry of North Cascades National Park and ultimately to the face of Sahale Glacier. Leaving from a trailhead on the Cascade River Road, the hike begins on the popular Cascade Pass Trail, a path through forests and wildflower-strewn meadows that leads to spectacular views of craggy peaks, Stehekin Valley, and—on a clear day—even Mount Rainier. At the pass (3.7 miles), most day hikers turn around, so the next six miles hiked on the trail to the glacier and the trail camp at its base afford wonderful solitude. Many hikers lay over here and savor the glorious high-alpine scenery, but the strong and ambitious use the day to proceed up the glacier (with ice ax and crampons) and summit Sahale Peak. It's a Class 4 climb, requiring the use of hands at times, to reach the 8,700-foot summit.

HORSEBACK RIDING

Although horses often travel to places you wouldn't be able to reach without their help, these mighty mammals are much more than transport. Horses can be extra eyes and ears, better than our own, their actions alerting riders to pay attention. They can also mingle with wildlife, getting closer than a human could do on foot. Apply for the proper permits and saddle up on these horseback routes.

1 | Olympic
NATIONAL PARK
Washington

Towering trees, cold rivers, damp trails through ferns. In this green-shadowed world of the Pacific Northwest temperate rainforest, Sitka spruce and western hemlocks grow 250 feet high. In their shade, maples wear shaggy beards of lichen and thick carpets of epiphytic moss and ferns. The forest is rich and layered with life, while high overhead mountain peaks gleam with glaciers, essentially sterile and seeming as remote as the moon. The contrast is particularly strong along the Hoh River, a beautiful place to linger among the great trunks. Yet it's a natural desire to get above it, up In the blue sky for a view from the top. That's exactly where the **Hoh Lake Trail** goes: 4,000 vertical feet to a series of high subalpine meadows. Wildflowers abound. Wide views of Mount Olympus dominate the southern quarter. The trail winds among sparkling lakes, sometimes tiptoeing along narrow ridges, before plunging back down into the forest and Sol Duc River to complete a magnificent two- or three-night trip. Olympic maintains 365 miles of trails for horse use. Note that special commonsense conditions apply to some areas (for example, no horse camping above 3,500 feet elevation, and camping only in designated stock camps on sensitive trails) along with the usual minimum-impact camping regulations.

2 | Rocky Mountain
NATIONAL PARK
Colorado

The Rockies reach their greatest elevation in Colorado. The park's popular **Trail Ridge Road** offers an opportunity to experience this realm of thin air, climbing above 12,000 feet on a dizzying traverse of the Continental Divide. A less crowded, yet truly memorable, way to feel that high is on horseback through the wilderness. Concessionaires run stables within the park, and many more are located outside its boundaries. After high snowdrifts have melted, you can choose from a variety of trip lengths, a majority of which set off from Sprague Lake. Beginners commit to two hours, while more advanced, saddle-hardened riders may be up for the challenge of a daylong ride. Follow **Glacier Creek Trail** to pass crystal lake waters, spotting wildflowers and wildlife along the way.

3 | Mammoth Cave
NATIONAL PARK
Kentucky

Mammoth Cave National Park has one of the world's largest cave systems. But for equine enthusiasts, the wonders above ground offer great appeal, too. More than 60 miles of backcountry trails north of the Green River provide diverse rides through rolling hills and river valleys, from wide, gravel paths perfect for beginners to sheer cliffs, narrow ledges, and muddy creek bottoms for the more experienced. One interesting path, the 12-mile **First Creek Trail,** follows ancient buffalo traces, historical routes that led animals and humans alike to the nearby river waters. As your route takes you from one creek to another, watch for Indigenous shelters, settler farmsteads, and views of Sugarcamp Hollow. The developed Maple Springs Campground has sites designed to accommodate horses, and many private horse-camping facilities operate just beyond the park's boundaries.

4 | Theodore Roosevelt
NATIONAL PARK
North Dakota

Theodore Roosevelt, America's great conservationist president, said his lifelong passion for wild country and its creatures began in the hills of North Dakota. "I heartily enjoy this life," he wrote, "and there are few sensations I prefer to that of galloping over these rolling limitless prairies." Whether riders stay on marked trails or take off across the country, there is reason to believe that horses were made for this sort of terrain. The 96-mile-long **Maah**

▶ *The call of the wild extends to horses as well, and feral breeds can be spotted at Theodore Roosevelt National Park, Assateague Island National Seashore, Cape Lookout National Seashore, and Bighorn Canyon National Recreation Area, among others.*

Daah Hey Trail connects the North and South units of the national park named after Roosevelt. For about half the distance, it parallels the Little Missouri River. Then, after visiting the undeveloped site of Roosevelt's Elkhorn Ranch, it crosses the river and loops through badlands and undulating prairie to the northern terminus. Early summer and autumn are the best seasons.

5 | Yellowstone
NATIONAL PARK
Wyoming

Thorofare sounds like a busy place, but it's the farthest you can get from a road in the lower 48 states. The area is named for Thorofare Creek, which offers a relatively easy way through northwest Wyoming's

Absaroka Range. The creek rises on one side of Thorofare Mountain in Teton Wilderness; the Yellowstone River rises on the other. The two streams take different directions from there but come together just inside the park on their way to a lush delta on Yellowstone Lake. The broad, flat-floored valley is rimmed by 10,000-foot mountains and blessed with a hearty complement of those things that make the Wyoming wilderness sing: bears, elk, moose, wolves, eagles, wildflower meadows, pine forests, a big sky, and not a motor or electric light to spoil the tranquility. It's country that makes you want to breathe deep and keep some portion of it in your heart forever. Safe to say, most visitors do. Go online for a list of backcountry campsites that allow stock, as well as information on reservations and permits.

6 | Capitol Reef
NATIONAL PARK
Utah

If John Ford hadn't been able to film in Monument Valley, he might have come here, and Western movies from then on would have had a different classic look. Capitol Reef National Park has enough varying backdrops to keep riders thinking they've passed through a series of extravagant film sets, most of which seem a bit exaggerated, as if the scenery designers got carried

Riders on horseback cross the remote Bechler River, which flows entirely within Yellowstone's boundaries.

away. Ride the **Halls Creek** drainage south of the Post Corral for a look at the reef—not a former ocean reef but a geological warp, more specifically a monocline, called the Waterpocket Fold. Wind and water have turned this great wall of rock into a strangely sculpted fantasy, poked full of holes and gussied up with windows, arches, and domes. Head for more open, rugged country via the Hartnet Road into Cathedral Valley, a fine place to imagine galloping cavalry regiments. On the park's west side, a recommended loop is to follow the **South Draw Road** (at the end of Scenic Drive) and return by way of **Pleasant Creek.** The park has no developed facilities for overnight horse use except at the Post Corral, but Capitol Reef does allow backcountry camping provided campers observe park regulations and restrictions; many trails and hiking routes are closed to horse use.

7 | Bryce Canyon
NATIONAL PARK
Utah

Horses have been a means for exploring Bryce Canyon for generations, and in 1931 the Park Service completed trails for this very purpose. When weather permits—typically May through October—there are two options for stock rides: guided or private excursions. Bryce Canyon's concessionaire Canyon Trail Rides leads two- and three-hour horse and mule tours into **Bryce Amphitheater,** a jovial, all-levels (ages 7+) look at the rock formations that make this park so distinct. Private rides (which must be scheduled prior to entering the park) begin at the Mixing Circle, then descend into the canyon on a dedicated trail before meeting up with the **Peek-a-Boo Loop,** a steep and stunning trail where hoodoos are sure to wow. Although sunrise and sunset are specifically spectacular times to view the park's dramatic colorations, stock camping isn't offered. Those interested in overnighting don't have to look far, though, as the U.S. Forest Service has adjacent campgrounds.

World-Famous Mules

Arguably the world's most famous riding trail, Bright Angel in Grand Canyon National Park is traversed atop mules, not horses. Head down, down, down into the wonderland of Arizona's Grand Canyon to spend a night or two in the bottom of the canyon at Phantom Ranch. Reservations are by lottery more than a year ahead, and for good reason. As breathtaking as the view from the rim is, experiencing the canyon from within is the stuff of bucket lists. Sure-footed mules have been carrying canyon riders for more than 100 years. The 10.5-mile excursion is strenuous and challenging, taking more than five hours each way, but the historic Phantom Ranch provides a steak dinner, comfy cabin, and a hearty breakfast before the return trip. Simply put, this trail ride—and its sturdy stock—are unforgettable.

8 | Great Smoky Mountains
NATIONAL PARK
North Carolina & Tennessee

Straddling the Tennessee–North Carolina border, Great Smoky Mountains is one of the most biodiverse locations on the planet, playing host to an estimated 100,000 species. It's also quite horse-friendly, with about 550 miles of trails, stock-approved backcountry campsites, and concessionaire stables. Riding trip options abound, including several sections of the **Appalachian Trail.** Start at one of the park's five camps set up specially for horses (reservations required). To get a full transect on a day ride, leave Cades Cove on the Anthony Creek Trail, which climbs through deep forest to top out on the Appalachian Trail. Return to Cades Cove on the Russell Field Trail for a loop of about 12 miles or continue to Gregory Bald and come down via the Gregory Trail (22 miles with a shuttle at Forge Creek Road). For the complete transit, consider the **Benton MacKaye Trail,** which runs nearly 100 miles; not heavily used, it is an adventure.

9 | Natchez Trace
NATIONAL SCENIC TRAIL
Alabama, Mississippi & Tennessee

Running from Natchez, Mississippi, to Nashville, Tennessee, the Natchez Trace was a 500-mile footpath established by Native Americans long before either city existed. It remained an important route through the period of European settlement. The footpath is mostly gone today, but five stretches of trail, totaling approximately 65 miles running near or atop the original, have been designated a national scenic trail for hikers and riders, cutting through habitats such as hardwood and pine forests, creeks, and swamp wetlands. The most scenic part is perhaps the 20-mile **Highland Rim** section near the Nashville end of the trail.

The Bradley Fork Trail in Great Smoky Mountains connects to the Appalachian Trail.

10 | Glacier
NATIONAL PARK
Montana

The unmatched splendor of Glacier is only hinted at from the roads, and the urge to venture deeper into its beauty strikes many. Park trails are generally open to stock, with limitations to minimize impact in sensitive areas. Guided rides are available at Many Glacier, Lake McDonald, and Apgar. Swan Mountain Outfitters, the park horse concessionaire, also offers a variety of route lengths. This leaves a lot of options, including one of the country's finest alpine trails, the **Highline,** which hugs the Continental Divide from Logan Pass north to Waterton Lake. Birds don't get better views of the mountains than hikers and riders

on this trail. The first few miles from Logan Pass are closed to stock, but horseback riders can join the route via the **Swiftcurrent Trail** from Many Glacier, a superb trail that climbs high over the divide before dropping a short distance to Granite Park Chalet, a classic mountain lodge set near timberline. For a longer journey, the Highline continues northward to Goat Haunt on gorgeous Waterton Lake. From there, one trail heads north to Canada (special border regulations apply to horses); another climbs west over Brown and Boulder Passes to a trailhead at Kintla Lake. A third trail turns off short of Goat Haunt, crosses Stoney Indian Pass, and heads down the valley toward Belly River. You can turn this into a loop that returns to Many Glacier.

DRAMATIC DAY HIKES

The best day hikes have an element of a quest or mission, which is why
so many of them climb to the top of a mountain or guide you to a significant landmark.
The point of each of these daylong hikes is the adventure of getting there—and the
inspiring views gained along the way.

1 | Mount Rainier
NATIONAL PARK
Washington

You might expect a lot from a place called Paradise. And when it comes to Mount Rainier, the national park's lauded historical area won't disappoint. Begin the **Skyline Loop Trail** at the Henry M. Jackson Visitor Center, where you'll cross (and likely photograph) picturesque steps inscribed with an apropos quotation from conservationist John Muir. The network of trails here offers something suitable for every ability, and while more strenuous than most, the 5.5-mile loop certainly rewards all effort at Panorama Point. Beginning on pavement, Skyline can be hiked clockwise or counterclockwise. Clockwise has an early elevation climb, while counterclockwise brings a more even start but a steeper descent. Along the way, subalpine wildflower meadows, cascading waterfalls, and many glaciers alternate with close-up looks

at Mount Rainier, Mount Adams, and the Tatoosh Range. If the clouds have cleared, you might even spot Mount Hood, 100 miles away in Oregon. Peak time for Paradise is July and August, with expansive wildflowers and extraordinary summit views.

2 | Acadia
NATIONAL PARK
Maine

Can a single hike distill all the beauties of coastal Maine's Acadia? The 5.5-mile **Jordan Cliffs Loop** comes close, offering a pleasant walk in the woods, a challenging ascent, and a glorious view from 1,373-foot Sargent Mountain, the second highest summit in the park. As a bonus, the trail provides timely assists in the form of some beautifully crafted stone steps and thoughtfully placed iron rungs. Start at the Jordan Pond House, follow the Spring Trail across the Jordan Cliffs Trail, and take the Sargent

Mountain East Cliffs Trail to the summit. Once there, just turn slowly and start identifying. On a clear day Maine's highest mountains, Baxter Peak and Katahdin, are visible far to the north. To the south are the Cranberry Isles, to the west Somes Sound, and to the east the granite domes of Pemetic and Cadillac Mountains. Worked up a sweat? Take a dip in Sargent Pond before returning via the Penobscot Mountain Trail.

3 | Wind Cave
NATIONAL PARK
South Dakota

The aboveground scenery is just as compelling as the subterranean at South Dakota's Wind Cave National Park—and what a remarkable contrast. After their stygian stint, visitors emerge to encounter big skies, wide-open spaces, abundant wildlife, and primal American prairies that once covered a third of North

Relax after a day of hiking at picturesque Sand Beach in Acadia National Park.

America. The 7.3-mile **Highland Creek/Centennial Trail Loop** starts on the Highland Creek Trail, which later intersects the Centennial Trail; the trailhead for the Centennial is the destination. The hike begins on mixed-grass prairie with distant views of the Black Hills and close-up views of a huge prairie dog colony. The trail then edges some lovely riparian stretches along Beaver Creek and Highland Creek and traverses ponderosa pine forests. Watch for bison wallows or, better yet, the beasts themselves—the park is home to more than 400 of them. But beware: Their docile appearance is deceptive; they can charge at surprising speed. Elk, coyotes, deer, and pronghorns are also plentiful.

4 Everglades
NATIONAL PARK
Florida

There's no better way to understand the Everglades as the so-called River of Grass than to step right into that river on a slough slog—a toes-on experience of the essence of the glades. Rangers lead slogs into **Shark Valley Sloughs** several times a week during the winter. Required equipment includes long trousers, socks, and a pair of old running shoes (sandals and boots will get sucked into the muck). Slog participants slosh their way through crystal clear water and dark black earth, out of which grow vast stands of saw grass dotted with little carnivorous bladderworts. In water sometimes greater than knee-deep, sloggers will make their way to a cypress dome, where cypress trees grow bright with bromeliads and orchids. Rangers will point out the odd gator hole or two along the way, but their creators aren't likely to be home during that time of year. Once a hiker gets the hang of slough slogging (like knowing where the snakes lurk), the world of the sloughs opens for exploration. The footing is dicey, the saw grass sharp, the going slow, but these are the Everglades, up close and very personal.

5 | Grand Canyon
NATIONAL PARK
Arizona

People come to Arizona to gape into the abyss of the Grand Canyon, to see the play of sunlight on rock, clouds on canyon, maybe to watch a rainstorm rush through, leaving a rainbow in its wake. Add to that the chance to eye the endless buttes and temples and side canyons and amphitheaters of rock. The best way to see it all, while avoiding the

▶ *A maximum of 300 hikers are allowed each day on the Half Dome Trail beyond the base of the subdome, a policy that reduces crowding, improves safety, and protects natural resources at Yosemite National Park.*

pounding of a canyon descent, is to hike the **Rim Trail** from Maricopa Point (where it becomes a dirt path) west to Hermits Rest—a distance of 6.7 miles. Take the park shuttle to the trailhead and pick it up at trail's end, or at one of six spots along the way. As the path undulates along the precipice, versus plummeting to the bottom, it reveals the park's (maybe the world's) most stunning panoramas—canyons within canyons, cauldrons of rapids far below. It isn't the most challenging hike, but it is one of the more memorable.

6 | Yosemite
NATIONAL PARK
California

It's mind-boggling to contemplate ascending any of the iconic rock faces in Yosemite National Park. When it comes to El Capitan, most of us will get no closer than sitting on a blanket in the meadow with

binoculars watching world-class rock climbers at work. But with will, determination, and proper footwear, many fit hikers can climb Half Dome. The challenging 14-mile round-trip hike begins at the Happy Isles Nature Center on the Mist Trail before joining the John Muir Trail and finally the **Half Dome Trail,** climbing 4,800 feet in the process. The last 900 vertical feet are on thin-air Half Dome itself. The crux is the final 400 feet, where strongly secured cables provide much needed assistance. Hang on and keep going. The view from the top is amazing, but for many hikers, besting the physical challenge of the hike and, perhaps, overcoming fears of exposure offer a greater reward. Permits, available by lottery for day hikers, are required seven days a week.

7 | Great Smoky Mountains
NATIONAL PARK
Tennessee

Few peaks anywhere are as beloved as Mount Le Conte in Tennessee. People return to it year after year. Le Conte lovers are slightly chagrined that it's only the third highest peak (6,593 feet) in the Smokies, so they're endeavoring to raise it—hence the growing pile of rocks at its summit. The 5.5-mile (one way) **Alum Cave Trail** is the shortest and most interesting path to the top. It features more open ridges than other routes, so the views are greater—though always best in the haze-free shoulder seasons. Its wonders include two varieties of rhododendron (rosebay and Catawba), peregrine falcons doing aerobatics at Inspiration Point, and forest habitats that include old-growth hardwoods. The cable handholds along the ledge just below Cliff Top might seem unnecessary on a dry day, but when the rock is slick or icy, you're grateful to have them. Alum Cave Bluffs, at mile 2.3, is a great slate overhang and site of a historic salt mine. The true summit is High Top with its can't-miss cairn. But proceed to Myrtle Point for the Great Smoky view that draws visitors back to Le Conte time and time again.

8 | Zion
NATIONAL PARK
Utah

Trail? What trail? Hiking the **Narrows**—the 2,000-foot-deep defile of fiery orange-red sandstone in this southern Utah park—means sloshing right through the North Fork of the Virgin River in water that can be waist deep, making it a thoroughly sensual experience. (Late summer thunderstorms can create flash floods, so be sure to check with the visitor center before setting out.) Wear old tennis shoes, seal valuables in a waterproof bag, and savor an entry into the deep and silent viscera of Mother Earth. The hike proceeds upstream from the Temple of Sinawava and continues as far as the heart and feet desire; every twist of the river reveals new subtleties of the vertical walls that frame it: sunset colors, striations of sedimentation, and plays of light. In places, the slot canyon constricts to 20 feet. Look high above and see tributary streams tumbling down

---

Trekking Poles

If you've never hiked with trekking poles, you have no idea what you're missing. Trekking poles are performance tools that can help any hiker move faster, more safely, and with better balance. With trekking poles in your hands (two are much better than one), you're suddenly a quadruped, a mountain goat. The *clack-clack* assist of two extra limbs on the ground helps you negotiate even extremely steep slopes without slipping. On side slopes they help keep you hugged snug against the mountain. A little slip that might otherwise be a cause for panic requires only the placement of a pole.

Poles also take pressure off the knees and hips. That adds up on a long hike, especially if you're carrying a pack—you can go faster or longer without aching joints. They particularly shine on long downhill hikes. By placing the poles out in front of you as you descend, you easily reduce the shock that normally gets delivered straight to your knees. They're nearly indispensable on tricky stream crossings, where any sane person goes looking for a pole anyway.

Trekking poles are adjustable to suit your height; that also means you can pack them away if you need your hands for technical scrambling. They come with padded handles and wrist straps, and some have shock absorbent tips. Those can be nice for long hikes on hard rock, but even basic poles will serve you well. Soon enough you won't want to hike without them.

Yellowstone isn't just for hydrothermal wonders. You'll also find fascinating petrified forests.

from side canyons. Hiking poles are highly recommended because the river—with its sandy banks, rocky stretches that require boulder hopping, and cool, at times swift, water—is a force to be reckoned with.

9 | Yellowstone
NATIONAL PARK
Wyoming

In northwest Wyoming, the **Specimen Ridge Trail** off Yellowstone Park's Northeast Entrance Road rises 1,600 feet in 3.5 miles without relief, so only hiking gluttons attempt this challenging hike, let alone reach the ridge top and the spectacular views it affords.

When standing high on the ridge, the hordes at Old Faithful become a cloudy memory. After the trail passes through some grassy sagebrush meadows and clears a forest of Douglas fir, it emerges into vast open country that's like a secluded grandstand overlooking Yellowstone's northeast quadrant. To the north is Lamar Valley—wolf country. To the south, Mount Washburn (10,243 feet) rises over central Yellowstone. In between are expanses of meadows, and the foreground is special, too. The specimens in question are petrified stumps of oak, redwood, birch, and maple trees, as well as conifers—this is one of the world's largest petrified forests.

Late summer wildflowers round out the show. As the trail continues, the views magnify. The ridge summit, 9,614-foot Amethyst Mountain, is at mile 11 of the hike, and if you've arranged a car shuttle, you can proceed seven more miles down to the Lamar River Trailhead. Or just turn around anywhere, anytime. The view is just as good going down.

10 | Haleakalā
NATIONAL PARK
Hawaii

Rangers at this Maui park probably get a little tired of the "like hiking on Mars" descriptions of the **Halemauʻu Trail,** which features a 1,400-foot plunge to the floor of Haleakalā's volcanic crater. The crater is bereft of vegetation, striated with red and black cinder cones, and certainly has an otherworldly look. But there's a lot more life here than on Mars. The early part of the hike meanders through native scrub forest that looks exactly as it did 2,000 years ago—a rarity in Hawaii, which has been overrun by non-native species. Just before the big plunge into the volcano, stunning views of the Big Island and Maui's north shore open up. Deep inside the crater awaits something nearly as alien as the landscape: astounding silence. Hōlua Cabin makes a reasonable turnaround point for this 7.4-mile, six- to eight-hour hike.

BICYCLE RIDES

National parks are blessed with opportunities for unparalleled biking: trips that can traverse a scenic desert, circle the rim of an ancient volcano, follow the path of a historic canal, or see the waters of the Atlantic beyond the sands of Cape Cod. The diversity and grandeur of the surroundings is awe-inspiring.

1 Crater Lake
NATIONAL PARK
Oregon

Perhaps because the window of opportunity is so very small, the desire to bicycle at Crater Lake is so very large. With an average annual snowfall of more than 43 feet, paved **Rim Drive** is usually closed until June. When the snow is cleared, the road reopens in stages, until around October when weather begins closing the road again. Riders who slip in through that gap will cycle beside America's deepest lake (1,943 feet), which dazzles guests with its rich blue waters shimmering within a six-mile-wide caldera (volcanic crater) created when Mount Mazama erupted and collapsed about 7,700 years ago. The 33-mile rim road, edged with fir and pines, creates a wonderfully appealing tour for riders who can handle some steep grades, high altitude, and numerous curves.

Beginners are pleased to finish a single circuit; experienced cyclists push themselves to pedal three laps—a complete century ride. On two Saturdays each September, the popular Ride the Rim event closes the park's East Rim Drive to motor vehicles, affording cyclists with 24 car-free miles.

2 Glacier
NATIONAL PARK
Montana

Considered one of the nation's most stunning *drives*, **Going-to-the-Sun Road,** an engineering marvel, is also one of the most impressive *rides*. Tackling the one road that bisects the park is demanding, and cyclists straining to propel themselves up and over 6,646-foot Logan Pass will wish their bicycle was equipped with cruise control. The payoff is in the sights: the crisp blue sky, the

Matterhorn-shaped peak of Mount Reynolds, the alpine flowers and fir trees, and especially a glance back to see the recently conquered thin black ribbon of road snaking around the steep hills. From border to border, the ride is roughly 50 miles, so plan on several hours for skilled cyclists to make the complete stretch. Remember that planning is imperative as peak season restrictions and weather closures limit route availability.

3 Acadia
NATIONAL PARK
Maine

Thanks to the vision and fortune of John D. Rockefeller, Jr., and the extraordinary efforts of road crews, access around Acadia was made far easier and much lovelier with the addition of 45 miles of stone and gravel **carriage roads** between 1913 and 1940. Designed for pedestrians, horses, and cyclists to pass through the island's most idyllic areas, the route came with a ban on motorized traffic that remains enforced today. Alternatively, the most popular activity in Acadia is traveling the 27-mile **Park Loop Road,** a route that tracks the island coast before slicing into the center of the park with a separate road leaping up toward 1,530-foot Cadillac Mountain. The loop road is relatively level while the ascent to reach the peak—and the commanding views of Maine's coast—presents steep grades. Since there is no shoulder, though, the road is only suggested for cyclists quite early or late in the day. Depending on the route selected, you may be sprayed by mist from the sea, dodging pedestrians, or giving wide berth to equestrian traffic. That's part of the beauty of Acadia—there's something for everyone.

4 Mississippi
NATIONAL RIVER & RECREATION AREA
Minnesota

Similar to hiking the Appalachian Trail from Maine to Georgia, it's actually possible to bike the length of the Mississippi River from its headwaters at Lake Itasca to the Gulf of Mexico via the **Mississippi River Trail.** Determined riders could tackle the 3,000-mile series of independent trails across 10 states on paved roads, atop levees, through the woods, and beside the water, but that would require quite a lot of effort. Instead, in southeast Minnesota, more than two dozen communities have come together with the national recreation area and a series of smaller regional parks to create a 72-mile bike path along the famous waterway. Beginning in Anoka, north of Minneapolis–St. Paul, the ride mirrors the mood of the river on a trail that is more remote and rural and natural. As riders pedal south, the landscape changes and so does the experience. The feel of the ride and dynamic of the river adopt a more urban appearance as it flows through residential and business areas near Minneapolis, introducing city sights like the Stone Arch Bridge and its view of St. Anthony Falls. After following the river's meander through the Twin Cities, the trail continues south to Hastings along a stretch of river that blends industry and nature. This ride should please cyclists looking for a country ride in the heart of the city, and a city tour that rolls across the country, and prepare them for the longer bike ride downriver.

5 Cuyahoga Valley
NATIONAL PARK
Ohio

Cuyahoga Valley has a number of superb trails, but for bicyclists the standout is a path once reserved for barge-towing mules. Dubbed

Rules of the Road

With bicycling comes a sense of freedom and independence that's only magnified on a ride in a national park. In reality, the rules of the road in most national parks are likely more stringent than when pedaling around town. Before heading out, check with rangers. While rules may vary, generally riders are expected to wear a helmet and high-visibility clothing, ride single file, have visible headlights and taillights if riding after dark, and stay to the right and ride with traffic. Recognize, too, that along with risks such as snow, ice, and sand, national parks can present unusual hazards, such as deer, elk, moose, prairie dogs, and an assortment of other wildlife blocking your way. If you see bison or bears on the road, rangers suggest you turn around. Ride safe, ride happy.

the **Ohio & Erie Canal Towpath Trail,** this 20-miler rolls beside the famous waterway, while additional miles of trails extend beyond the park into the Ohio & Erie Canalway. In addition to abundant shade and a wide berth, what makes this family-friendly ride so pleasing is its ease and the chance to enjoy a variety of sights and sounds—from marshes and covered bridges to boardwalks, woods, and historic structures like the circa 1836 Boston Store. For riders with waning energy, simply find a Cuyahoga Valley Scenic Railroad station and flag down a passing train (yes, train) that, for a few dollars, will take you and your mount back to the start.

6 | Theodore Roosevelt
NATIONAL PARK
North Dakota

Established to honor the area's importance to conservationist Theodore Roosevelt, this park in southwestern North Dakota combines the rugged beauty of badlands with the rich diversity of animal life. The park's South Unit is its most popular, with a visitor center, hiking trails, and the town of Medora drawing guests. For full immersion, choose the paved 48-mile **Scenic Loop Drive,** which winds through the region's rolling, colorful terrain and features many pull-outs with interpretive signage designating significant natural and

The Province Lands Bike Trail (p. 90) takes riders to Cape Cod's Herring Cove Beach.

historical parts of the park. Keep your eyes peeled for prairie dogs, wild horses, and, indeed, bison (the South Unit herd count has been known to top 400). Although the Scenic Loop is open to motor vehicles, it's easy to imagine Roosevelt's enthusiastic approval of the unencumbered panoramic views gained from behind handlebars.

7 | George Washington
MEMORIAL PARKWAY
District of Columbia, Maryland & Virginia

For 18 miles the multiuse **Mount Vernon Trail** meanders along the Potomac River from Theodore Roosevelt Island to George Washington's Virginia home at Mount Vernon, and even with more than one million annual users, it's well

worth contending with walkers, runners, and skaters just to experience the history and scenery it delivers. Depart from Theodore Roosevelt Island and a mile later ride toward Arlington National Cemetery by crossing below the Memorial Bridge, the symbolic dividing line of the Union and Confederacy. Shortly after, at Lady Bird Johnson Park is the LBJ Memorial Grove on the Potomac and a panoramic view of Washington, D.C. Clock on more miles and pedal past memorials to service personnel, Daingerfield Island, the charm (and conveniences) of Old Town Alexandria's shops and restaurants, the wetlands and hiking of Dyke Marsh, the remains of the Civil War–era Fort Hunt, and, three miles later, historic Mount Vernon. In other words, in just 18 miles riders can travel from America's present through its undeniably storied past.

8 | Everglades
NATIONAL PARK
Florida

Despite the name, you won't spy sharks as you tour the Shark Valley area of Everglades National Park. Teeming with alligators, turtles, deer, snakes, and birds of all types, this region was named for the Shark River, and for many it is the gateway to exploring the Everglades. Begin at the visitor center, convenient to Miami, and opt for the popular 15-mile **Shark Valley Tram Road.** Closed to private vehicles, the wide, flat loop has regular tram trips led by park rangers and naturalists. But for a self-paced ecotour, set out on your bike (or rent one from the area concessionaire) and enjoy the quite easy, not-quite-breezy ride. At the halfway point of the loop, the 70-foot Shark Valley Observation Tower provides sweeping views of the River of Grass marsh, as well as restrooms and water-filling stations. Keep in mind that the Florida climate can be stifling, so hydrate before venturing out and pack plenty of fluids, too.

9 | Cape Cod
NATIONAL SEASHORE
Massachusetts

Flexing its long arm 60 miles into the Atlantic and wrapping itself around the eastern coast of mainland Massachusetts, Cape Cod National Seashore continues to

▶ *The new Biscayne-Everglades Greenway, the first of its kind to connect two national parks, is a 42-mile paved loop tour through South Florida's natural, cultural, and historical resources.*

guard against development and vigilantly preserve its shoreline and sand dunes. Generally low and long and level, the terrain makes pedaling a pleasure, and its three short, but enjoyable, biking trails give riders exactly the kind of carefree riding any visitor expects to find on the cape. The **Nauset Trail** begins at the Salt Pond Visitor Center, which is the prime spot for park maps and information on programs. The entire ride is only 1.6 miles (double that for a round-trip). It leads through the woods and sometimes beneath a canopy of trees on a paved trail that slowly dips and twists like a low-grade roller coaster until, just after a long bridge, it arrives at Coast Guard Beach and the Atlantic Ocean. The easy two-mile **Head of the Meadow Trail** winds from the town of Truro to East Harbor. The 5.4-mile **Province Lands Bike Trail** features steep hills, sharp curves, and low tunnels in a loop that takes

in Beech Forest and Pasture Pond. Caution is called for along certain stretches of the path. The loop has several access points and spur trails.

10 | Saguaro
NATIONAL PARK
Arizona

Tucson is one of the country's most cycling-friendly cities. And it's here, surrounded by the Sonoran Desert's wildly distinctive landscape, that riders flock to the eight-mile **Cactus Forest Scenic Loop Drive.** Begin at the Rincon Mountain District's visitor center for a comprehensive immersion in desert ecology before you take off on the paved path. A steep downhill at the start gets the adrenaline pumping, which you'll need for the many hairpin turns and the challenging climb at the halfway point. Along the narrow route, rolling desert, several washes, and the wayfinding Rincon Mountains keep riders in awe. The wildlife can do that, too, but it's more likely you'll cross paths in the early morning hours. Almost bisecting the loop, the 2.5-mile Cactus Forest Trail can be added to the route by either the north or south trailhead. The trail is thin with sections prone to erosion, so bikers are encouraged, as always, to use discretion. Scenic overlooks and well-earned picnic areas add to the desert drive's appeal.

The Cactus Forest Scenic Loop Drive in Saguaro National Park is perfect for cars and bikes.

WHERE KIDS ARE WOWED

National parks make for the ultimate field trip. From cave exploration and a hands-on paleontology lab to witnessing a hatchling release and hiking a trail through history, the following outings are kid-approved for all ages, balancing safety and ease with a healthy dose of thrill. You're sure to stir up a sense of curiosity, opening the doors to experiential education.

1 Great Smoky Mountains
NATIONAL PARK
North Carolina & Tennessee

Strengthen the ties that bind while you eat, sleep, play, learn, and live in America's most visited national park. That's the goal of the popular six-day, five-night **Family Camp** at Great Smoky Mountains Institute at Tremont. Happy campers choose from a thrilling list of guided activities each day—hiking, biking, swimming, crafts, kids club, musical entertainment, and more—all with the backdrop of the Smoky Mountains. Registration includes activities, meals (many of which are themed!), and comfortable dorm lodging. It's common for Family Camp to sell out quickly, but you can still build your own adventure by staying at one of the national park's 10 easy-access, frontcountry campgrounds and participating in some of the many ranger-led programs offered each year.

2 Cuyahoga Valley
NATIONAL PARK
Ohio

All aboard! Follow the Cuyahoga River, named after a Native American word for "crooked," as a vintage **Cuyahoga Valley Scenic Railroad** railcar travels through the valley on a two-hour excursion. In addition to the chance of spotting eagles, deer, beavers, and more, passengers get a unique view of the rural and urban national park landscape between northeast Ohio's Akron and Peninsula communities. Have young explorers? Themed family-friendly trips, such as Riding With a Ranger and Trains, Tracks, and Trails, are short, interactive rides ideal for wee riders. From November through December, families can also book a train to the North Pole, which includes hot chocolate, elves, carols, and even a visit from Santa himself.

3 Mammoth Cave
NATIONAL PARK
Kentucky

Some cave tours are too long for children or require climbing lots of stairs or squeezing through tight (possibly scary) spaces. The **Frozen Niagara Tour** at Mammoth Cave is short (a quarter-mile round-trip, taking just over an hour) and easy enough for kids, yet offers unparalleled cave scenery such as Rainbow Dome (a spectacular display in size and color), Crystal Lake (a small natural pond visitors can also boat on), the Frozen Niagara flowstone (a rock formation, not a waterfall), and the Drapery Room (for more awe-inspiring formations).

Young folks can let their imaginations take over here, as formation shapes suggest animals, cartoon characters, and anything else their creativity can dream up. With this introduction to underground wonders, a child just might turn into a dedicated caver.

4 Yellowstone
NATIONAL PARK
Idaho, Montana & Wyoming

For many, a multigenerational trip to Yellowstone National Park is a family rite of passage. At 2.2 million acres, the country's first national park is **a natural playground** so rich

Mammoth Cave's Booth's Amphitheatre is named for actor Edwin Booth, who performed here in the 19th century.

with wonders that there's never a dull moment. Take, for instance, the park's countless active geysers, springs, and hydrothermal features, including, of course, iconic Old Faithful. Lake Yellowstone can be explored via kayak, rented boat, or guided tour, while anglers of all skill levels can enjoy a permitted fishing outing. Play the Quiet Game for ultimate wildlife viewing, which is quite popular in the Lamar and Hayden Valleys, where bison, bears, and wolves roam. In the northeast, Tower-Roosevelt Junction hosts a Pleasant Valley chuckwagon cookout with steak, beans, fruit crisp, and more. And while kids won't soon forget the singing cowboys—or any of the park's many adventures—a stop by Yellowstone General Stores will solidify the memories with a souvenir.

5 Badlands
NATIONAL PARK
South Dakota

What do you want to be when you grow up? After a trip to the **Fossil Preparation Lab** at Badlands National Park, you might just answer: "paleontologist." Here, at the Ben Reifel Visitor Center, watch and ask questions as park paleontologists and interns perform the tasks of identifying species, removing rock from fossils, and cataloging specimens in real time. The lab, which is open daily from mid-June to mid-September, has plenty of fossils to investigate—the Badlands is home to one of the world's richest fossil mammal beds from the late Eocene and early Oligocene epochs. After that, hike the family-friendly Fossil Exhibit Trail for a tactile experience outdoors.

6 Mount Rushmore
NATIONAL MEMORIAL
South Dakota

Kids have seen them in schoolbooks, on television, and in movies: four gigantic (we're talking 60 feet high!) presidential heads—George Washington, Thomas Jefferson, Theodore Roosevelt, and Abraham Lincoln—way up on the mountain, looking just as distinguished as they have for decades. Nobody can fail to be impressed by the first sight of Mount Rushmore through the Avenue of Flags (try to arrive in the morning for best light). To get a closer view, walk the half-mile **Presidential Trail,** which approaches the base of the mountain and passes the studio (open seasonally) where sculptor Gutzon Borglum worked. Consider renting audio wands or multimedia devices for a narrated tour of dozens of park stops, including fascinating details about the monument's history and creation.

Every Kid Outdoors

Families of fourth graders, get ready to get out! From September 1 through August 31 of a child's fourth-grade school year, the Every Kid Outdoors program offers free access to hundreds of national parks and federal recreational lands. (Homeschooled learners can register, too.) The process is simple: Print the coded pass from the site online, plan your trip, and hit the road. No matter where students live within the United States, they're within two hours of an included site— from spotting puffins in Acadia National Park to a forested trail walk in Portland, Oregon's Marquam Nature Park. To register and learn more about covered destinations, visit *everykid outdoors.gov.*

Ninety percent of Mount Rushmore's iconic granite sculpture was carved using dynamite.

7 | Great Sand Dunes
NATIONAL PARK & PRESERVE
Colorado

The tallest **dunes** in the United States are found in, of all places, Colorado, tucked against the east side of the Sangre de Cristo Mountains. Between the westerly winds and some unique geological circumstances, the main dune field has spread across about 30 square miles, with dune heights reaching above 700 feet. The dune field is not so much a hiking place as a "go out and have fun" spot. Kids (adults, too) are welcome to run up the sand dunes and slide or roll down. Certain precautions should be heeded, especially when the sand is scorching hot in summer, but no one is likely to hurt themselves if they fall down. To sand sled, or sandboard, consider renting slick-based boards from retailers in the San Luis Valley. Snow sleds and skis work only on very wet sand; sections of cardboard and saucers will barely budge. If the weather conditions have been too dry, the sand may be too soft for sledding, but the dunes are great to

Red foxes call both Yellow-stone and Indiana Dunes National Parks home.

explore any time. People are free to wander, provided they don't impact the vegetation. The summit of High Dune, the most common destination in the dune field, is a long mile-plus climb through the shifting sands. If it's flowing, nearby Medano Creek offers a great place to cool off afterward.

8 | Padre Island
NATIONAL SEASHORE
Texas

Summer is high time at Padre Island National Seashore, and it's not only because of the undeveloped beaches prime for swimming, camping, and shelling. From mid-June through August, when given the "go," hundreds of visitors will gather in the wee morning hours to watch **sea turtle hatchlings** make the brave journey into the Gulf of Mexico. The Division of Sea Turtle Science and Recovery at Padre Island (the only unit of its kind in the park system) works to monitor, rehabilitate, and protect five species of turtles, the smallest and most endangered of which is the Kemp's ridley. Since sea turtle numbers are too low to risk losses, all nests found in the park are moved to protected places. The nests are then given approximate "due dates," and the Hatchling Hotline (361-949-7163) keeps callers updated on open releases. Not all releases are public and public releases don't occur on

▶ *Park rangers happily field questions atop Gateway Arch National Park's 630-foot focal point. One frequent answer is that, yes, the arch is designed to sway and can withstand an earthquake, but under normal conditions you won't feel it budge.*

a regular schedule, but if you're able to catch one, the experience is extraordinary.

9 | Gateway Arch
NATIONAL PARK
Missouri

Cities captivate kids, and any child would jump at the chance to overlook St. Louis from a height of 630 feet. Gateway Arch National Park, known for decades as the Jefferson National Expansion Memorial, honors the country's westward growth. Its stainless-steel **arch,** the tallest monument in the nation, is a must-do, and all ages will enjoy the tram to the top. (Keep in mind that tickets sell out quickly.) The arch is just the start of all the park offers. Once you've been sky-high, head

underground for the free **Museum at Gateway Arch National Park,** showcasing 200 years of Missouri history. Grab a quick bite in the Arch Café or reserve the family-friendly Sunday brunch aboard the park partner riverboat cruise. You can also make a day of it by exploring the five miles of **paved paths** around the park, with a special stop at the 1839 Old Courthouse.

10 | Indiana Dunes
NATIONAL PARK
Indiana

On the southern shore of Lake Michigan, Indiana Dunes National Park is the site of many Midwesterner beach days. The vast size of the lake gives it all the perks of the ocean without the jellyfish to contend with. But there's plenty else to do here beyond the park's 15 miles of shoreline. More than 1,100 flowering plant and fern species may not thrill the wee ones, but five dozen butterfly species will certainly turn heads. Clocking just shy of a mile, the **Dunes Succession Trail** loop highlights the four stages of dune development, plus lessons and lore from West Beach dunes. The reward for climbing 270 stairs? Sweeping views of the Chicago skyline. Geocaching is also a popular pastime at Indiana Dunes, and you can learn more about the four types of treasure hunts offered on the park's website.

SUMMITS

Our national parks are full of peaks that can be reached by nearly anyone in good physical condition, given they have the proper gear and preparation. Whether you're walking up or mountaineering, challenge blends with beauty at these summits worth conquering.

1 | Denali
NATIONAL PARK & PRESERVE
Alaska

The Indigenous Athabaskans have always called the mountain that dominates Alaska **Denali**—"high one." Given the name Mount McKinley in 1896, this 20,320-foot summit, the tallest in North America, was officially renamed Denali in 2016. It's a formidable mountain, with much greater vertical rise (17,000 feet) than Mount Everest. More than 1,000 climbers attempt it annually (a special use permit is required), nearly 95 percent of them via the West Buttress route; about 70 percent succeed. It's a serious challenge that can take up to three weeks to accomplish. Climbers generally fly by ski-plane from Talkeetna—where the park ranger station is the hub for all climbing activity—to Kahiltna Glacier, where they camp at 7,800 feet, acclimate to the altitude, and begin to ferry loads to camps higher on the mountain. Most use lightweight plastic sleds to haul supplies up the mountain. Cold temperatures, strong winds, thin air, and glacier crevasses all present dangers—in winter, Denali is one of the coldest places on Earth. Even in May, prime time for climbing, it's commonly minus 50°F at the camp, which sits at 17,200 feet above sea level. The summit is spectacular for its degree of exposure and steep drop-offs, and, of course, for the view from the roof of North America.

2 | Olympic
NATIONAL PARK
Washington

Mount Olympus (7,980 feet) has all the attributes of a trophy summit but one—extreme elevation, which is good news for anyone who has difficulty adjusting to lofty heights. One of the world's most challenging sub-8,000-foot peaks, craggy Olympus serves up a long approach, climber-swallowing crevasses, snowfields that require crampons, and cold, unpredictable weather. The approach is 17.4 mostly gentle miles on the Hoh River Trail through pristine old-growth rainforest. Climbers then set up camp at Glacier Meadows, just below the alpine zone. The big day requires a headlamp-illuminated departure in the wee hours. Climbers ascend the icy moraine of Blue Glacier before traversing the glacier itself and climbing Snow Dome. Then it's a return to the upper part of the glacier, a narrow col, and a scramble up an exposed ridge to the summit, 3,765 feet above camp. In clear weather, the view is awesome: the Strait of Juan de Fuca, Vancouver Island, Mount Rainier, the distant snowy peaks of the Cascades, and a thousand surrounding shades of green.

Climbers trek to the summit of Denali.

3 | Rocky Mountain
NATIONAL PARK
Colorado

Sure, anyone can drive to the Continental Divide on Trail Ridge Road, but a person gets extra satisfaction by hiking up to the backbone of our continent. In north-central Colorado's Rocky Mountain National Park, this hike to the broad summit of 12,324-foot **Flattop Mountain** begins at picturesque Bear Lake and requires no climbing skills at all. Just keep putting one foot in front of the other for 4.4 miles while ascending 2,849 feet. But stop often along the hike to enjoy spectacular views of Longs Peak (the park's highest mountain at 14,259 feet) in the distance, with forested slopes, glacier-carved valleys, and alpine lakes nestled in between. Wildlife is plentiful. The rugged, pointed summit of 12,713-foot **Hallett Peak,** the next Continental Divide mountain south of Flattop, is also easily seen. To reach that peak means continuing another 0.6 mile and ascending 400 feet or so. Getting to the top of Hallett does require a bit of boulder scrambling, but nothing too difficult. On its summit, a hiker will probably feel more like a "mountain climber" than on the summit of Flattop, which is well described by its name. Beware of the high altitude if not acclimated to the Rocky Mountains, though; it can take an unexpected toll on a person.

4 | Crater Lake
NATIONAL PARK
Oregon

Scenic views of Crater Lake are abundantly available to you in this Oregon park—an incredibly deep-blue lake set within the caldera (collapsed summit) of an ancient volcano that exploded 7,700 years ago. Quite naturally,

The First Climb Is the Hardest

Climbers who make the first documented ascent of a major mountain—especially without the benefit of a guide or route information—are often honored by history as pioneers, explorers of the unknown.

Consider Hazard Stevens's account of the first ascent of Mount Rainier published in the *Atlantic Monthly* in 1876: "We gained a narrow ledge ... and creeping along it, hugging close to the main rock on our right, laboriously and cautiously continued the ascent. The wind was blowing violently. We were now crawling along the face of the precipice almost in mid-air. On the right the rock towered far above us perpendicularly. On the left it fell sheer off, 2,000 feet, into a vast abyss. A great glacier ... stretched away for several miles, all seamed or wrinkled across with countless crevasses." At the summit, the climbers enjoyed a glorious view, but one tinged with apprehension: "On every side of the mountain were deep gorges falling off precipitously thousands of feet, and from these the thunderous sound of avalanches would rise occasionally."

Some first ascents are recounted in more understated fashion. In 1871, John Boies Tileston became the first man to summit Mount Lyell in Yosemite: "I was up early the next morning, toasted some bacon, boiled my tea, and was off at six. I climbed the mountain, and reached the top of the highest pinnacle ['inaccessible,' according to the State Geological Survey at the time], before eight. I came down the mountain, and reached camp before one, pretty tired."

one of the finest panoramas can be enjoyed from the 8,929-foot summit of **Mount Scott,** the park's highest point and a popular hiking destination. Not only does Crater Lake lie in the foreground below but the long-distance vista takes in Mount Shasta in California and, in Oregon, Mount Thielsen and the Three Sisters. The alpine wildflower display below the summit can be spectacular in mid- to late-July. The moderately strenuous trail to the top gradually climbs 1,250 feet in 2.2 miles, beginning at a trailhead on the eastern part of Rim Drive, which circles Crater Lake. Keep in mind that Crater Lake National Park is quite snowy and may not be fully open until mid-July, while the high elevations may already be covered with snow by October.

5 | Zion
NATIONAL PARK
Utah

Every year, thousands of people visiting this southern Utah park successfully make the 5.4-mile round-trip hike to **Angels Landing** to enjoy what may be one of the most spectacular views in one of America's most beautiful places. Because of the trail's popularity, a permitting system grants lucky spots through either a seasonal or day-before lottery. This adventure comes with serious cautions, though: The last half mile traverses a very narrow

▶ *Stopping at Scout Lookout in Zion National Park, rather than continuing to Angels Landing, is still an admirable feat. The last hurdle before this scenic spot is a series of 21 steep switchbacks called Walter's Wiggles, named for the park's first superintendent, who created them.*

ridge (in geological terms, a sandstone fin) with sheer drops on both sides. Even though chains have been installed for safety, several people have had fatal falls. The Angels Landing Trail begins at the Grotto and ascends 1,488 feet to its 5,785-foot end point, a small summit once described as so isolated that "only an angel could land on it." After a steep two-mile start, the trail reaches Scout Lookout, where many hikers decide to stop; the last section is no place for those afraid of heights or for small children, and Scout Lookout itself has a fine vista of Zion Canyon. Those who proceed are indeed rewarded for their efforts, with many hikers recalling Angels Landing as one of their most thrilling outdoor experiences.

6 | Grand Teton
NATIONAL PARK
Wyoming

There's a reason **Grand Teton** is a symbol of North American mountaineering. The 13,770-foot summit is archetypally majestic—a sawtooth pinnacle overlooking the valley of the Snake River—and topping it requires climbing 13 highly exposed pitches. Still, with the commensurate technical knowledge, fitness, and preparation, almost anyone can climb it. The mountain can be ascended in one long day, or from a base camp high on the mountain and an alpine (early) start. The most popular route is by Exum Ridge, which, when first climbed by Glen Exum in 1931, required a leap across an exposed chimney ... without a rope. It and other routes require technical climbing—not difficult, but quite exposed. The reward is a summit view of the entire Greater Yellowstone ecosystem and 14 mountain ranges in four states.

7 | Kings Canyon
NATIONAL PARK
California

Many of the Sierra Nevada summits in this eastern California park are reached by either very long and strenuous day hikes (as is needed to summit nearby 14,494-foot Mount Whitney) or overnight backpacking trips.

Breathtaking alpine views await climbers on Mount Whitney in Sequoia National Park.

Lookout Peak, however, offers a wonderful vista for far less effort and planning. The most scenic approach to the 8,531-foot top is via the Don Cecil Trail, which ascends the north-facing slope of Kings Canyon beginning in the Cedar Grove area. The hike climbs about 4,000 feet in seven miles, passing a lovely cascade on Sheep Creek along the way. The reward for this workout is an eye-popping panorama east toward the head of Kings Canyon, which shows the U-shaped form of a typical glacier-carved chasm. It's truly one of the best views in the combined Sequoia and Kings Canyon National Parks.

8 Sequoia
NATIONAL PARK
California

At 14,494 feet, **Mount Whitney,** the highest peak in the lower 48, looks like a castle tower of bare granite when viewed from below in the Owens Valley—and only gets more regal up close. Most climbers take the short way from the east, choosing one of two routes: one a walk-up, the other a semitechnical way called the Mountaineers Route. The walk-up is an extremely popular challenge that can be done in one long day (with a predawn start). That option permits a light pack, but also means 6,100 feet of trudging with little chance for acclimating, and a 22-mile round-trip from the trailhead in Whitney Portal (8,361 feet). Trail camps along the way can be used to break up the journey. Most of the hike is in Inyo National Forest, which issues a daily quota of permits. The Mountaineers Route is the way John Muir soloed Whitney in 1873. Though far less busy, this way calls for some scrambling and route finding, and is best done with a guide. Really serious climbers can ascend technical routes on Whitney's East Buttress. The reward for everyone is a soaring view of the Great Western Divide highlands and the Owens Valley lowlands.

9 | Mount Rainier
NATIONAL PARK
Washington

Mount Rainier lords over Washington State with such aloof dominance that it simply demands to be climbed, and 9,500 people or so per year attempt it. The active volcano's 14,410 feet of perpetual snow and ice make it a proving

ground for any mountaineer harboring Himalayan dreams, but it is also a trophy unto itself. Most climbers go with a guide service and spend a day learning the necessary basics of mountaineering—cramponing, ice-ax arrest, rope travel, rest stepping, and pressure breathing—before a

two-day ascent of the mountain. The standard Rainier route is up Muir Snowfield out of Paradise (5,400 feet) to Muir Camp at 10,000 feet. After dinner and a nap, you rise around midnight for a headlamp ascent of Cowlitz and Ingraham Glaciers, proceed up Disappointment Cleaver as the sun dawns, and dodge crevasses and bulges the last 1,200 feet. Then it's a short walk across the crater and up the crest.

There's no doubt why Mount Rainier is said to be the hardest climb in the continental United States.

10 | North Cascades
NATIONAL PARK
Washington

Renowned for their ruggedness and challenging approaches that often require glacier travel, the North Cascades have been called "America's Alps." At 9,200 feet, Goode Mountain is the highest peak in the park, but North Cascades' quintessential trophy is 8,815-foot **Forbidden Peak,** regarded as one of the 50 classic summits in North America. It and most of the other tallest mountains in the Washington park are in the South Unit. Access it from Cascade River Road by way of an old miner trail. In typical North Cascades fashion, the approach is a bit of a grind, with detours and bushwhacks, but once above tree line you'll enjoy views of the gorgeous landscape. After some glacial moraine, the usual route proceeds to the stunning west ridge, with views of lakes, meadows, deep valleys, and jagged peaks and some enjoyable Class 5.6 climbing to the summit. Not enough of a challenge? Consider the skill-demanding Torment to Forbidden traverse, bagging Mount Torment (8,120 feet) first before continuing over to Forbidden.

CANYON HIKES

For most, the great canyons of the national parks are drive-to jaw-droppers that lie below scenic rides and rim trails leading to stunning overlooks. For others, canyons beg to be explored on foot. These hikes can be quite an undertaking—almost always a steep, dry descent and a climb back— but the sensory experience is worth it.

1 Grand Canyon
NATIONAL PARK
Arizona

An out-and-back hike on the South Rim's **Hermit Trail** delivers the visceral impact of an epic Grand Canyon rim-to-rim trek without having to share it with hordes of fellow pilgrims. Intended for experienced desert hikers only, this lesser used trail provides awesome canyon views as it switchbacks down towering red-wall faces. Start at Hermits Rest off West Rim Drive and descend gradually as far as the heart desires through stands of piñon and juniper into a red-rock abyss. Backpackers can plunge 3,800 feet amid shale slopes and sandstone cliffs to lovely, cool Hermit Creek and a trail camp. Along the way stands evidence of this route's origins: a long-abandoned train track that served a popular tourist camp built by the Santa Fe Railroad. Set up camp and take a day hike down to Hermit Rapids on the Colorado River or on nearby portions of the Tonto Trail, which has superb river and canyon views. To make a day of it, follow the Hermit Trail to the Boucher Trail to Dripping Springs, hiking beneath sheer 1,200-foot cliffs to a lovely spring nestled within an alcove of Coconino sandstone. The seven-mile round-trip drops "only" 1,700 feet and represents a grand cross section of canyon formations.

2 Yellowstone
NATIONAL PARK
Wyoming

Yellowstone National Park seems to contain every type of natural wonder on Earth, so it's no surprise that a magnificent canyon is among them. The 20-mile-long, 1,000-foot-deep, V-shaped **Grand Canyon of the Yellowstone** in Wyoming is both a thermal area and a river (the Yellowstone) canyon. It was at one time covered by rhyolite lava flows, and some geysers and hot springs are still evident here. An assortment of trails explore the canyon's expanse, such as the North Rim Trail from Inspiration Point to Grandview Point. The Brink of the Lower Falls Trail is a strenuous showcase of the river's dramatic 308-foot drop. Distinct colors in the canyon walls indicate thermal spots where the rock has essentially rusted. The park's nominal yellow stone indicates the presence of iron.

3 Zion
NATIONAL PARK
Utah

The greatest hikes aren't only beautiful—they're also an adventure. **Angels Landing** in Zion National Park is definitely both. It ascends a dramatic red sandstone formation that rises 1,488 feet from the middle of Zion Canyon, and hikers have no clue how

*A hiker pauses to take in the
enormity of the Grand Canyon.*

A stream cuts along the border of the U.S. and Mexico in Big Bend.

they'll get to the top until they do it. It begins innocently enough, skirting the Virgin River from a trailhead near the Grotto picnic area and slowly climbing through slopes dotted with piñon and juniper trees. Portions of it are paved. But soon enough, the real action begins in the form of Walter's Wiggles—21 steep, supertight, zigzagging switchbacks, like a spiral staircase carved right into the huge block of sandstone. The final leg to Scout Lookout is as steep as a trail can be, and the drop-offs are sheer, so hikers can clutch handhold chains bolted into the cliff. The view of the river and the Zion Canyon floor below looks like an aerial photograph, only there's no airplane. Plan ahead: Angels Landing has a strict permitting process.

 **Glen Canyon**
NATIONAL RECREATION AREA
Arizona

True canyoneering is a technical craft that entails climbing, rappelling, and often swimming through slot canyons that can narrow down to a body squeeze, then drop off 40 or 50 feet into another slot. **Cathedral Wash** gives a sense of that adventure without the technical demands, though this three-mile round-trip hike does require a bit of scrambling. The trailhead is on the Lees Ferry access road in Arizona, which curves around a prominent formation called Cathedral Rock. The hike leads through narrow passageways lined by cliffs of limestone and sandstone, smoothed and eroded into all manner of formations—arches, alcoves, overhangs, muddy pools, and dry waterfalls that are anything but dry when a flash flood courses through. This is obviously not a hike to make during or after a rain in the vicinity. In the heart of the canyon, hikers must make their way along ledges and ease themselves down drop-offs. Finally, the trail opens up and reaches the Colorado River, which signals an about-face for the return hike.

5 | **Big Bend**
NATIONAL PARK
Texas

The **Santa Elena Canyon Trail** is Big Bend National Park's form of instant gratification. Short and arguably easy, the 1.6-mile path arrives at the mouth of the Santa Elena Canyon, where you'll have an up-close experience with the Rio Grande. Begin at the end point of the Ross Maxwell Scenic Drive. After crossing Terlingua Creek, the trail ascends concrete stairs before descending into the canyon, meandering along the edge of the river. Look up: These canyon walls (predominantly limestone) rise a dramatic 1,500 feet above the Rio Grande. It's important to be aware of recent rainfall, as Terlingua Creek can flood, making the trail impassable. Park staff warn guests that trying to cross the creek at off-trail locations is not only dangerous but can also cause severe erosion, so stick to the trail and listen to weather reports to ensure safe passage and a memorable experience.

6 | Cedar Breaks
NATIONAL MONUMENT
Utah

Utah's early Indigenous population called Cedar Breaks the "circle of painted cliffs." The circle is really a giant amphitheater with its floor a maze of multicolored pinnacles, spires, fins, columns, and arches. It's a big commitment to walk to the depths of Cedar Breaks, but you can do it via **Rattlesnake Creek Trail,** a strenuous, nine-mile route that flirts with the northern boundary of the park and passes through Ashdown Gorge Wilderness Area. The trail drops 2,500 feet in four miles before it connects with Ashdown Creek, which can be followed into Cedar Breaks. Or, with a car shuttle, a hiker can continue five miles to Utah 14. Along the way stands Flanigan Arch, which rises 100 feet and spans 50 feet. This is a hike only for the fit and adventurous, although summer heat is far less a concern than in the Grand Canyon because Cedar Break's rim is above 10,000 feet. Less intrepid visitors can enjoy two short rim hikes—the **Alpine Pond Trail** leads to a great view at Chessman Overlook and to a spring-fed pond; the **Spectra Point/Ramparts Overlook Trail** follows the south rim to a viewpoint before it passes through a stand of ancient bristlecone pines.

Drink Early, Drink Often

Most canyon hikes are in arid Southwest parks where rangers warn hikers to carry plenty of water—a minimum of a gallon per person per day is typically a good rule of thumb for wilderness hiking. Nevertheless, it's common for hikers to compromise a bit, carrying less than what is recommended or not drinking enough even if they are carrying plenty. Remember that there are limited water stations inside national parks, so always plan to pack in your own supply.

In a paper published in the early 20th century called "Desert Thirst as Disease," researcher W. J. McGee recounts the tale of prospector Pablo Valencia, who went without water for nearly a week in the Arizona desert. McGee came upon Valencia shortly before he would have died and describes the phases of (near) death by thirst in vivid detail.

First comes the cotton-mouth stage—"best relieved by water," McGee sagely opines. But soon enough, "saliva ceases, and membrane-mucus dies into a collodion-like film which compresses and retracts the lips, tightens on the tongue until it numbs and deadens, shrivels the gums and starts them from the teeth, shrinks linings of nostrils and eyelids." Around this time, Valencia discarded his hat and shoes—clearly not a good idea, but apparently not uncommon. During this phase, the sufferer also often begins to babble incessantly.

McGee goes on to describe "a progressive mummification of the initially living body, beginning with the extremities and slowly approaching the vital organs." Eventually the shrunken tongue swells and forces its way through the jaws, becoming "a reeking fungus on which flies ... love to gather and dig busily." Luckily Valencia survived, and so will you if you always carry the amount of water that park rangers suggest.

A rainbow arcs over White House Overlook in Canyon de Chelly National Monument.

7 | Canyon de Chelly
NATIONAL MONUMENT
Arizona

Canyon de Chelly National Monument lies within the Navajo Nation, some of whose people live and farm in the canyon. Hiking here must be in the company of a professional Navajo guide. That's good, because local knowledge is necessary to navigate its hundred or so different trails, most of which are ancient, and many of which are extremely steep, "hand-and-toe" trails. Some are dangerous. Some require a notched log (ladder) to reach a goal. The guides know the safe routes and, of course, the cultural history of the canyon, which has been settled for more than 2,000 years. Hikes are always customized to suit the interest and fitness of the hikers. A typical hike would be along **Bare Trail to Tunnel Trail,** starting in Canyon del Muerto, ending in Canyon de Chelly, and taking in at least a half dozen ancestral Puebloan ruins and caves along the way. Any Canyon de Chelly hike is a walk through time. The park's website has a list of certified local guides.

8 | Bryce Canyon
NATIONAL PARK
Utah

Only a thin red line divides fantasy from geology in Utah's Bryce Canyon. A hike amid the park's rococo hoodoos—sandstone spires arrayed in amazing mazes, fantastically animated forms shimmering in colors that Revlon can only covet—lends itself to true flights of fancy. The hoodoos rule. The **Queen's Garden/Peekaboo figure-eight loop** (6.5 miles) is Bryce's signature hoodoo hike. Working clockwise from Sunrise Point, descend into Bryce Amphitheater on the Queen's Garden Trail into hoodooland. Next, pick up the Peekaboo Trail to crisscross a ridge rife with hoodoos and climb out via Wall Street's towering sandstone spires. That's all a warm-up for the eight-mile **Fairyland Loop,** which loses and gains 2,309 cumulative feet as it navigates first a hoodoo graveyard of stumpy towers, then a forest of tall hoodoos that rise to the canyon rim. The

▶ *Don't be deceived by its name: Bryce Canyon is not an actual canyon. Rather, it is a series of more than a dozen limestone amphitheaters cut into the Paunsaugunt Plateau.*

park's popular full-moon hikes—no flashlights permitted—drop into this same hoodoo fairyland.

9 | Death Valley
NATIONAL PARK
California

California's Death Valley is really a park of canyons—slots and chasms and fluted corridors piercing the mountains that frame the valley. The one not to be missed is **Fall Canyon,** which can be reached from the Titus Canyon (a jeep road) trailhead: Proceed three miles up a wash surrounded by twisted striations of metamorphosed marble and dolomite to a dry waterfall, then another three miles through a narrow slot. Mosaic Canyon, near Stovepipe Wells, leads past walls of polished marble and ends at a dry waterfall two miles up. Golden Canyon is probably the most popular—an interpretive trail leads through the mile-long canyon and walls show tilted and twisted layers of rock, revealing the valley's faulting action, as well as mudstone deposits and ripple patterns that indicate an ancient lakeshore. At the head of the canyon is Red Cathedral's steep, rust-colored fluted cliffs.

10 | Colorado
NATIONAL MONUMENT
Colorado

As stunning as this Colorado monument's Rim Rock Drive is, it pales in comparison to experiencing the depths of Monument Canyon firsthand. Here steep-walled gorges and naturally sculpted rock formations reveal 1.7 billion years of geological history. Solitude and the singing of birds add to its pleasures. The signature hike is the **Monument Canyon Trail,** which descends 600 feet in the first mile, then flattens out and continues another five, passing some of the park's most striking landforms along the way. Note the layers of rock, from variegated sandstone and mudstone to the dark reds of Kayenta and Wingate sandstones. About halfway down are three major formations—Pipe Organ, Praying Hands, and Independence Monument, the last a remnant of a wall that once divided Monument Canyon from Wedding Canyon. But the hike isn't entirely made of stone—it also leads through piñon pine and juniper trees, and bighorn sheep might be spotted along the way.

10 BEST PARKS FOR
WINTER SPORTS

National parks near peak capacity in summer—but what about winter? With some park services limited or closed and many roads blocked by snow, fewer visitors head into these white, open spaces, so those who do have easier access to snowshoe hikes, ice fishing, and alpine and cross-country skiing in wondrous landscapes.

1 Yosemite
NATIONAL PARK
California

Following the first snowfalls, Yosemite slowly transforms into a winter wonderland of outdoor activities—from **snowshoeing** to **cross-country and alpine skiing**—that would look at home in a snow globe. With the exception of ice-skating on a classic rink at Curry Village, the focus of the park shifts to surrounding areas where, on ranger-led hikes, the sound of frozen terrain crunching beneath snowshoes is echoed by heavy pants of breath. And on a variety of cross-country winter trails, snow swishes beneath skis at Crane Flat, the Mariposa Grove of giant sequoias, and along Glacier Point Road, where heroic trees frosted with a cloak of snow and ice frame the paths. More inclined toward inclines? Slide into the Badger Pass Ski Area to find equipment, lessons, and whole-family thrills.

2 Olympic
NATIONAL PARK
Washington

Named for the forceful winds that blow here, 5,242-foot Hurricane Ridge is Olympic's hub for winter activities. Typically open Friday through Sunday during the season, the site is an alpine destination for **snowshoeing, cross-country and downhill skiing, snowboarding,** and **tubing.** More than 15 miles of ski routes cover everything from flat terrain to backcountry slopes, and rangers regularly lead snowshoeing outings on Saturdays. *(Call 360-565-3131 for information about road accessibility.)* As action-packed as weekends can be at Hurricane Ridge, Olympic's other diverse ecosystems are also worth wandering. While the mountains are blanketed in snow, the temperate rainforests drip with precipitation and lush greenery and the Pacific coast offers swell-watching from its snow-free shores. Just be sure to check forecasts ahead of time: What starts as a sunny day in Olympic can quickly shift to blinding blizzards.

3 Yellowstone
NATIONAL PARK
Idaho, Montana & Wyoming

With a riot of geothermal activity taking place beneath the surface, Yellowstone is one of the more spectacular parks to see in winter. Working overtime underground are 60 percent of the world's active geysers, creating hundreds of fumaroles, mud pots, hot pools, limestone terraces, and hot springs. So after a veil of white blankets the mountains and trees, some areas are punctured by heated steam and water that percolate through the ground. Where snow does manage to find a permanent home between mid-December and mid-March, you can tackle miles of trails on **skis**

A thick blanket of snow in Olympic National Park makes for a winter wonderland trek.

Rocky Mountain's snowshoe trails are also suitable for cross-country skiing.

or **snowshoes.** Casual skiers will be pleased with groomed trails near services; bolder outdoors enthusiasts disappear into backcountry treks that heighten the impact of the experience, knowing they'll need to be aware of changing weather, limited daylight, potentially aggressive wildlife, and the very real possibility of triggering an avalanche. Exploring by **snowmobile** or **snowcoach** is also permitted, but only when accompanied by a commercial guide (the park has a list of authorized operators that can provide services and rentals if necessary). All in all, the winter sports here are as wild as Yellowstone.

4 | Rocky Mountain
NATIONAL PARK
Colorado

The natural crease created by the north–south Rocky Mountains range leads to two distinct settings. To the west is an area of fluffy, powdery snow and, to the sheltered east side, harsh swatches of patchy frost, ice, and snow. No matter the conditions, the craggy rocks and towering peaks create a great American landscape worthy of winter sports like **hiking, sledding, snowshoeing,** and **cross-country skiing.** The park's primary winter destination is Hidden Valley, where kids of all ages can get some action

out of their saucers and sleds (no metal runners allowed)—beware, conditions can be fast and icy. Cross-country trails snake across the park, mostly west of the Continental Divide. Sans skis, snowshoes can get you onto trails, with the easier paths on the east side of the park. Meanwhile, snowmobiles give riders a chance to rev it up on a two-mile stretch of the North Supply Access Trail in the park's southwest corner. Even hikers can get into the action: A good pair of waterproof boots is all you need to enjoy the lower elevation trails where snowfall is less.

5 | Glacier
NATIONAL PARK
Montana

As climate change continues to impact the remaining glaciers, winter activities at Glacier National Park focus more on **cross-country skiing** and **snowshoeing** rather than ice climbing and rappelling into a crevasse. Despite the towering presence of the north Rocky Mountains, many trails are low and level and are perfectly carved for cross-country skiing. Take some time to familiarize yourself with avalanche conditions, then strap on the sticks and follow a glide path past meadows and woods of aspen and mixed conifers and around the shoreline of frozen lakes. Of course, this is a wilderness land, which means alternate trails can display a more aggressive personality—and many do. In some cases, the trails stretch as far north as the Canadian border, may remain unplowed and icy, and be far removed from the safety and shelter of park conveniences. The Going-to-the-Sun Road—the park's main east–west thoroughfare—is closed to vehicles (but open to cross-country skiers and snowshoers) due to snowfall beyond a certain point on each side of the park, so with the exception of the Lake MacDonald/Avalanche region, the established cross-country trails are concentrated along the edges of the park in the Apgar/West Glacier, Marias Pass, St. Mary, and Two Medicine areas.

6 | Mount Rainier
NATIONAL PARK
Washington

Where there's a hill, there's a way. In winter at Mount Rainier, the hills are alive in the Paradise section of the park, where kids tackle the slopes on **inner tubes** and **slippery flexible saucers.** As they go slip-slidin' away, rangers lace up their **snowshoes** and trudge onto trails to lead you (in rented snowshoes) on educational outings that explore the secrets of survival in subzero temperatures. One active life-form that thrives on frigid temperatures is the **snowmobiler,** who is granted access to a 6.5-mile

> *In the winter of 1971–72, the snowiest of all our national parks, Mount Rainier, received an outrageous 1,122 inches of snowfall.*

stretch of snow-covered roadway in the southwest section of Rainier; a 12-mile section of Wash. 410 in the north, from the park boundary to White River Campground; and all road loops of Cougar Rock Campground. Overall, a generous snowfall averaging 708 inches a year is more than enough to set the stage for these and other winter activities at Mount Rainier.

7 | Cuyahoga Valley
NATIONAL PARK
Ohio

More than 125 miles of trails, ranging from simple to challenging, run through this Ohio park—across open fields, through coniferous tree stands, past ponds, down steep hills, and even along a stretch of the Ohio & Erie Canal Towpath Trail. Cuyahoga offers a lineup of winter activities similar to other snowy national parks—**sledding** and **snow tubing** (particularly in the Kendall Hills), **snowshoeing,** and **cross-country skiing** (including a trek that leads along the Ledges

Trail to the park's ice-covered rocks). There's even **downhill skiing** at the Boston Mills and Brandywine ski resorts, which together host 17 trails. What adds a unique twist to Cuyahoga's lineup is **ice fishing** on Kendall Lake. When the ice is at least seven inches thick (you'll have to be the judge of that), anglers can drop a line for largemouth bass, crappie, and bluegill.

8 | Acadia
NATIONAL PARK
Maine

Although the Northeast is one of the nation's most frigid regions, the winter activities of coastal Maine's Acadia are relatively limited. Limited, but not disappointing. The majority of the island's famed carriage roads, 45 miles of which were created for nonmotorized vehicles, remain true to the cause as **cross-country skiers** hit the road. Sections groomed by volunteers are, of course, the most popular, and online maps and park guides direct you to the most popular areas. Skiers are also permitted to make tracks on unplowed park roads, which can be a bit more challenging since these same roads are also open to **snowmobiles.** Just steer clear of the high-revving vehicles and have a good time. Anglers willing to brave the cold can also **ice fish** in some of the park's lakes, weather permitting.

9 | Great Basin
NATIONAL PARK
Nevada

Great Basin may not be the first park people think of when planning a winter excursion, and perhaps for good reason. During the park's long winter season, most guests are long gone and so are many basic services, such as food. Still, the desert park does provide one of the more interesting slates of activities, because the people who come here really want to be here. **Skiers** won't just ski—in many cases they'll venture off marked routes, relying on orienteering skills to find their way on trails that loop through woods and around canyons. Trailheads that start at around 7,000 feet go up from there, so the challenge isn't downhill skiing, it's knowing how to get up these unplowed, ungroomed trails, just as it takes skill to come down on trails that can be narrow, steep, and icy. Additionally, the Lehman Creek Trail is a popular **snowshoeing** destination; free rentals are available at the visitor center. And for the bravest of the bunch, Lower Lehman Creek remains open for **backcountry winter camping.**

Wintertime Wildlife

When snow covers the country's national parks, some of their wildest residents stir into action, escaping the harshness of high-altitude winter environments by moving to lower valleys and meadows. While bears hibernate in Yellowstone, the park's wolves, bison, and coyotes are easily spotted out and about, their coats a bold contrast against the frozen landscape. Glacier's mammals, too—white-tailed deer, elk, and moose—have trouble hiding as they forage for limited food supplies. Park roads remain open at Rocky Mountain National Park to indulge wildlife enthusiasts' efforts to see snowshoe hares, ptarmigans, mule deer, and coyotes, as well as Steller's jays, long-tailed black-billed magpies, and Clark's nutcrackers. In Maine, Acadia welcomes migrating snowy owls, which travel thousands of miles from the arctic to winter in the park. As always, rangers urge visitors to keep their distance to protect the well-being of these majestic creatures.

10 | Sleeping Bear Dunes
NATIONAL LAKESHORE
Michigan

The towering sand dunes of Sleeping Bear doze on the shore of Lake Michigan. Already impressive in summer, this curving 35-mile lakeshore transforms in winter, when the snapping cold hurtles across the water and decks the dunes with snow and ice. With the scenic hiking trails covered in white, the land is primed and ready for winter activities including **snowshoe hikes** (taken with or without a ranger) across dunes, fields, and forests. Many of the park's hiking trails are recycled for **cross-country skiing,** including the loops of the Old Indian and Platte Plains Trails. Imagine a **sledding** excursion across the Sahara and you'll have an idea of what it's like when visitors take on the slopes at Dune Climb—the park's designated sledding area—after the powder coats the sand.

ACCESSIBLE PARKS

While the inherent nature of wild places means that not all visitors will be able to experience all areas, prioritizing accessibility allows people of all abilities to enjoy our natural spaces. The National Park Service is stepping up efforts to make at least portions of every park accessible to all, and here are some of the best so far.

1 | Everglades
NATIONAL PARK
Florida

Encompassing the largest wilderness area in the United States east of the Mississippi River, Everglades National Park ranks as a World Heritage site, international biosphere reserve, and wetland of international importance for its nine distinct ecosystems and great biodiversity. Relatively few of this park's most interesting features can be seen through a windshield—you need to experience the area on a more personal level—but don't let that stop you. Seven accessible trails are available, one of the easiest being the 0.8-mile **Anhinga Trail,** a flat, wheelchair-accessible loop that begins at the Royal Palm Visitor Center, four miles from the main park entrance. Named for a large, black, fish-eating bird, the Anhinga Trail has close views of alligators, turtles, and a variety of birds such as herons, egrets, gallinules, and—yes—anhingas. Part of the trail runs along land; the other is a boardwalk perched above a marshy wetland. All the animals are accustomed to humans, so photography opportunities are excellent, even for normally skittish birds. For another accessible way to enjoy the Everglades, check out the two-hour **Shark Valley Tram Tour** and the **Gulf Coast boat tour,** each of which offer assistive learning devices.

2 | Rocky Mountain
NATIONAL PARK
Colorado

There aren't many places in the United States where travelers can drive up to a true alpine mountain lake, and Bear Lake in Rocky Mountain National Park is one of them. That fact, however, has good and bad points: good in that such a gorgeous site is accessible, and bad because even the large parking lot at Bear Lake can fill up at times. Even when the lot is full, though, a park-and-ride shuttle provides access from other parking areas back down Bear Lake Road. The 0.6-mile **Bear Lake Trail** encircling the small lake isn't fully flat and is more challenging. Nonetheless, wheelchairs can make it around much of the lakeshore and traverse the full trail with assistance. The view over Bear Lake includes the stark profile of Hallett Peak looming on the Continental Divide, part of a scene of unforgettable beauty. Other accessible trails include **Coyote Valley, Sprague Lake, Lily Lake,** and **Holzwarth Historic Site.**

3 | Wind Cave
NATIONAL PARK
South Dakota

The world's first park created to protect a cave, Wind Cave may have received its official designation in 1903, but Indigenous groups have regarded the "hole that breathes cool air" as sacred for generations. Currently ranked the seventh longest cave on the planet, Wind Cave continues to expand its size as cavers explore more passages each year. The only way to see the underground world of Wind Cave is via one of the many ranger-guided tours. Of particular note is the **Accessibility Cave Tour,** designed to give those with limited mobility, vision, and hearing the chance to experience the cave. The half-hour tour *(call 605-745-4600 to schedule)* provides an introduction to Wind Cave, with an emphasis on its signature feature, boxwork, an unusual formation composed of thin calcite fins that can resemble honeycombs.

A few cave tour scripts can be accessed online, and with enough notice, sign-language programming is also available.

4 Redwood
NATIONAL & STATE PARKS
California

Among the tallest trees on Earth, coast redwoods once covered two million acres of the Pacific coast, but 96 percent of the original old-growth forest has been lost to logging. Newton B. Drury Scenic Parkway, about 40 miles north of Eureka, California, takes drivers on a 10-mile route through a magnificent stand of old-growth redwoods that escaped the loggers' saws. (Watch for Roosevelt elk along the way.) At milepost 128 of the drive is the Big Tree Wayside, a parking area with a very short trail leading to the aptly named Big Tree, a redwood more than 300 feet tall, 66 feet in circumference, and an estimated 1,500 years old. Beginning at the wayside, the 0.3-mile wheelchair-accessible **Circle Trail** provides an easy way to experience a mature grove of redwoods, one of the world's most distinctive ecosystems and an environment that can justly be called awe-inspiring. This site is actually located in Prairie Creek Redwoods State Park, part of an unusual partnership in which federal and California state lands are jointly administered. Few trails anywhere provide so much grandeur for so little effort.

5 White Sands
NATIONAL PARK
New Mexico

Ranking among the most otherworldly landscapes in the United States, this New Mexico park protects a 115-square-mile area of sand dunes formed from the bright white mineral called gypsum, a type of calcium sulfate. Strong southwesterly winds continually reshape the dunes, moving them as much as 40 feet a year and creating a harsh environment where a surprising number of plants and animals manage to survive. Beginning at the wheelchair-accessible visitor center, the eight-mile (one way) Dunes Drive leads into the heart of the park. A

Easier Access

Thanks to an interagency partnership, the free, lifetime Access Pass is available to U.S. citizens or permanent residents of any age who have been medically determined to have a permanent disability. More than 2,000 recreation sites (managed by five federal agencies) are covered by the pass, which permits the owner and everyone in their vehicle into areas where per-person fees are typically charged. Visitors can learn more, find a location where they can apply for a pass, and start planning their next trip at *nps.gov.*

trailhead 4.5 miles along the route marks the start of the fully accessible **Interdune Boardwalk Trail,** a 600-yard elevated path with interpretive signs explaining the natural history of White Sands. The trail is an excellent way to see wildflowers and other plants that have adapted to life in the lower, slightly wetter spots between the dunes. The boardwalk protects the interdune area, which is easily damaged by footprints. At times, animal tracks are visible in the sand beside the path. The trail features benches for rest stops and ends atop a dune with a scenic view of this unique terrain.

Iconic bridges traverse Acadia's automobile-free roads.

6 Acadia
NATIONAL PARK
Maine

In 1913, John D. Rockefeller began overseeing a network of carriage roads through the most scenic sections of Mount Desert Island. Roadways were graded to smooth their passage, 17 stone-faced bridges were erected, and automobiles were (and still are) prohibited. In 2022, concessionaire Carriages of Acadia added a custom-built **wheelchair-ready carriage** to its fleet, providing comfortable outings for all. Another way to see the park is the complimentary, **wheelchair-accessible Island Explorer bus,** which is available from late June through early October. Since parking is especially limited in popular Acadia, the shuttle helps guests avoid long waits. Visitors with vision impairments will want to arrange a stop at the Hulls Cove Visitor Center, which stocks materials like braille brochures and a CD of the 56-mile audio tour that takes in the Park Loop Road, Cadillac Summit, and Somes Sound.

7 Yellowstone
NATIONAL PARK
Wyoming

Quite a few of the highlights of this celebrated park can be seen from roads and viewpoints in northwestern Wyoming, including the Upper and Lower Falls, parts of Mammoth Hot Springs, and some of Yellowstone's famed wildlife. Perhaps the best experience for persons with limited mobility can be found at the park's **Upper Geyser Basin,** which includes the iconic Old Faithful geyser. The visitor center and historic Old Faithful Inn are fully accessible, and a short trail leads to a viewing area of Old Faithful itself. This much publicized geyser is not actually that faithful: It varies in eruption frequency from 45 to 110 minutes. From Old Faithful, a hike/bike path accessible for wheelchairs heads 1.5 miles north to Morning Glory Pool, named for its evocative colors. Along the way the path passes features such as Castle Geyser, so-called for the tall cone that has built up around the vent. Visitors with limited mobility will also enjoy nearby Firehole Lake Drive, with accessible viewpoints of thermal features like Great Fountain Geyser, which sometimes sends steam and water up to 200 feet into the air.

8 | Carlsbad Caverns
NATIONAL PARK
New Mexico

Entering the underground wonderland of southeastern New Mexico's Carlsbad Caverns via the natural entrance requires a very steep walk of 1.25 miles, descending 750 feet. Persons with limited mobility have another option, however: taking the elevator from the visitor center down to the **Big Room.** (The visitor center, including a theater and restaurant, is easily accessible from the parking lot.) The one-mile Big Room Route is reached from the cave's underground rest area, and about two-thirds of that length is accessible to persons in wheelchairs with assistance; some sections are closed to wheelchairs because of steepness or narrowness of the trail. Visitors can rent the park's self-guiding audio program, which provides interpretation of some of the sights along the way. Among Carlsbad's famed cave formations and attractions visible along the wheelchair-accessible portion of the Big Room Route are the Hall of Giants, Temple of the Sun, the Caveman, Lower Cave Lookout, Rock of Ages, and Crystal Springs Dome.

▶ *Dogs that are trained to do work or perform tasks for people with disabilities are classified as service dogs. Those that meet this definition are allowed on-leash in National Park Service facilities and on park shuttles.*

9 | Grand Canyon
NATIONAL PARK
Arizona

The mostly paved, 13-mile trail along the South Rim of the Grand Canyon features magnificent views into one of the planet's most famous geological wonders. Wheelchair-accessible buses serve the entire length of the linear **Rim Trail.** Use the park shuttles, which stop at many points along the trail, to customize a visit to individual abilities and desires. The Rim Trail accesses Grand Canyon National Park's most popular area, the South Rim Village, which includes the main visitor center at Mather Point, the bookstore, the Market Plaza (with grocery store, bank, and post office), and many of the historic lodges. Ask at the visitor center for a Scenic Drive Accessibility Permit, which grants visitors with mobility issues access to **Hermit Road** and

Yaki Point Road, two scenic roads otherwise closed to the public. Also inquire about the variety of interpretive ranger programs available most days of the week.

10 | Yosemite
NATIONAL PARK
California

You don't have to travel far from the car to spot iconic Yosemite points of interest like El Capitan, Half Dome, and Yosemite Falls. Scenic roadways and overlooks abound. And yet continuing to improve accessibility—from designing new construction to renovating historic structures—remains a leading focus at this California national park. With 24-hour advance scheduling, open-air (seasonally) Valley Floor Tours gives **guided immersions** into Yosemite's history, wildlife, and natural wonders. The **Deaf Services Program** supplies trip planning and interpreting services for tours, ranger programs, and theater presentations. **Tactile park maps and exhibits** make learning easier, such as at the Yosemite Valley Visitor Center, where hands-on displays present geological and historical features of the park. Plus, Junior Rangers with diverse communication, learning, motor, or sensory needs can dive into an adapted activities booklet that helps all young explorers fully engage with Yosemite.

Committed to accessibility, Yosemite features paved trails, elevated walkways, and special programming.

Hooting in the Olympic Woods

In 2000, author Robert Earle Howells talked his way on to a spotted-owl census mission in Olympic National Park. Spotted owls no longer call the sites he visited home, and technology has changed the shape of avian monitoring, but his adventure is a testament to the efforts to preserve and document the wilderness within the National Park System.

The Bogachiel is one of five major river systems in the western lowlands of Olympic National Park. A constant, Pacific-born westward procession of super-saturated clouds, unimpeded until they collide with Mount Olympus, bestows staggering volumes of moisture upon this part of the park every year—something like two billion gallons of water per square mile. A little muck on the valley floor comes as no surprise.

THE LURE OF OLD GROWTH

It was raining steadily the morning I met wildlife biologist Scott Gremel at park headquarters in Port Angeles. But a high-pressure ridge moved in to quell the deluge just as we set off on the trail. Side streams flowed with spring-freshet exuberance. Every fern I brushed dumped a gallon or so of water onto my lower body. It was like walking through a car wash.

The first 1.5 miles of the Bogachiel trail run through second-growth spruce forest on U.S. Forest Service land. "Barred-owl habitat," sniffed the biologist, referring to a feathered competitor that has been displacing spotted owls in the region.

It was a desire to delve into some of these old-growth stands that drew me to Olympic. Since spotted owls are symbols of old-growth preservation—where old-growth habitat declines, so do the birds—I figured an owl researcher could lead me into some prime forest stands. Gremel has been participating in annual spotted-owl counts since 1994.

After four or five gymnastic stream crossings over slick concatenations of stones, we set up camp between the river and a mélange of spruce, western hemlock, and vine maples swagged with moss. While we ate lunch, we watched an otter surface downstream from us, then dive and begin swimming upstream, only to emerge exactly at our feet. The critter showed its startled face for two seconds, then torpedoed back down the Bogie.

GETTING A BIRD

How do you spot spotted owls? You hoot for them. As we bushwhacked up a 70-degree slope on Indian Pass, picking our way over and around huge fallen logs, Gremel would stop, swallow, and emit a staccato, four-note, *"Ooo. Oo-oo. Ooo."* I finally dared a hoot series of my own. Gremel cautioned me to "Get the 'who' out of your hoot." Owls don't cotton to consonants.

Only by picking your way through its chaos do you get close to grasping the nature of an old-growth northwestern rainforest. It's characterized by a thousand shades of green and infinite degrees of rot. Between 200-foot western hemlock trees and 300-foot Douglas firs are impassable thickets of huge ferns and innumerable fallen logs, and everything's so thickly furred with moss that shapes are just murky suggestions rather than clear outlines that would suggest a course of travel.

After a few minutes of hooting, we heard a soft whistle from downslope. "Contact whistle," Gremel whispered. "She's nearby." From his pack, Scott pulled a one-gallon

A northern spotted owl at the Kalaloch stand site in Olympic National Park

coffee can riddled with breathing holes for the white mouse within. Almost immediately, an owl soundlessly swooped to a branch 50 yards away. Gremel released the mouse on a log, and in seconds, the bird was on it. A red band on one leg identified her as a female. On a perch just 20 feet away, she gulped the rodent, then froze, eyes fixed on us.

Amazing. Preposterous. Gremel estimates the million-acre park holds just 229 pairs of the endangered owls, and here was one. In owl-research parlance, we'd gotten a bird.

EVER ONWARD

We repeated the process for two more days. No more birds. Perhaps this was consistent with the 3 to 4 percent annual decline in spotted-owl population since the mid-1990s. On the way out, we scared a herd of elk whose barnyard scent we'd caught a few minutes before.

TECH BOOM

As in so many other fields, technology has impacted the way biologists study and interact with the environment of Olympic National Park. They now use a system called remote acoustic monitoring to record owl calls at night, and software to analyze the recordings for trends, giving them the data needed to help support dwindling owl populations. Those so inclined can participate in citizen scientist programs (page 288) to contribute to similar efforts at various parks. If this seems to lack a certain connectedness, that's understandable. But maybe it's enough to let technology work its own sort of magic, while we stand in the mist, the forest towering overhead, and breathe in the peace that comes with being outdoors.

10 BEST PARKS FOR

ROCK CLIMBING

Our national parks are home to some of the most fabled rock-climbing sites in the world. Even if you're not a climber, you can enjoy the theatrical thrill of watching athletes scale seemingly impossible rock faces. Intrigued? You can arrange for lessons or a guide at nearly every park that has great climbing.

1 | New River Gorge
NATIONAL PARK & PRESERVE
West Virginia

The New River's superhard **Nuttall sandstone cliffs** are so revered by climbers that they'd just as soon call this West Virginia park New River Gorge National Cliffs. The cliffs stretch for 12 or so miles, range from 30 to 120 feet high, and are riddled with slabs, cracks, corners, steep overhangs, and vertical walls to the tune of more than 1,400 identified routes. That bomber rock—harder than most granite—and scores of elegant climbs make the New River very popular. Many climbers favor **Bridge Buttress,** which stretches for two miles downstream from the New River Bridge. It's where New River climbing was pioneered in the 1970s, and where most of the park's moderate climbs are, mainly in the 35- to 75-foot range. Parkwide, most routes are 5.9-plus—but newbies can find plenty of easy routes—and

sport climbs slightly outnumber trad routes. For sheer number of routes, look to Endless Wall—four miles of cliffs, 675 routes, some as high as 150 feet. Across the river from Endless, the Kaymoor and South Nuttall areas feature sport climbs that require a longer hike in than other sites. Climbing can be year-round, but arguably the best months are April to mid-June and mid-September through the end of October.

2 | Joshua Tree
NATIONAL PARK
California

Anyone who has scrambled for a line at a popular climbing site or queued up at the local rock gym to grab artificial nubs will celebrate the overwhelming abundance of superb rock in Joshua Tree National Park. It's amazing, grippier-than-Velcro stuff called quartz monzonite, so abrasive that a person can walk up

or down a 70-degree slab in their tennies. Downside: It masticates city-slicker fingers within a couple of hours. Among the many delights of Joshua Tree is the bouldering. Opportunities are everywhere, and of all difficulty levels, so anyone can get into the action. Included in more than 8,000 climbs are Class 4 scrambles, scads of intermediate top-rope walk-off routes, and multipitch climbs that'll thump anyone's heart; many are found near campgrounds. The hotbeds are in the north, such as **Real Hidden Valley,** across the road from Hidden Valley Campground. A mile-long trail winds among some of the best climbs in the park, including three rewarding 5.8 and 5.9 routes up **Hidden Tower,** a thick spire that dominates the valley—exposed, vertical, but laced with the holds. **Indian Cove** is a great mixed-difficulty venue comprising crags, slabs, and cracks, with overtones of theater. From the

In the New River Gorge, a rock climber scales to new heights on the Endless Wall via the 5.11 Leave It to Jesus route.

comfort of a campsite chair, climbers can eyeball the day's challenges over coffee and celebrate them over a cool beverage.

3 | Yosemite
NATIONAL PARK
California

Some of the most beautiful walls and spires in the world tower above eastern California's Yosemite Valley, the place where big-wall climbing was born, and where it has advanced beyond all possible dreams of the sport's pioneers. The accomplishments of climbers here are as humbling as the granite faces themselves. It used to take weeks to scale the famous Nose of **El Capitan,** but now it has been ascended in less than three hours. It has even been conquered by blind climbers (with sighted assistants) and soloed. El Cap remains the holy grail of Yosemite climbers, but the park has many other renowned routes, virtually all of them traditional, including Royal Arches, the Glacier Point Apron, Sentinel Rock, Snake Dike on Half Dome, and Cathedral Spire. Many are "only" 5.7 to 5.9, but these are long routes, and only one look at the tall, polished granite walls is needed to gauge their exposure. Climbers looking for less daunting climbs should head to **Tuolumne Meadows,** which has a wealth of smaller slabs and domes— one to four pitches versus 39 on the

▶ *Practice conscious climbing ethics by leaving no trace, which includes using as little chalk as possible, choosing earth-toned webbing, packing out human waste, and avoiding cliff edges, cracks, and ledges that are prone to erosion.*

Nose. Yosemite also has dozens of bouldering areas, including several around Camp 4, the traditional hangout of Yosemite climbers.

4 | Little River Canyon
NATIONAL PRESERVE
Alabama

Little River in northeastern Alabama is the great secret of Southeast sport climbers, at least those with the mettle to tackle its predominantly difficult routes. The cleft's sandstone cliffs tower as high as 400 feet above the river, and most of them overhang. That means two things: First, challenge—most of the canyon's 400 or so routes are 5.11 to 5.14 in the 80-foot range, though there are some 5.8s and 5.10s. Second, many walls are climbable even when it's raining—overhangs like **Lizard Wall** work as a sandstone

umbrella. Speaking of bonuses, a little-regarded climbing option are the thousands of boulders in the river itself. Climbers can challenge themselves and hop off into the cool river. You won't readily find guides or outfitters in the area; this is very much a word-of-mouth destination.

5 | Pictured Rocks
NATIONAL LAKESHORE
Michigan

Come winter, America's first national lakeshore is also known for its frozen fortresses. Flocking to the rapidly growing sport of ice climbing, enthusiasts swing axes into ice, grip on tight, and work their way up frosty "ladders." Here in the Upper Peninsula of Michigan, low temperatures, abundant waterfalls, and sandstone cliffs combine with lake-effect snow to form impressive ice columns. In fact, the Michigan Ice Fest, held here every February, is the longest-running ice-climbing festival in the country, drawing upwards of 1,000 climbers annually. Although the area's namesake Pictured Rocks cliffs do have ice, they aren't recommended for climbing. Rather, **Sand Point** and **Miners Falls** are among preferred sites. Sand Point is known for its easy proximity and blue ice curtains, while a three-mile ski or snowshoe trek reveals the 40-foot column of Miners Falls. Learn or sharpen your technique with a course led by the Michigan Ice Fest Guides (*downwindsports.com*).

6 | Grand Teton
NATIONAL PARK
Wyoming

Alpine climbing—climbing to reach a summit—has long ruled in northwest Wyoming's Teton Range, where the serrated peaks are beautiful and irresistible. But pure rock climbing on boulders, faces, slabs, and pinnacles is also superb in Grand Teton. Often the two will merge, as many of the best summiting routes entail rock climbing, and many students in Grand Teton's rock-climbing courses are there to learn enough rock climbing and rope handling technique to scale the big mountains. The heart of the scene for either type of climber is **Jenny Lake,** where the ranger station serves as an information source for rock climbing in the park. Jenny Lake is also the site of three legendary boulders—Cutfinger Rock, Falling Ant Slab, and Red Cross. Symmetry Spire above Jenny Lake has great rock routes with a dash of alpine flair; an ice ax and crampons might be needed to reach its 5.6 to 5.9 climbs. **Cascade Canyon** is a rock-climbing hotbed, including the highly regarded Guide's Wall and its many variations in the 5.7 to 5.9 range.

7 | City of Rocks
NATIONAL RESERVE
Idaho

Few places anywhere receive the kind of whispered reverence that City of Rocks inspires among climbers. The City is a collection of abruptly rising spires ranging from 30 to 300 feet in the Albion Mountains of south-central Idaho. The granite is sticky and the choices nearly endless, with hundreds of trad routes from 5.4 to 5.12, and plenty of 5.11-ish sport climbs, too. Formations tend to be highly featured, with lots of cracks, pockets, edges, and knobs, and many have unusual protruding flakes that are surprisingly solid. A classic example

Climbing Argot

Aid climbing: Employing artificial devices to directly assist with upward progress

Bouldering: Climbing on a large rock, usually without rope

Crack climbing: Following a crack in a rock by jamming a finger, fist, or body into it and moving upward

Free-climbing: Climbing with ropes and hardware to protect in case of a fall, but not to help ascend

Pitch: A segment of a climb that can be accomplished with a single length of rope

Protection: Devices such as nuts and cams, placed in rock cracks to protect a climber in the event of a fall

Rack: A climber's collection of protective devices, slung around a shoulder on a climb

Sport climbing: Climbing using bolts fixed to the rock

Top rope: A climb that is anchored from above

Trad (traditional) climbing: Climbing by placing own protection, rather than using bolted-in protection

Yosemite Decimal System: Common difficulty-rating system in the United States. Class 5 climbing entails exposure, requires technical moves, and a fall results in severe injury or death. Ratings for technical climbs begin at 5.1 and progress to 5.15.

of the latter is **Rye Crisp on Elephant Rock.** Virtually every wall and feature has a mix of easy and difficult climbs, so it's almost always possible to find a choice line. Climbing starts as early as April and runs into October; be aware of private lands within the reserve.

8 | Pinnacles
NATIONAL PARK
California

At Pinnacles National Park, a *tap tap tap* can tell you all you need to know about the permanence of a hold. When you hard-tap it with your fingertips or knuckles, a hollow sound means the hold is more likely to fail. In fact, many holds and some bolts—which are not maintained by the National Park Service—can be old or damaged here, so climbers are urged to check and double-check before relying on them. That warning doesn't squelch the popularity of Pinnacles, where hundreds of high-thrills routes have been developed. The difference is that here rock is volcanic rhyolite breccia, essentially fragments cemented together to form a much weaker rock than granite. Park staff notes that it is common for climbers who lead 5.9 routes in other areas to determine that 5.6 routes at Pinnacles are unacceptably dangerous. The **East District rock** is more solid and tends to draw more climbers to sites like the cheekily named

A climber scales Devils Tower in the Black Hills of Wyoming.

Tourist Trap and **Discovery Wall.** From January through July, check for posted raptor advisories to help you avoid sensitive areas.

9 | Devils Tower
NATIONAL MONUMENT
Wyoming

There may be no more enticing climbing destination than the 867-foot granite Devils Tower monolith rising above the northeast Wyoming range. The routes tend to be long, 500 to 600 feet, and they require a serious rack; there's virtually no bolting on the tower. Trad rules here. The most popular way to the summit is the 5.7 **Durrance Route,** a six-pitch traditional crack climb (as most routes are here) of nonstop fun and awesome views—famed climber

Todd Skinner free-soloed it in 18 minutes. The lower rock has plenty of sustained pitches, too. Climbing can be fine year-round, but the park encourages a moratorium in June, when Native Americans—who consider the tower sacred—often hold ceremonies. Also, some routes are closed in spring to protect nesting prairie falcons.

10 | Rocky Mountain
NATIONAL PARK
Colorado

Colorado's Rocky Mountain features great climbs on both sides of the Continental Divide, but the most popular climbing areas are on the east side, namely the **Longs Peak cirque, Glacier Gorge, Loch Vale, Tyndall Gorge, Odessa,** and **Mummy Range.** Routes tend to be clean, on firm granite, and with wonderful views. Most are multipitch, which climbers appreciate, given that they need to hike two to six miles just to start. At Loch Vale is **Petit Grepon,** listed as one of the 50 classic climbs in North America in the book of the same name. The south face is the most famous way up—a 5.8 climb with eight exposed pitches to an extraordinarily small summit. **Spearhead** at Glacier Gorge is every bit as classic and just one of several stunning formations there, where crack climbs range from 5.6 to impossible. And the famous Diamond face of **Longs Peak** is gorgeous, if difficult; even the so-called **Casual Route** is 5.10.

EPIC BACKPACKING TRIPS

For independent spirits who prefer to trek for days and cover tens, if not hundreds, of miles, the parks offer a bonanza of boundless opportunities. From the Appalachians to the Continental Divide, these trails allow you to immerse yourself in nature and do something many only aspire to: experience America as you've never seen it before.

1 Yosemite
NATIONAL PARK
California

The time invested on California's **John Muir Trail** rewards hikers with what may be the most memorable month of their lives. With the trailhead within Yosemite National Park, the 215-mile trek ties in the Ansel Adams Wilderness and Sequoia and Kings Canyon National Parks before ending at Mount Whitney, the highest peak in the lower 48. Backed by plenty of time and loads of stamina, journeys can be accented with an ascent. The route, which requires a permit, follows the Sierra Nevada, ascending and descending some 50,000 feet as hikers pass, ford, and scale over, around, or near legendary sites such as subalpine Tuolumne Meadows, the Merced River, Half Dome, Cathedral Lake, and Lyell Canyon and Glacier. Keep walking and the scenery includes Jeffrey pines, fir trees, blue spruce, azaleas, ferns, violets, and fields of wildflowers. There are creeks and peaks, butterflies and waterfalls, cascading rivers, narrow canyons, grouse, deer, bears, and lakes.

2 Rocky Mountain
NATIONAL PARK
Colorado

Since the challenging **Colorado Grand Loop** clears the Continental Divide, consider it the rare cross-country hike that'll take only a week, traveling west to east and back again in just 43 miles. Kicking off with an ascent from the Bear Lake Trailhead to Flattop Mountain (elevation 12,324 feet), short but intense stretches lead to much appreciated campsites. After crossing the 12,061-foot Boulder-Grand Pass, the trail climbs below the south face of Longs Peak and eventually reaches the base of the Palisades cliffs and then affords great views from the summit of Longs Peak; afterward, a well-deserved downhill hike drops nearly one mile over the course of 10. Hikers get a bird's-eye view of the American West with panoramas of wildflowers, purple mountains majesty, and the occasional elk. If planning to hike in the fall or late spring shoulder seasons, though, hikers should add a little more to the pack: specifically an ice ax and crampons.

Mount Rainier's Skyline Trail can be looped into a longer trek via the Mazama Ridge and Wonderland Trails.

3 | Grand Canyon
NATIONAL PARK
Arizona

It's possible to hike from one soaring rim of the Grand Canyon to the other quicker than the suggested four days and three nights on the 24-mile **Rim to Rim Trail,** but why would you want to? Doing so rushes one of the most epochal adventures in the National Park System. Via the North Kaibab Trail at the North Rim, descend 14.2 miles and nearly 5,850 feet to the base of the canyon, the storied Colorado River. This in itself is a feat: Of the more than five million people who visit the canyon annually, only one percent hikes all the way to the bottom. Campgrounds dot the route, or you can try snagging a lottery spot at the famed Phantom Ranch, a circa 1922 lodge that books 13 months in advance. When it's time to ascend, you'll take the Bright Angel Trail out, climbing almost 4,500 feet and 9.6 gratifying miles to crest the South Rim. May and October are arguably ideal for the journey, as summer temperatures can top 120 degrees and winter can be icy. But no matter your timing, training and planning are required as your body and pack will need to be prepared for strenuous conditions. You'll want to stage a car at the South Rim, or book a service like the Trans-Canyon Shuttle for your triumphant return to the North Kaibab Trailhead.

▶ *On August 14, 1937, the last mile of the Appalachian Trail opened on Mount Sugarloaf in Maine. The footpath was officially designated the Appalachian National Scenic Trail in 1968.*

4 | Mount Rainier
NATIONAL PARK
Washington

Pacific Northwest scenery such as displayed on postcards, calendars, and bookmarks is what inspires hikers to take on the **Wonderland Trail.** Clocking in at just under a century, the 93-mile trail is resplendent with glaciers, volcanoes, and swatches of high-altitude wildflowers and towering cedar and fir trees, all of which will be revealed from a repeated series of crests slicing across Washington State's Mount Rainier. Planning is key here, and rangers encourage hikers to consider elevation and skill before applying for a permit. A range of wildlife includes elk, deer, and bears (beware), but perhaps even more challenging are sections of the trails that, when not conveniently flat, rise and fall a collective 23,000 feet and areas where hikers are expected to ford streams that may or may not be traversed by a bridge or fallen tree.

5 | Chesapeake & Ohio Canal
NATIONAL HISTORICAL PARK
District of Columbia, Maryland & West Virginia

When is an epic hike not an epic hike? When it's a **184.5-mile stroll** through a national park beside the remnants of a historic canal. Actually, this may be a good way to test the waters of hiking by keeping one foot on the trail and one close to conveniences between Point A (Georgetown) and Point B (Cumberland) along a towpath that parallels the north bank of the fabled Potomac River. Out of commission for nearly a century, the canal now serves a greater purpose: providing Washingtonians with a convenient path for jogging, biking, and walking. While most of the trail's three million recorded annual visitors stick relatively close to D.C., others opt for a "thru-hike" that passes towns followed by long stretches of solitude and areas where the once flowing canal has been reclaimed by vegetation. It's adventure without the drama, and an opportunity to hike past locks, dams, tunnels, and historic sites like Harpers Ferry.

6 | Glen Canyon
NATIONAL RECREATION AREA
Arizona & Utah

The spectacular **Rainbow Trail** winds for nearly 30 miles around the footprint of Navajo Mountain; because much of the trail falls within the Navajo Reservation, this five-day hike will require a permit issued from the Navajo Nation Parks and Recreation Department *(928-871-6647, navajonationparks.org).* Setting off through a scrub forest, the path slips into First Canyon then crosses the state line from Arizona into Utah. After following a dry creek bed in Horse Canyon, the trail starts its drop of nearly 2,000 feet into Cliff Canyon. Sunset Pass provides spectacular views, and later, hikers encounter a Navajo hogan and have the chance to camp out beside Rainbow Bridge, the world's largest natural bridge. A sacred site to the Navajo (don't walk beneath it), the bridge stands 290 feet high and spans an incredible 275 feet. The trail next snakes past the crimson shades of Super Dome, campsites near Owl Bridge and Bald Rock Canyon, and glimpses of centuries-old rock art created by ancient tribes. The entire journey? A masterpiece.

7 | Big Bend
NATIONAL PARK
Texas

More than 200 miles of trails meander through rugged and remarkable Big Bend National Park. From the Rio Grande and Chihuahuan Desert to the Chisos Mountains, 6,000 feet of elevation change reveal a wide diversity of plants, animals, and landscapes.

Backpackers settle in for the night at Maine's Bald Mountain Pond on the Appalachian Trail.

Backpackers can shape multiday routes by piecing together short, easy walks with longer day hikes. The result is a DIY adventure through the vast Texas expanse. And we mean DIY—oftentimes hikers will need to rely on topographic maps and a compass to navigate Big Bend's vague trails. (Rangers will need to know your detailed itinerary before issuing a permit.) The mountain-sliced 5.6-mile **Window Trail** is quite popular, as is the 12.6-mile **South Rim Trail,** where a steady climb from the Chisos Basin to the South Rim reveals expansive views of the desert. From there, you can reach Emory Peak (the highest

point of the Chisos Mountains) by adding just three short, worthwhile miles to the journey. Since extreme heat and dry desert conditions can derail a trip quickly, you can make the most of this Texas park by visiting from October to April.

8 | Voyageurs
NATIONAL PARK
Minnesota

In the remote and untrammeled Minnesota North Woods, the **Kab-Ash Trail** is a popular year-round draw. Although it is less than 30 miles east to west, it still requires a few days to complete. In wintertime, cross-country skiers take to the trail for adventure in the snow-hushed woods. When the snows melt (in the few brief summer months), the rough country trail provides firm footing for hikers, meandering through a mix of backcountry forests of pines and rock ridges and wetlands that connects the Kabetogama and Ash River communities on the southern edge of the national park. On the west, the trail literally begins where the road (Salmi Road) ends and the wilderness begins, home to bald eagles, white-tailed deer, and black bears. From here the path snakes east and south and then north and, midway through, spins into loop trails and eventually hooks into new paths that venture to the shores of Kabetogama Lake, Sullivan Bay, and the Ash River. Where the trail crosses the road to the Ash River Visitor Center, stop at the Beaver Pond Overlook to see the creatures at work.

9 | Wrangell–St. Elias
NATIONAL PARK & PRESERVE
Alaska

According to experienced hikers, the epic 45-mile Alaska trek from the wreckage of the gold rush town of **Bremner** to the **Tebay Lakes** is tougher than most, and the fact that it can take nearly two weeks to trudge attests to its difficulty and may even encourage participants to consider an easier journey instead. Hikers can tell they are making progress by checking off what they'll encounter: a few dozen frigid streams that need to be waded across; glaciers (which, hopefully, will still be there); a half dozen ice fields; and nine icy and challenging passes. The trail is steep, overgrown, and the footing can be tricky, but it places the adventurer in the middle of the world's largest protected swath of wilderness. Plenty of planning and prior Alaskan wilderness experience is a must—or go ahead and hire a guide to help you tough it out. Doing so makes you one of the few people on Earth to return home with visions of rivers, lakes, glaciers, and heroic mountains rarely seen by human eyes.

10 | Appalachian
NATIONAL SCENIC TRAIL
Georgia to Maine

Along with "thru-hikes" on the long-distance Pacific Crest Trail and Continental Divide Trail, accomplishing a similarly straight shot on the fabled **Appalachian Trail** completes what's lauded as the Triple Crown of American hiking. Although some extreme hikers have completed the 2,190 miles from Georgia to Maine in as little as 41 days, allot at least six months for the expedition—while adding a few more years just to plan and train for it. It's no exaggeration. By the time someone's walked from Springer Mountain, Georgia, to Mount Katahdin, Maine, they've hiked through 14 states and balanced along the ridge crests and ventured into the valleys of the Appalachian Mountains to see national parks, national forests, state parks, historic sites like Harpers Ferry and the Bear Mountain Bridge, then the Adirondacks, Blue Ridge, Catoctin, Piedmont ... the list and the views go on.

Pacific Crest National Scenic Trail

Hikers who have the fortitude and the time (about six months) may be tempted to tackle the entire 2,650 miles of the Pacific Crest Trail, a path that runs from Canada to Mexico or vice versa via California, Oregon, and Washington. If this seems too much (and it more than likely is), slice off a 200-mile portion of the trail—Section 20—which leads into and out of three national parks: Sequoia, Kings Canyon, and Yosemite. While on the move for several weeks, hikers make their way to or near Mount Whitney via the adjoining John Muir Trail, into Tuolumne Meadows in Yosemite's eastern side, and to nice views of the high-elevation desert of the Owens Valley, located within distance of a hitch into the town of Lone Pine for supplies. In addition to views of canyons and meadows, other rewards include a refreshing encounter with Chicken Springs Lake or beautiful alpine-like Guitar Lake, and much appreciated campgrounds featuring small diners and well-stocked stores. It's wild. It's the West.

4×4 DRIVES

National parks preserve a unique beauty that ignites a sense of freedom—and that feeling is magnified when drivers discover the roads less traveled. Namely the thousands of miles of unpaved dirt, rock, and gravel traversable only by 4×4 vehicles, freeing people to explore remote stretches of backcountry that lead to rarely seen vistas.

1 Death Valley
NATIONAL PARK
California & Nevada

Just crossing Death Valley via paved Calif. 190 is a demanding experience, requiring scaling the Inyo Mountains before tackling the challenge of a long and lonely trek across the valley floor. Driving off-road is illegal in Death Valley, but threading through its three-million-plus acres is a potpourri of 4×4 roads. In total, there are more than **1,000 miles of roads** to explore, of which the majority are unpaved. These wilderness avenues test driving skills and nerves with sheer drop-offs, river crossings, and washboarded roads leading to sights the vast majority of park visitors have never seen, nor ever will: strange geological formations, hidden canyons, and the ruins of mining camps and cabins. **Titus Canyon Road** is the most popular backcountry road (a high-clearance two-wheel-drive vehicle is sufficient). Tips: Bring spare gas and water, don't rely on your GPS, and come between October and April to avoid summer temperatures that can clear 120°F.

2 Canyonlands
NATIONAL PARK
Utah

Considering that just a few paved roads run through Canyonlands, it's pretty clear that this destination is prime for 4×4 adventures. In fact, the central Utah park has so much remote backcountry that deciding where to go will be the hardest part of the journey. In the **Needles** and **Maze Districts** in the southeast and southwest portions of the park, extremely technical and treacherous routes are best tackled by skilled drivers. For everyone else, a somewhat easier alternative is sticking with the 100 miles of unpaved fun on the **White Rim Road** loop, which encircles the canyon in the Islands of the Sky section of the park. It can be rugged, rocky, and, at times, spookily dangerous with sheer drop-offs that, with a slip of the wheel, could turn a truck into tinfoil. All around are spectacular views of buttes and mesas and wide-open skies, while far, far, far below, the Green and Colorado Rivers wind sinuously through their respective canyons. Campgrounds dot the loop, so it is possible to turn a day trip into a leisurely exploration, allowing time to soak in the majesty of the red-rock country.

3 Padre Island
NATIONAL SEASHORE
Texas

Arched like a bow sandwiched between the Gulf of Mexico and the Intracoastal Waterway, the longest stretch of undeveloped barrier island in the world is custom

Brave the bumps in a 4×4 to explore less traveled park regions, like these unpaved roads in Death Valley.

Get a permit before driving onto the sand at Cape Point in Cape Hatteras.

designed for four-wheel touring. From the park's north entrance, a paved road parallels Malaquite Beach for about five miles then terminates at the beach vehicle barrier, after which **60 miles of beach** stretch south, which in Texas means 60 miles of public highway, complete with posted speed limits. While regular vehicles can travel down the first five miles or so along South Beach, a 4×4 is necessary to continue down Little Shell and then Big Shell Beaches to the end point at Mansfield Channel, within sight of South Padre Island. From about milepost 20 to 40 vehicles will encounter soft and/or deep sands. And some stretches of beach can become particularly tricky when affected by weather, wind, and

tides. The farther south one goes, the more remote and tranquil the scenery, with Gulf waters lapping the shore to the east and dunes, grasslands, and mudflats (all three habitats off-limits to vehicles) to the west, backing the beach. The park has plenty of primitive campsites for those who wish to camp out on a piece of land that hardly ever sees a human.

4 | Cape Hatteras
NATIONAL SEASHORE
North Carolina

The **70 miles of beautiful Atlantic shoreline** in North Carolina's Cape Hatteras are a magnet for off-road aficionados. However, national parks must frequently balance the

desires of visitors with the needs of nature, and at Cape Hatteras that means periodically imposing beach closures or driving restrictions in order to protect the nesting grounds of endangered sea turtles and a wide range of birds, especially between mid-March and early November. That may sound limiting, but the amount of protected shoreline usually fluctuates, so a generous number of miles should always be accessible. The barrier island beaches can be challenging—savvy four-wheelers deflate their tires to about 18–25 psi to get a wider grip and reduce demand on the transmission. In addition to the sea oats and sand and wonderfully picturesque Atlantic Ocean and Pamlico Sound (accessible by off-road vehicle, too), the seashore is also home to the tallest brick beacon in the United States: 210-foot-tall Cape Hatteras Lighthouse.

5 | Great Sand Dunes
NATIONAL PARK & PRESERVE
Colorado

Drivers could motor from Maine to California and never spy sand dunes as majestic as the ones within this park (some surpass 700 feet), and that's reason enough for 4×4 owners to make tracks for Colorado. Even though the dunes are off-limits to motorized vehicles, the **Medano Pass Primitive Road** more than provides an eventful experience in the shadow of the dunes. The road begins at the ominously named Point of No Return parking area, beyond which high clearance is required. The road threads the land between the dunes and follows Medano Creek up into the Sangre de Cristo Mountains, stretching 22 miles from Great Sand Dunes to Colo. 69 on the east side of the mountains. Over the dozen miles to 9,982-foot Medano Pass, drivers must tackle soft, deep sand for a couple miles; inch up the mountains, fording nine streams; and contend with a rocky roadbed on the approach to the pass. At the pass, the road exits the national park and preserve and heads down 10 miles through San Isabel National Forest. Now named Forest Service Road 559, the road enters the Wet Mountain Valley and intersects with Colo. 69, a paved road that can turn this into a daylong loop ride. Drivers should expect to travel only five to 10 miles an hour. The road is usually open between May and November, weather permitting.

6 | Joshua Tree
NATIONAL PARK
California

Stark and spare, the landscape in Joshua Tree National Park is well suited for 4×4s. There's a limited but far-reaching series of dead-end and loop wilderness roads that follow historic paths down old mining roads, across dry washes, and into canyons; others bump along rocks and around boulders while demanding extra traction to summit steep hills leading to panoramic views. The dirt roads in the **Covington Flats** area access the most verdant parts of the park and some of its largest Joshua trees; an overlook at Eureka Peak provides a view of the neighboring community of Palm Springs and the Morongo Basin. The 18-mile **Geology Tour Road,** on the other hand, delivers adventure in the form of rock piles, boulders, parched ravines, and a dam created by ranchers more than a century ago to water their cattle. There will also be remnants and ruins of a long-gone industry seen in an assortment of abandoned vats and mines and dangerous shafts that should pique anybody's curiosity—and sense of caution.

4×4 Adventures: Be Prepared

Outside many national parks where 4×4s are permitted, a fleet of vendors are usually geared up to lead off-road tours. In addition to having the right equipment, their unique understanding of the area's history and terrain will help guide you to the region's most beautiful and intriguing destinations. Not only that, if a jagged boulder rips open the undercarriage of their truck, they'll call their insurance company—not yours. But if you insist on traveling solo in your own 4×4, do so with the understanding that while it can be fun, it's not a game. Often you'll be traveling outside cell phone range and beyond the reach of park services and rangers, so if your vehicle gets stuck, breaks down, or someone is injured, you're essentially on your own.

Take precautions. Have your vehicle inspected before setting out, talk with rangers about road closures, pack current maps and a GPS, and carry plenty of water, extra fuel, and all manner of vehicle fluids. Stow a tool kit, spare tire(s), towrope, shovel, and high-lift jack. If this seems like too much preparation, consider the rewards that await you: rarely seen vistas away from the maddening crowds.

Canyonlands' White Rim Road (p. 132) provides one of the most stunning 4×4 adventures in the United States.

7 | Rocky Mountain
NATIONAL PARK
Colorado

Most visitors to Rocky Mountain National Park are content with the miles of paved roads tangled up in the park's 266,000 acres, and in fact the park is off-limits to 4×4s. But those who want a more intimate experience with nature head for the manageable dirt and gravel **Old Fall River Road,** a seasonal one-way tract on the east side of the park. Open from roughly early July to when the snow arrives in September or October, Old Fall River Road is an entry-level off-road excursion. In the 11 miles between the start of the road at Horseshoe Park and the Alpine Visitor Center at 11,796 feet, drivers proceeding at a cautious 15 miles an hour will pass large boulders, tall trees, and deep valleys and have a chance to stop at pullouts for scenic views and perhaps even spy elk, deer, and bighorn sheep. One of the most impressive sights is the road itself, the first route to clear the Continental Divide, in 1920. The narrow road does feature some tricky switchbacks with zero guardrails, but those moments are more than compensated for with views of alpine tundra, wildflowers, valleys, and mountains.

8 | Grand Canyon
NATIONAL PARK
Arizona

Unfortunately for four-wheel drivers in Grand Canyon National Park, mules have the advantage in navigating the depths of the canyon. Where drivers do win out is on roads where they're given free rein to explore portions of the backcountry that frame the chasm, primarily near the North Rim. Those who travel with a sense of adventure will find miles of unpaved roads at the end of Ariz. 67 on the subalpine **Kaibab Plateau,** although North Rim services are open only between snow's departure and its return, typically mid-May through mid-October. Among the routes for 4×4s are primitive roads leading through canyons and to lookout points. One of the most popular routes (a relative term here) is the 17-mile tract of dirt and gravel that leads west through a lush forest of spruce, pine, fir, and oak. Along the way, rocks crush and grind under tires as the road alternates between roughly smooth and seriously bumpy. The payoff is the end point, Point Sublime. Here the view is as

▶ *Did you know? In the state of Texas, virtually all beaches are considered public highways, meaning that all on-road laws apply.*

majestic as any afforded by the South Rim—except here you could likely have the Grand Canyon all to yourself.

9 | Capitol Reef
NATIONAL PARK
Utah

The 58-mile **Cathedral Valley Loop** in Utah's Capitol Reef is a feast for the eyes and a challenge to drive. Two-wheel drives with high clearance can travel this road, but the extra traction of a 4×4 can be much more comforting when weather tangles with this unpaved trail of rock, sand, and dirt that snakes through the mountains. The access point is on Utah 24, about 12 miles east of the park's visitor center (with another entry point found about eight miles past that), outside the park. The loop is created by stitching together Utah 24 and the dirt Hartnet and Cathedral Roads. Most visitors begin the loop on Hartnet Road, which requires a ford of the Fremont River shortly after turning off Utah 24. The road leads to remote regions and geological

curiosities, including spectacular multicolored mesas, spires, and other rock formations. The road is usually closed from mid-November to mid-March, when not even the most skilled drivers and sturdiest 4×4s find it navigable.

10 | Big Cypress
NATIONAL PRESERVE
Florida

Everglades National Park doesn't allow off-roading, but Big Cypress National Preserve, which is adjacent to Everglades and shares its same ecosystem, has many roads and trails that invite off-road vehicles. With a valid permit, you can explore sections of the 720,000-acre preserve that are nearly impassable on foot in order to enjoy hunting, fishing, frogging, camping, wildlife observation, and other recreation. Approved trails are available in the **Bear Island, Corn Dance, Stairsteps,** and **Turner River** units of the park (secondary backcountry roads remain closed pending an environmental review). When applying for your permit, be sure to check for annual 60-day closures, which are intended to ease visitor pressure on the preserve's natural resources. Big Cypress has both tropical and temperate plant communities, but by far its most captivating resident is the protected Florida panther, an elusive subspecies of cougar.

TRAIL RUNS

From the heart of a high desert to a shaded New England trail, routes vary as widely as runners. And that's the beauty of America's national parks: There's a jogging path to match them all. Lace up a pair of trail shoes and set out to pound the ground—your new favorite running route may just be the next one.

1 | Cuyahoga Valley
NATIONAL PARK
Ohio

It's easy to picture the past when you're traveling the **Ohio & Erie Canalway Towpath Trail.** For decades in the 19th century mules towed canalboats loaded with goods and passengers, leaving worn indentions with each step and every mile. Today's runners can journey in these proverbial hoof-steps, which are now paved or covered in crushed limestone to make up the 101-mile Towpath Trail. The northernmost trailhead is Canal Basin Park in downtown Cleveland, while the southernmost trailhead is Canal Lands Park in New Philadelphia. Twenty miles lie within Cuyahoga Valley National Park (and this sole part is open 24/7), while other sections are managed by state and local parks divisions. Overall, more than 2.5 million visitors traverse the towpath annually, oftentimes enjoying diversions to natural and historic sites, such as the Canal Exploration Center, 63-foot Brandywine Falls, and the charming Peninsula community.

2 | Valley Forge
NATIONAL HISTORICAL PARK
Pennsylvania

Rich with history, Valley Forge also offers an opportunity for a pleasing run along a wonderful loop road. The scenery protected within the park's boundaries—from the cannon and monuments to the restored headquarters of George Washington—is so impressive and this land so bucolic that it's a challenge to imagine the grounds as they were in the winter of 1777–78, when George Washington and his Continental Army toughed it out for several harsh months. Starting near the visitor center, the six-mile **Joseph Plumb Martin Trail,** a hilly, paved multipurpose path, parallels the scenic North and South Outer Line Drives, providing a perfect avenue for joggers. Sweeping across rolling hills and past re-creations of cabins constructed by troops to shelter themselves from the snow and sleet, the path stretches from the visitor center past the magnificent National Memorial Arch and on to Knox's Quarters. The trail connects with others to make a loop of the park's windswept fields, filled with deer and birdsong.

3 | Rock Creek
PARK
District of Columbia

Washington, D.C., isn't only for politicians. Runners can find much to enjoy, including several choice trails in Rock Creek Park, one of the oldest federal parks (established 1890). At just over 1,700 acres (twice that of Central Park), Rock Creek Park

You can find charming Boulder Bridge, built in 1902, upstream from the Blagden Mill site in Rock Creek Park.

packs in an impressive 20 miles of dirt trails that usher runners into an assortment of tree-lined paths along wooded hillsides, ridges, and valleys that are perfect for a gentle jog. The blaze-marked single-track **Western Ridge** and **Valley Trails** are both nicely suited for runners ready for a moderate workout, not a marathon. For a little more elbow room, trot onto one of the wider paths that double as equestrian trails (so watch your step) or drop down to the paved multipurpose route that parallels the park's eponymous stream. By using connector trails and roads, it is possible to make short and large loop runs. In addition, many trails branch out of the park into nearby D.C. neighborhoods, such as Dumbarton Oaks and Palisades Park, as well as into the Maryland suburbs.

4 | Muir Woods
NATIONAL MONUMENT
California

When even motorcyclists find this coast region's paved roads challenging, imagine what joggers feel as they take on the jittery single-track trails of Muir Woods. Snaking in and out of the national monument, the **Dipsea Trail** is one of the most challenging in the nation. Each year the single-track terror attracts hundreds of contestants to a century-old race famed for its utter punishment of runners. The trail requires a vertical ascent of more than 2,200 feet over its 7.4-mile length from the town of Mill Valley to Stinson Beach, so prepare for demanding climbs—including three sets of stairs, together numbering more than 685 steps—and the obstacles of rocks, roots, branches, and slopes that are alternately grueling and, when conquered, rewarding. Runners should pace themselves and enjoy the scenery (walk if need be; many runners do in places); focus on the cooling pleasures of the old-growth coast redwood forest (the only one in the San Francisco Bay Area), and soak in the awe-inspiring Pacific Ocean views at Stinson Beach.

5 | Golden Gate
NATIONAL RECREATION AREA
California

With the twin icons of San Francisco and the Golden Gate Bridge providing inspiration (and a water bottle providing hydration), the maze of trails in the **Marin Headlands** is a runner's delight. One option is to scale a solid section of the Coastal Trail from Rodeo Beach, climbing into the headlands and cresting out along Wolf Ridge, then descend on the Miwok Trail to find a way through the Tennessee Valley. Another option from Rodeo Beach is to head south around Rodeo Lagoon and capture sections of the coast en route to water's edge at the Point Bonita Lighthouse. Both runs are out-and-backs, but GPS can help you map a variety of loops with varying distances. If your body's running

Rail to Trail

Throughout the country, long-abandoned railways are bustling once again, this time with bikers, runners, walkers, and other outdoor enthusiasts. In fact, it's estimated that more than 25,000 miles of rail corridors have been converted to trails nationwide. By design, rail trails travel through some of the country's most iconic locations, including historical monuments, seashores, and parks. Here are just a few of note:

George S. Mickelson Trail (South Dakota): Venturing 109 miles and passing through 100 converted railroad bridges, this rail trail is just a short jaunt from sites like Wind Cave National Park and Mount Rushmore National Memorial.

Mammoth Cave Railroad Hike and Bike Trail (Kentucky): A nine-mile portion of the route formerly used by the Mammoth Cave Railway (which made its last run in 1931) has been converted to highlight scenic overlooks and interpretive waysides.

Greater Yellowstone Trail (Idaho and Wyoming): Once completed, the 180-mile-long trail will connect Grand Teton and Yellowstone National Parks in a world-class regional trail system.

Moab Canyon Pathway (Utah): Just 13 miles long, this trail features surreal rock formations, including arches and cliffs, as it traces the border of Arches National Park.

on autopilot, take in the setting's remoteness, including the hills and fields that accent windswept cliffs overlooking the Pacific Ocean. It's worth the time spent training.

6 | Cape Hatteras
NATIONAL SEASHORE
North Carolina

The coastline of North Carolina's Outer Banks is a natural jogging trail for runners. Finding the optimal stretch of strand to sample would be difficult were it not for **Old Lighthouse Beach** and the Cape Hatteras Lighthouse. At 198 feet, this beacon, one of three within the park boundaries, is the tallest lighthouse in the United States. The chance to run with this iconic landmark in sight is the "OBX" (Outer Banks) equivalent of a Paris jog alongside the Eiffel Tower. Although eroding sands necessitated the lighthouse's move a half-mile inland in 1999, the shoreline is still close enough to start the day with a sunrise beach run here ... or perhaps return at dusk for a brisk jog in the pulsing glow of its powerful light.

7 | Acadia
NATIONAL PARK
Maine

In 1919 a group of "Rusticators"—wealthy families who summered on Mount Desert Island—donated a substantial 30,000-plus acres of mountains, lakes, and seashore to

▶ *First held in 1905—and taking place the second Sunday every June—the annual Dipsea Race at Muir Woods National Monument is the oldest trail race in the country.*

the federal government to help create Acadia in Maine, the first national park east of the Mississippi. In addition to this generous gesture, John D. Rockefeller, Jr., advanced the design of 45 miles of **carriage roads** that encircle the isle. These roads now serve in part as excellent running trails. One particularly pleasing stretch of about three miles wraps around Witch Hole Pond and offers a splendid path through eastern deciduous trees, including oak, maple, beech, and other hardwoods. Try to come during the fall, because in a region known for its fiery burst of changing leaves, autumn trail running in Acadia can seem, in the words of one runner, "as if you're running through a painting."

8 | Rocky Mountain
NATIONAL PARK
Colorado

The thrill of running in the Colorado Rockies is not only in the sights, but in the challenge. The six-mile round-trip run around the **Loch Vale**

begins at the already thin-air altitude of 9,240 feet before scampering as high as 10,210 feet. The incentive, of course, is the pure joy of being here, where the waters of Andrews Creek and Icy Brook flow from Andrews Glacier and Taylor Glacier into the Loch, a subalpine lake as picturesque as one can imagine, with emerald green aspens and evergreens framing the sapphire blue lake. From the Glacier Gorge Trailhead the moderately difficult trail climbs past Alberta Falls through a gorge to Glacier Junction where, incredibly, two glacial valleys have converged. The trail then clears Icy Brook and springs ahead through a series of switchbacks that climb up through the Loch Vale toward the waters of the Loch. The ultimate reward: even more spectacular views of an unspoiled American wilderness.

9 | Lewis and Clark
NATIONAL HISTORICAL PARK
Oregon & Washington

It was quite the feat for 28-year-old Meriwether Lewis and 32-year-old William Clark to cross the country and reach the Pacific in December 1805. A run in this historic park that honors their achievement is nowhere near as exhausting. The park shares the accomplishments of the explorers with nearby state parks and preserves in both Oregon and Washington. In Oregon, the **Fort to Sea Trail** runs between Fort Clatsop, the winter encampment for the Corps

The Sand Point Trail's raised boardwalk leads runners through Olympic's lush vegetation.

of Discovery, and Sunset Beach on the Pacific coast. Appreciate the beauty of the Pacific Northwest along the 6.5-mile trail—the misty forests, streams, dunes, and endless sea appear untouched by time. The **Lewis and Clark River Trail** connects to the Fort to Sea Trail at Fort Clatsop. It runs 1.5 miles to Netul Landing, the expedition's canoe landing on today's Lewis and Clark River, which the Corps paddled up to reach their winter campsite.

10 | Olympic
NATIONAL PARK
Washington

The eight-mile round-trip **Spruce Railroad Trail,** on the shores of Lake Crescent in Olympic National Park, is a tale of timing. The railroad was initially built to transport harvested spruce timber intended for World War I aircraft, but it wasn't completed until 1919, which was too late for its planned purpose. Used as a common carrier and for logging after the war, the railroad was eventually abandoned in 1951. Following extensive upgrades, including accessibility expansion and the renovation of the McFee and Daley-Rankin Tunnels, the Spruce Railroad Trail opened as a beloved segment of the grand 135-mile **Olympic Discovery Trail.** Runners of all skill levels rejoice at the route's level grade and scenery, such as the strikingly blue Devil's Punchbowl and bridge.

MOUNTAIN BIKE RIDES

Sporting scraped elbows, bloody knees, and big smiles, mountain bikers appear as rugged as the trails they tackle. Steep passes and deep woods are no obstacle in their pursuit of fire roads, back paths, and coveted single-tracks—narrow routes that slice across deserts, rocks, and forests—set against the backdrop of beautiful national parks.

1 Mammoth Cave
NATIONAL PARK
Kentucky

Although the main attraction at this Kentucky park is a massive hole in the ground, there are plenty of aboveground thrills as well, including a set of trails and loop mountain bike rides that either dive into the forest, follow a paved path, or provide a grab bag of dirt and gravel. For a surf-and-turf ride, mountain bikers have dubbed one route the **Ferry Loop,** a 32-mile long-distance run that includes a ferry trip across the Green River; the route sticks to roads, including six miles of gravel. Kicking off from the Maple Springs Trailhead, the nine-mile one-way **Big Hollow Trail** cuts through the park, featuring dense woodlands and rocky outcroppings. South of the Green River, the **Mammoth Cave Railroad Bike and Hike Trail** is another nine-mile run that parallels a historic railroad bed from the visitor center to the park's southern boundary at Park City. Along the route, find historic sites, interpretive displays, and scenic overlooks.

2 Indiana Dunes
NATIONAL PARK
Indiana

The eponymous dunes are off-limits to bikes, of course, but with more than 37 miles of interconnected trails winding through Indiana Dunes National Park, cyclists are sure to find plenty of awe as they tackle their treks. The park's most rugged ride is perhaps the gravel **Calumet Bike Trail.** At 19 miles round-trip, the route follows the South Shore Railroad tracks before connecting to other trail systems. Although Calumet is flat, prepare to slog through standing water, which can pool up to six inches in spots much of the year. The reward is more water, but this time in the form of sandy beaches, as the park's 15 miles of Lake Michigan shore give sweet relief after a day of rides. Tip: The South Shore Railroad, which has railcars designated as bike-friendly, has multiple stations within Indiana Dunes.

3 Big South Fork
NATIONAL RIVER & RECREATION AREA
Kentucky & Tennessee

In Big South Fork, mountain bikers have the best of all worlds: access to horse trails, backcountry roads, and single-track trails like the 8.4-mile **Collier Ridge Loop** with its repeated climbs and exciting descents. The **Grand Gap Loop** is also now open daily to bikes, so riders can take on a single-track that rips into some serious wilderness and challenges riders for 6.4 miles. Expect dense southern forest and drop-offs, ridges and creeks, and a track that includes tight switchbacks then swirls across the land past a lumberyard of felled trees mingled with rocks, dirt, and mud. Bikes bounce off small boulders, ricochet over roots, and negotiate narrow board bridges that make this a challenging run for any rider.

4 | Big Bend
NATIONAL PARK
Texas

Although Big Bend is located in one of the hardest to reach sections of the nation—southwest Texas, on the Rio Grande—when riders reach the middle of mile after mile of mountain-biking roads, they don't mind the drive. The relatively untouched roads make this a prime destination for mountain bikers. For those who like things slow and smooth, there are roughly **100 miles of paved roads,** while riders that prefer a fast and furious outdoor adventure can choose from more than **150 miles of backcountry roads** that

▶ *Boy Scouts of America provided 78,544 volunteer hours to build the Arrowhead Trails at New River Gorge National Park. The endeavor is one of the largest youth service projects in national park history.*

provide a grab bag of challenges atop dirt and gravel trails. Rangers and riders praise the merits of the **Old Ore Road,** a bumpy and rocky 26-mile one-way unpaved 4×4 road that reveals views of the Chisos Mountains. Another recommended ride is the 20-mile out-and-back from **Panther Junction to Chisos Basin,** which requires peak physical condition. The good news is that after conquering the narrow, steep, and curvy road—often working through 15-degree grades—once a rider hits the heights, the downhill run is a blast.

5 | Point Reyes
NATIONAL SEASHORE
California

A sampler of coastal habitats and terrains helps deliver a splendid range of mountain-biking options in this park, with trails that rip through

evergreen forests and skim across coastal scrubs and beside estuaries. While mountain bikers can't enter protected wilderness areas, they do have access to all **paved roads** as well as **fire roads, horse trails** (equines have the right of way), and a handful of **single-track trails.** The off-road trails are generally narrow and winding and grant riders admission to beaches and bays and atop mountain crests. Concentrate on a flurry of trails in the heart of the park or cross the border to pedal outside of Point Reyes and into the north unit of neighboring Golden Gate National Recreational Area.

6 | New River Gorge
NATIONAL PARK & PRESERVE
West Virginia

BASE jumping commands the lion's share of daredevil attention at New River Gorge; however, more practical outdoor adventurers are content to make use of the dynamic and dramatic hills and single-track trails cleaved into this densely wooded and highly challenging landscape in southern West Virginia. In 2011, more than 1,000 members of the Boy Scouts of America created the **Arrowhead Trails** stacked-loop system, which features four mountain bike loops. Also popular is the **Long Point Trail,** a three-mile round-trip path that unveils panoramic views of the New River Gorge and its soaring bridge. Arrowhead Bike Farm,

Canyonlands National Park is a mecca for mountain bikers, enticed by its challenging terrain and gorgeous scenery.

located just outside the park, features year-round camping, lessons, and guided adventures through the park. Bring your own bike or rent one from Arrowhead.

7 Canyonlands
NATIONAL PARK
Utah

It may not be trailblazing, but a ride along the **White Rim Road** in Canyonlands will be trail *amazing*.

Jeeps have already made this a popular ride, and even though it's not single-track, mountain bikers still enjoy the wide road's rough tread and the chance to pedal around the rim of the sandstone cliffs and canyons. Located in the Island in the Sky District, the road kicks off with a sharp descent down a cattle trail turned road called the Shaffer Trail, switchbacking down to the White Rim Road for the majority of the ride; at the end of the White Rim, riders

pick up Mineral Road, which leads to Utah 313 and back to the starting point at Red Sea Flat. Riders can stop to catch their breath at photo ops of White Rim Canyon. While some claim that hardcore riders can conquer the 100-mile loop in one long, bone-bouncing day, an ordinary person who has a goal to complete the circuit should allot at least three or four days. Campgrounds and rustic restrooms are tacked around the entire loop; bring plenty of water. If

Pass through ancient forests on Old Highway 101 in Jedediah Smith Redwoods State Park, part of Redwood National and State Parks.

the route is ridden counterclockwise, riders face a 1,000-foot ascent up switchbacks, so a vehicle stationed near the finish is a convenient way to get bailed out of the park.

8 | Redwood
NATIONAL & STATE PARKS
California

Aware that conservationists had to work overtime to protect the redwoods that remained after many had been felled by axmen, rangers at Redwood have been intent on safeguarding the environment by limiting the access of mountain bikers to backcountry trails. Designated cycling routes, often on rehabilitated logging roads, range from three to 11 miles in length and from easy and level to steep and difficult. Many of the trails interconnect with each other and with **U.S. 101,** allowing for lengthier loop rides and providing a variety of habitats in one ride, from coastal scrub and ocean views to ancient and second-growth coast redwoods, conifers, and Sitka spruce forest to open prairie. Some terrific single-tracks and vegetation seemingly plucked from the set of *Jurassic Park* make this park one of the finest for riders.

9 | Whiskeytown
NATIONAL RECREATION AREA
California

Nestled in the Klamath Mountains, Whiskeytown welcomes bicycles on nearly all trails. Add in the numerous routes found in the surrounding national forest lands, and you can be assured of enough trails and tracks for all levels of mountain bikers. At times riders have to put on the brakes and clutch their bike when fording a stream, but in the wilds of Whiskeytown that's part of the appeal. The **Oak Bottom Water Ditch Trail** is a relatively flat gravel-and-dirt single-track that follows the lakeshore for 2.7 miles. (It's also the best birding trail in the park, if you slow down for a brief break.) The three-mile **Mount Shasta Mine Loop** winds past an abandoned mine that dates back to the early 1900s, then climbs steadily to views of surrounding peaks before descending via a forest access road. Note: Entrance passes are required at the park.

10 | Hawai'i Volcanoes
NATIONAL PARK
Hawaii

Perhaps no park serves as the perfect commercial for the joys of mountain biking more than Hawai'i Volcanoes on the Big Island. The fact that motorists dominate the two-lane paved roads makes mountain bikers more desirous to reach the dirt and gravel paths that release them to more remote regions of the park, where adventure and vegetation are found in abundance. Beyond the obvious appeal of volcanic landscapes, the park also

frees riders to immerse themselves in deserts and rainforests, but when the craters call, a few paths command the most attention. Launching from just below 4,000 feet at the Thurston Lava Tubes, the **Escape Road** is really that—an escape route should the lava start to spew. According to a ranger, the gravel and rock trail is a "crazy insane steep downhill" and a "way challenging uphill" enveloped within a forest of ferns. Near the park entrance,

Escape Road intersects with the park's premier route, **Crater Rim Drive.** Here cars are restricted to the pavement, while mountain bikers can look for the drive's parallel gravel and cinder path perched above the steaming caldera floor. Cooled lava has closed the drive to a complete circle tour, so at some point riders have to double back. Exercise caution on the narrow drive, not only looking out for cars but also steering clear of the crater.

The Right to Bike

While national parks face more serious threats than outdoor enthusiasts who like cycling, it took significant time for the International Mountain Bicycling Association (IMBA) to convince the National Park Service that mountain bikers deserved a place to pedal. Founded in 1988, the IMBA lobbied for years that cyclists would be good stewards of the land. In 2002, it displayed an admirable level of persistence and prescience when it partnered with the Park Service's Rivers, Trails, and Conservation Assistance program, affirming that both organizations were committed to encouraging active exploration of national parks. In 2005, the Park Service agreed and, in the words of the IMBA, "formally recognized mountain biking as a positive activity, compatible with the values of our National Park System." The Park Service began to open mountain bike trails in its parks, starting with pilot projects on dirt roads and introducing single-track trails. The IMBA pledged to assist in the care, maintenance, and development of the trails.

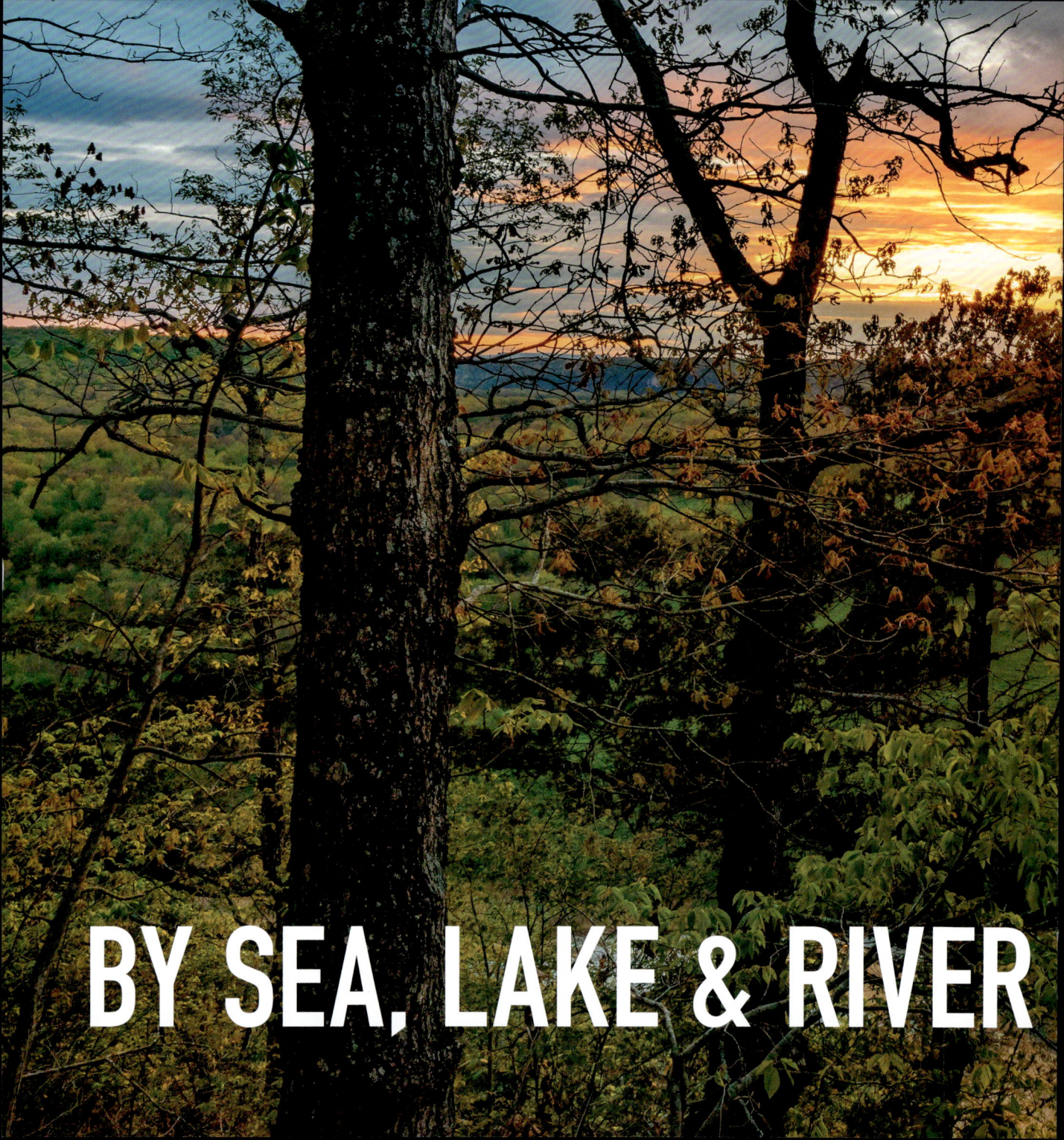

BY SEA, LAKE & RIVER

Sunset illuminates a portion of the Buffalo National River in Arkansas.

10 BEST PARKS FOR
FLY-FISHING

Properties in the National Park System protect entire functioning ecosystems. That includes fish. Although many parks allow fishing with a permit, regulations are strict and the sport is primarily catch and release. To that end, these spots are for people who love where they fish more than what they catch.

1 Yellowstone
NATIONAL PARK
Wyoming

Northwest Wyoming's **Firehole River**—the length of which is entirely within Yellowstone National Park—must be the strangest fly-fishing stream in the world, but in the best of ways. If you fish in the fall (the best season) in the early morning (the best time of day), you might think you've stepped into a different world. Gone down the rabbit hole, as it were. Mists from hot springs fill the valley. You can hear the thumping and wheezing of fumaroles and boiling water, perhaps the hiss of a geyser. Hulking bison grunt from the meadows, or loom ominously in the sulfur-smelling fog. The air is cold but the river is warm, heated a few degrees by the biggest concentration of geysers and hot springs in the world. During summer, the Firehole can be too warm for trout—they escape into cooler side streams. But in autumn and spring the river is one of the world's premier fly-fishing waters for **brown and rainbow trout.** Yellowstone was among the first to institute low-impact fishing regulations. The Firehole has been a fly-fishing-only stream since 1968 and for rainbow and brown trout it's catch and release only.

2 Great Smoky Mountains
NATIONAL PARK
North Carolina & Tennessee

Headwater streams above 3,000 feet in elevation are home to tiny **brook trout** as beautiful as warblers, and a big surprise if you think American freshwater fish can't put on the flash of their bright cousins in tropical waters. Red speckles ringed with blue spangle their green backs. Their bellies are orange, and their fins sport jaunty black and white stripes. These are gorgeous native fish living in their native place. The story of brookies in the Great Smoky Mountains is an unexpectedly happy one. As in many other places around the country, the introduction of non-native trout (in this case rainbows), along with habitat destruction before the park was established, decimated the native population. But a restoration effort came to the rescue. By removing non-natives from sections of streams protected by natural obstacles like waterfalls, the park gave brookies a chance to thrive without competition. In 2006, after 30 years of strict limitation, the park opened virtually all streams for brook-trout fishing. Today, most remain at fish capacity—a great opportunity for anglers. Among the best is Raven Fork, along the Enloe Creek Trail, and the East Prong of the Little River, along the trail from Elkmont Campground.

An angler casts for brook trout in a clear mountain stream in Great Smoky Mountains National Park.

3 | Grand Teton
NATIONAL PARK
Wyoming

The best fishing always seems to be in beautiful places. This is partly because natural habitat and clean water are the best protection for wild fish. But it might also depend on the meaning of the word "best." Catching a lot of fish (something that is almost never common) is less important than where you make your attempt. For many, wetting a line is merely an excuse to be outside in the happy embrace of nature. Beautiful water, good companions, a smoothly floating boat, and great craggy peaks in the background could be enough without the fish. But fish there are in this beautiful alpine setting in northwest Wyoming. The **Snake River** hurries out of Yellowstone, tarries for a while in Jackson Lake, and then runs fast over a stony bed through Jackson Hole. Most fishers float in rafts or drift boats. Some bank fishing is possible. You fish the river by drifting the bank, tossing flies as close to overhanging vegetation as possible, or by pulling out on a gravel bar and letting wet flies swing in the deep pools. The fish are Snake River **cutthroats,** which see an awful lot of fishers in a season and tend to be savvy. A good day for most people is to catch (and release, if it's November 1 through March 31) a couple fish, and to bask in the grand scenery.

Clear waters await in Rocky Mountain National Park.

4 | Rocky Mountain
NATIONAL PARK
Colorado

Fishing for **greenback trout** in the waters of Rocky Mountain National Park is a bit like searching for willow flycatchers in Arizona—a rare species, a special sighting. Greenbacks, a subspecies of cutthroat, sport prominent speckles, red lateral stripes, bright red cheek patches, and yes, a green back. Once widespread, they were displaced by planted exotics including brook, brown, rainbow, and Yellowstone cutthroat trout. In the 1930s, they were thought to be extinct, but in 1957 a small remnant population was found in the Big Thompson River. For several decades, the park and the state of Colorado have worked to restore greenbacks to their native habitat. The effort has seen some success. Put on the endangered list in 1978, they are now listed as threatened. In the park, restored populations of greenbacks (and their cousins, Colorado cutthroat) can be fished in specific lakes and streams using barbless hooks, but never kept. **Roaring River** and **Lawn Lake** are a center of interest for greenback fishers.

5 | Everglades
NATIONAL PARK
Florida

It's a far cry from little brook trout in mountain streams to 100-pound tarpon in the coastal waters of South Florida. The **Ten Thousand Islands** area near Everglades City on the Gulf Coast is prime water—a big place of open shallows, mangrove swamps, hidden backcountry inlets, lagoons, and beaches. Fishers typically use shallow-draft boats with motors to reach their fishing grounds but switch to quiet poling when they get close. Fishing from the shore is generally not practical. They call it sight fishing because you look for the fish before casting. You can see **tarpon** hanging in shallow water as if asleep, but they can be stirred to action by a well-placed fly tossed from a distance of about 60 feet. **Snook** prefer the cover of mangroves, where they lie waiting in ambush for smaller fish. **Redfish,** also called channel bass, red drum, and other names, are found in very shallow water, sometimes so shallow you see their dorsal fins. When taking a fly, they do it with speed and ferocity. This is not easy fishing. Searching is a challenge, and bugs do a lot of the biting, but those who experience Everglades fishing become passionate repeat visitors.

6 Yosemite
NATIONAL PARK
California

Nestled high in the Sierra Nevada peaks of eastern California, Yosemite National Park's Tuolumne Meadows is the upcountry opposite of the park's Yosemite Valley. Here, the mountains step back, the streams meander, and it's easy to understand what John Muir meant when he wrote "Oh, these vast, calm, measureless mountain days, inciting at once to work and rest!" The work is easy if anglers stay in the meadows, and not much harder if they walk up Lyell Fork beside the peacefully winding stream. It comes out of a steep-sided canyon with a nearly level floor—good waters for **brown trout** and for **brookies** higher up, if fishers backpack in and camp at least one night. The Dana Fork (which joins the Lyell to form the **Tuolumne River**) is smaller, more accessible, and faster moving. Think rainbows and, in the deeper pools, browns. The Tioga Pass Road parallels Dana Fork for about five miles. Road traffic might be heavy, but fishing pressure can be surprisingly light.

7 Chattahoochee River
NATIONAL RECREATION AREA
Georgia

Buford Dam holds back Lake Lanier just outside of Atlanta, Georgia. It also provides what some anglers consider to be the best **trout** fishing in the South. Purists, however, like to point out that almost all of the browns and rainbows along this 48-mile stretch of the Chattahoochee River below the dam come from hatcheries, and that the river needs continual habitat improvement to become a wild fishery. On the other hand, the river's proximity to Atlanta makes it a much loved urban asset, commonly referred to as **"The Hooch."** As with most hydropower tailwaters, river flow here can change dramatically—as low as 700 cubic feet per second to more than 10,000—which requires some planning for wading or boating. Fluctuation is less extreme during the cool months when peak power demands diminish. In addition to trout, a number

Let It Go

Preserve the thrill of the chase while limiting the impact on fish with these quick tips for catch and release:

1. Use barbless hooks (or use pliers to pinch down the barbs) to prevent unnecessary injuries.
2. Keep the tug of war brief. Prolonged fights can overstress a fish, causing it to produce a lactic acid that will reduce its chance of survival.
3. Cradle the catch without squeezing it and keep the fish in the water while removing the hook with wet hands.
4. Want photo evidence? Rather than lifting your fish into the air for a hero shot, cradle it in the water while someone else snaps a photo.
5. When releasing your fish, don't let it scrape or bruise itself, as rubbing off its mucous film can leave it susceptible to disease.

The Gunnison River is a popular location for fly-fishing.

enough for quick weekend visits, he specified that it be in Virginia's Blue Ridge Mountains near a first-rate trout stream. He chose a spot where two streams merge to form the scenic, boulder-strewn **Rapidan River,** in the newly authorized (but not yet formally established) Shenandoah National Park, saying, "Fishing is an excuse … for temporary retreat from our busy world." So it is, and the Rapidan remains a fine place to float dry flies for native **brook trout.** It's catch and release only, as are most waters in the park. There is no road access. Mill Prong Trail leads two miles from Skyline Drive to Hoover's Rapidan Camp on the river.

of other species can be found in the Hooch's cool waters, including **bass** and **catfish.** The park can provide a list of authorized fishing guide outfitters for both float and wade fishing as well as fly-fishing lessons.

8 | Black Canyon of the Gunnison
NATIONAL PARK
Colorado

Among the least visited of the country's national parks, Black Canyon of the Gunnison has minimal vehicle and pedestrian traffic, which means outstanding opportunities for anglers. But don't take our word for it—the **Gunnison River,** known as the Gunny, has earned Gold Medal Water and Wild Trout Water designations. Access doesn't come easy. The only route within the park that allows visitors to drive up to the river, East Portal Road, has 15 percent grades and hairpin turns. For those willing to hike down from the rim, a handful of steep, rigorous routes also lead to the water. The solitude, scenery, and self-sustaining trout populations make the effort worthwhile, though, as **rainbow trout** (catch and release only) and **brown trout** are abundant.

9 | Shenandoah
NATIONAL PARK
Virginia

When, in 1929, President Herbert Hoover decided to build a retreat far from the Washington scene but near

10 | Katmai
NATIONAL PARK & PRESERVE
Alaska

To catch wild **sockeye salmon** in the company of fish-catching brown bears holds a particular cachet for fly-fishers. There are few places in the park system where such an experience is possible—Katmai, at the head of the Alaska Peninsula, is one of them. Most venture here to watch the bears, who do their fishing without benefit of tackle. Bears concentrate at Brooks Falls but prowl the short length of the **Brooks River** between Brooks and Naknek Lakes. For humans, the river is fly-fishing only, and catch and release. Plan in advance—access is by floatplane, via Anchorage.

10 BEST PARKS FOR
LAKE FISHING

Drift along the shore, anchor in a bay, probe the inlets and narrow channels, or find a pleasant spot on land to relax while you keep an eye on a bobber. Using anything from flies to spinning lures to live bait, lake fishing can be as energetic or lazy as you like. Permits—from the state, park, or both—are required in all jurisdictions.

1 Glen Canyon
NATIONAL RECREATION AREA
Arizona & Utah

Imagine a gently sloping beach made not of sand but of smooth sandstone. Walk down to the water. You can see the stone slanting down into water of startling clarity before it gives way to a deep mid-ocean blue. You can get a sense of its scope by visiting the 710-foot-high Glen Canyon Dam. **Lake Powell** is on one side. On the other, the vertical walls of Marble Canyon drop away, far below, to the Colorado River. Then imagine walls that curve, that overhang like vast eggshells, pierced by arches and pinnacles and slot canyons, and you get a picture of the precipitous landscape that lies beneath the waves. With about 2,000 miles of shoreline (depending on water level), the lake surface is a delightful maze of channels and canyons. Stone beaches alternate with sand beaches and towering vertical cliffs. Fish love it, and so do campers, boaters, and anglers who come looking for **channel catfish, black crappie, walleye,** and **bass** (striped, largemouth, and smallmouth).

2 Amistad
NATIONAL RECREATION AREA
Texas

Amistad means friendship. The **reservoir** that carries its name, located at the confluence of the Rio Grande, Pecos, and Devils Rivers in Texas, shares its basin with Mexico. The lake has 850 miles of shoreline and strikingly clear water thanks to the quality of its contributing rivers, but also to its clean limestone bedrock and two large springs. This aspect of the lake attracts snorkelers and scuba divers, the most skilled of whom explore deep underwater caves in the limestone. No doubt they also come face to fin with the lake's famous **largemouth bass,** the fish most anglers here desire. Bass do well around the ledges and submerged cliffs of the long rocky shoreline. The lake also has **smallmouth, striped,** and **white bass,** as well as **channel** and **blue catfish.** A cliffy shoreline riddled with canyons, coves, and protected inlets provides good camping spots and enough variety to please even nonmotorized boaters.

3 Lake Meredith
NATIONAL RECREATION AREA
Texas

Formed by the Sanford Dam, and filled by the Canadian River, **Lake Meredith** seems all the more precious because of its arid, mostly treeless surroundings. **Walleyes** are the main objective for anglers in this Texas Panhandle lake, but also **bass, catfish, crappie,** and **trout.** The walleye population is naturally reproducing, with big fish running six to eight pounds. From April to June the lake provides the best walleye fishing in Texas. Fishers do best at night with minnows and night crawlers drifting slowly past cliffs and rocky outcrops. Summer and fall is the time for white bass, which, along with smallmouth bass, find good habitat in the lake's rocky basin.

4 Grand Teton
NATIONAL PARK
Wyoming

Carved out by glaciers, then enlarged by a dam before Grand Teton National Park was established, **Jackson Lake** can be an outstanding experience both for its location at the base of the spectacular Teton Range and for its excellent fishing. Jackson is a big lake when full, about 40 square miles and 438 feet deep at its greatest point. The water, coming out of Yellowstone National Park through the Snake River, or pouring directly off the Tetons, is as clear and enticing as mountain water can be. **Lake trout** (mackinaw) are the top draw for fishermen, but they share billing with **brook, brown,** and native Snake River **cutthroat trout.** While the best fishing may be June through September, the lake is open for fishing all year, with the exception of October when it protects lake trout during their spawning season. Ice fishing is the

▶ *Fish consumption advisories, often due to heightened levels of mercury or bacteria, have been in effect in recent years. Safe eating guidelines—many of which simply limit, not omit, certain fish—can be found at each park.*

local angler's secret, and it begins usually in January when the ice is thick enough to be safe. The other prime time happens in the spring, just after ice-out, when lake trout prowl shallow waters along the shore for prey, making them easy targets for bank fishing. After that they go deep, and anglers pursue them with trolling rigs. Access is easy, with many marinas, parking areas, and trailheads scattered around the lakeshore.

5 Voyageurs
NATIONAL PARK
Minnesota

The Canadian Shield makes for some of the finest lake country on the planet. Ancient Precambrian rock, smoothed by the great glaciers of the last ice age, creates a wonderland of **interconnecting lakes and waterways.** Beginning in the late 17th century, French-Canadian voyageurs traveled the water highways carrying trade goods to the interior and paddling back with furs. Voyageurs, in the heart of Minnesota's North Woods, protects a generous sample of the countryside the legendary canoeists traversed. The park is perfect for exploring with a boat and fishing gear. The piscine list is long and distinguished: **walleye, sauger, northern pike, smallmouth** and **largemouth bass, yellow perch, sturgeon, lake trout,** and, the champion of northern fish, **muskie.** Fishing here, especially at dawn or dusk, can be an idyllic experience. The plunk of spinning lures hitting the water, the gentle splash of ripples against the canoe, and the haunting cry of loons make powerful music.

6 Isle Royale
NATIONAL PARK
Michigan

Unlike lakes surrounded by a park, this is a park surrounded by a lake. **Lake Superior** is the world's biggest freshwater lake measured by surface area, and third largest if measured by volume. Michigan's Isle Royale is an archipelago consisting of one long narrow island and more than 450 small ones. Bays, inlets, open water, and **inland lakes** offer variety for the angler. In Lake Superior, **lake trout** and **salmon** take top honors. **Walleye,**

Cast a line with clear mountain views at Grand Teton National Park's Trapper Lake.

yellow perch, and northern pike swim the inland lakes. Visiting the island is a wilderness experience and not a casual trip. Ferry services operate from mainland ports in Minnesota and Michigan and will carry a limited number of canoes and kayaks. The *Ranger III,* sailing from Houghton, Michigan, six hours away, can carry boats up to 20 feet long. Canoes and kayaks can be rented on the island at Rock Harbor Lodge, on the park's south side, the only lodging other than camper cabins and campsites (reserve well in advance for all). Lake Superior can be a dangerous place for small craft; the park advises canoers and kayakers to stay on protected inland waters. Plan accordingly: Isle Royale and the surrounding islands are closed to visitors from November until mid-April each year.

7 | Yellowstone
NATIONAL PARK
Wyoming

For those who care about fish and natural habitats, the case of **Yellowstone Lake** represents a true challenge. Its native **cutthroat trout**—pure and unhybridized—are the core of a vibrant ecosystem that ties together the lives of fish, grizzly bears, osprey, cormorants, white pelicans, river otters, bald eagles, and a range of associated creatures. This northwest Wyoming lake and its tributaries have the largest remaining concentration of inland cutthroat trout in existence. **Lake trout,** fierce predators that feed on young cutthroat, were illegally introduced over several years beginning in the late 1980s. Since then, despite urgent efforts by management, they have proliferated and cutthroat numbers have fallen drastically. Spawning runs in tributary streams—once an important food source for bears and other fish-eating predators—have nearly disappeared. Lake trout, in contrast, spawn on the lake bottom. Because they live at depth, they are not a substitute food source for predators. It's a bad situation, but not a hopeless one. Gill nets, set deep, catch lake trout but not cutts. Efforts to target lake trout spawning areas show promise. If lake trout numbers can be kept in check, cutthroat are likely to survive, albeit in reduced numbers. For anglers, this is a chance to fish for a higher purpose. The park requires that all lakers caught must be kept or killed. Catch all you can. Do it for the ecosystem.

8 | Everglades
NATIONAL PARK
Florida

Everglades National Park is renowned for the saltwater fishing in its bays and mangrove swamps, but the park's inland waters—known as the **River of Grass**—deserve equal attention. **Largemouth bass** are king, with big ones commonly weighing in at around five pounds.

Transcontinental Trout

If bluebirds can fly over the rainbow, can trout swim over the Continental Divide? The seemingly impossible could happen at Two Ocean Plateau south of Yellowstone. Here, Two Ocean Creek splits at the edge of a marshy meadow: Pacific Creek heads west through the Snake River to the Pacific Ocean; the Atlantic Creek flows east through the Missouri River to the Atlantic. When water levels are high, as in the spring snowmelt, trout with a nomadic urge could make the crossing. Mountain man Osborne Russell wrote about the "parting of waters and fish crossing mountains" in the 1830s. "Here a trout of 12 inches in length may cross the mountains in safety."

Canoers lazily paddle down a calm stretch of the Rio Grande running through Big Bend National Park.

to check river levels, as canoes occasionally must replace rafts when levels are low.

3 | Glen Canyon
NATIONAL RECREATION AREA
Arizona & Utah

Although the recreation area revolves around Lake Powell, the park also includes the lower reaches of the wild and rugged **San Juan**

River. One of the most remote corners of the American Southwest, the stretch of river between Goosenecks State Park and Clay Hills Crossing skirts the northern edge of the Navajo Indian Reservation and is only reachable by boat. The river runs fairly gently, with nothing more than Class II rapids to raise your heartbeat. It's wilderness camping all the way, mostly on beaches tucked beneath soaring red-rock

walls. A few outfitters organize guided floats along the San Juan, and trips can range from two to six days in length.

4 | Lake Clark
NATIONAL PARK & PRESERVE
Alaska

Three national wild rivers shoot through this vast Alaska park: the **Tlikakila,** the **Chilikadrotna,** and the **Mulchatna.** Each has a distinct personality, and all of them are ideal for float trips of up to a week in length. The Tlikakila is probably the roughest of the three, a glacier-fed waterway that rages with Class III and IV rapids. It flows entirely inside the park and tumbles into the east end of Lake Clark. With float trips that can range up to 230 miles, the Mulchatna is the longest, although most of the river meanders outside the national park. The Chilikadrotna lies somewhere between the two in location and mood, a swift and sometimes treacherous river that intersects with the Mulchatna in the upland forest west of Lake Clark. Floatplane is the only way to reach the headwaters of all three rivers. The season generally runs June to September, although the Tlikakila, at a slightly higher elevation, might not be runnable until early July.

5 | Gates of the Arctic
NATIONAL PARK & PRESERVE
Alaska

It's not so much another state as a different planet—the endless tundra on the north slope of the Brooks Range is in a part of Alaska that even most state residents have never seen. The entire park lies north of the Arctic Circle, and within its boundaries are six national **wild and scenic rivers** and half a dozen other large waterways that are eminently floatable, including the

▶ *In order to receive the lauded designation of national river or wild and scenic riverway, a water source must flow naturally without dams or diversions. The National Park Service manages more than 3,500 miles of such rivers.*

Nigu and Kobuk. A float trip here is edge of the envelope boating: Take everything with you and hope nothing goes wrong, because help is far away. But the rush along these rivers is unparalleled: herds of caribou thundering across the frozen plains, grizzlies fishing downstream (and hopefully downwind) from your campsite, tramping across ancient Inuit hunting grounds. Want to leave the planning to the pros? Outfitters run six- to nine-day float trips from June to August on several of the park's rivers.

6 | Wrangell–St. Elias
NATIONAL PARK & PRESERVE
Alaska

America's largest national park offers plenty of big rivers fed by glaciers and spring rain, and damless along their entire length,

including the **Copper** and **Chitina.** The Copper forms the western boundary of the national park as it shoots through a mighty gap in the Chugach Range on its way to the Gulf of Alaska. Although it's a tributary of the Copper, the Chitina is just as mighty, a meandering river with many islands and channels that runs right through the heart of the giant park. Smaller rivers in the park like the Nabesna and White are also floatable. It's not unusual to go days along these rivers without seeing another boat or human being; however, waterfront wildlife is plentiful: grizzly and black bears, moose and elk, bald eagles and more game fish than you could catch in a lifetime. Outfitters run six-day guided floats along the Copper River as well as a four-day trip that includes the Kennicott, Nizina, and Chitina Rivers between McCarthy and Chitina Ranger Station.

7 | Saint Croix
NATIONAL SCENIC RIVERWAY
Wisconsin

The lower St. Croix between Wisconsin and Minnesota may be hard-core houseboat country, but the upper reaches of the river and its **Namekagon** tributary are more float-trip friendly, and near ideal for canoeists. It takes about six days to float the 100-odd miles of the Namekagon between the eponymous lake and Riverside

Floatplanes

Many floatable rivers can be reached only by floatplane. Small aircraft mounted with pontoons that can land on lakes, rivers, and even glaciers, floatplanes were originally developed for naval use during World War I. After the war, pilots realized the value of these planes to commercial aviation, especially in remote places like Alaska and northern Minnesota, where there was a lack of airstrips but abundant flat water.

Floatplanes began appearing as early as the 1920s in many places that now fall within the National Park System, especially Alaska units like Lake Clark, Katmai, and Gates of the Arctic, where they continue to perform vital passenger, cargo, and public safety duties. They are also a part of the landscape in parks in the lower 48 like North Cascades, Isle Royale, and Voyageurs.

The "muscle cars" of the floatplane world are Canadian-made de Havilland planes—the DHC-2 Beaver and the DHC-3 Otter—single engine, propeller-driven aircraft with short takeoff and landing capability. With their powerful engines and deep roar, both aircraft seem more like wild beasts than something man-made. The Beaver has been around since 1947, the Otter since 1951, and although manufacture of the original models stopped long ago, hundreds of the planes are still in service and much loved by their pilots.

Landing near its confluence with the St. Croix. Numerous primitive campsites and put-in points make it easy for floaters to chop the journey into shorter sections. The Namekagon is slender and serene, flanked by thick northern Wisconsin woods and interrupted every so often by minor rapids and small dams that require a short portage. Similar characteristics prevail along the 100 miles of **upper St. Croix** between Gordon Dam and Taylors Falls (home of the St. Croix River Visitor Center), a trip that takes six or seven days. Contact an outfitter for canoes and kayaks, as well as maps and shuttle service.

8 | Alagnak
WILD RIVER
Alaska

The **Alagnak River** packs a lot of punch into its 79 miles: white water and wildlife, seclusion and incredible tundra and spruce forest scenery. From its source in Katmai National Park and Preserve, the river flows west across the Alaska Peninsula before pouring into Bristol Bay. The incredible quantity and quality of game fish makes the Alagnak one of Alaska's most renowned sportfishing rivers and a popular venue for fly-fishing float trips. Rainbow trout, char, grayling, northern pike, and five different types of salmon inhabit the river. But boating or fishing the park is no easy matter: A complete lack of roads means you need to hop a floatplane to Kukaklek or Nonvianuk Lakes near the river's headwaters. You should ask your pilot to do a flyover of the river beforehand to

Rafts meander down the Colorado River past the soaring cliffs of the Grand Canyon.

check overall river conditions and potential hazards. Vast stretches of the Alagnak flow free and easy, but a narrow gorge in the upper section churns up Class III rapids. It takes about six glorious days to float the river at a leisurely pace.

9 | Grand Canyon
NATIONAL PARK
Arizona

Perhaps the granddaddy of all float trips follows the roller-coaster **Colorado River** through the bottom of Arizona's big ditch. John Wesley Powell famously pioneered the route during a daredevil 1869 expedition to survey the length of the river. Powell and his men ran

the Colorado in double-ribbed oak boats with watertight compartments to prevent sinking; nowadays the float is undertaken in neoprene rafts, plastic kayaks, and throwback wooden dories that come the closest to replicating Powell's wild ride through the canyon. Between the main put-in at Lee's Ferry and the westernmost haul-out at Diamond Creek, the river throws up 42 rapids rated Class V or higher. That's an awful lot of white water. But floating the Grand Canyon is more than adrenaline. Side hikes lead to fern-covered gullies and ancestral Puebloan ruins. Wildlife along the river ranges from rattlesnakes to desert bighorn sheep. There are quiet places to swim

or contemplate and even beach parties after dark. Outfitters offer guided float trips of anywhere from one to 18 days in length. Self-guided floats of the heart of the canyon are also allowed, but each group must have at least one member with river skills that match the demanding Park Service criteria. Permits for private trips are obtained via lottery.

10 | Buffalo
NATIONAL RIVER
Arkansas

Proving that it's not just wildlife that can be rare and endangered, the **Buffalo** is one of a few rivers in the lower 48 states without dams, reservoirs, or other impediments to its flow. This fact was recognized in 1972 when the Buffalo became the nation's first national river. It meanders west to east across northern Arkansas before a rendezvous with the White River near Buffalo City. About 135 miles of the river corridor is designated parkland, an Ozark wilderness that shelters myriad wonders, from waterfalls to sinkholes to elk herds and 300 different types of aquatic creatures. Rangers lead guided seasonal float trips in the lower district between Buffalo Point and Rush Landing, a journey that takes five to seven hours with a mid-river lunch stop. Independent outfitters organize much more substantial floats, including multiday trips between Ponca and Woolum.

WHITE-WATER THRILLS

Rafts and kayaks are the best ways to experience the big thrills and spills of our national park's white water. Most of these routes can be run in a half or single day, although some are a ripe two days of adventure—we're talking rapids, whirlpools, and risky currents—with a much needed breather in between.

1 Denali
NATIONAL PARK & PRESERVE
Alaska

Born of snow, sleet, and glaciers in the mighty Alaska Range, the Nenana River runs a swift course down the east side of Denali National Park and Preserve before pouring into the Tanana River near Fairbanks. It's a wild one, especially through the narrow **Nenana River Canyon** near the park visitor center, a stretch that features wicked Class III and IV rapids like the Coffee Grinder, Royal Flush, and Razorback. The dozen miles of river between the visitor center and the put-out point near the town of Healy takes around two hours. And don't try it without a dry suit; even on the hottest summer days, Nenana's waters are just above freezing. Outfitters are available to run white-water trips through the canyon and along milder stretches of the Nenana.

2 Glacier
NATIONAL PARK
Montana

The long and twisty Flathead River forms the western boundary of Montana's Glacier National Park, separating the national park from national forest on the opposite shore. The entire length of the watercourse alongside Glacier has been designated a national wild and scenic river. The **North Fork of the Flathead** takes three or four days to run. But parts of the **Middle Fork** between Walton and West Glacier can be rafted in a single day. The portion through narrow John Stevens Canyon features challenging Class III and IV rapids, especially hairy during the early summer when water flow on the Flathead peaks. But there's also plenty of time to eye the scenery—snowcapped peaks rising on either side, thick forest along the shoreline, and maybe even a glimpse of bears, moose, or wolves. The Middle Fork also has Hollywood cachet: Parts of the 1994 movie *The River Wild* with Meryl Streep and Kevin Bacon were shot on location here. Local companies offer single and multiday trips on both forks of the Flathead in rafts and inflatable kayaks. Those who wish to run the river themselves can also rent equipment and hire shuttle services.

Fall foliage puts on a vibrant show for New River Gorge rafters.

3 New River Gorge
NATIONAL PARK & PRESERVE
West Virginia

True, it's called the **New,** but this rugged river is actually one of the world's oldest. Flowing northward from western North Carolina through Virginia and into West Virginia, 53 miles of the New pass through New River Gorge National Park and Preserve. Within those miles are what rangers consider two very different characters of river, and which access point you choose determines which type of ride you'll contend with. The upper Southern section is tamer, with beautiful, runnable rapids that tend to peak at Class III. The lower Northern section is a wilder ride, with cutthroat currents and angry rapids ranging from Class III to Class V. Licensed outfitters lead trips daily; for those with advanced river experience, running it on your own is sure to challenge both skill and stamina. All routes give you an appreciation for moving water and how, over millions of years, its passage carved a gorge through the mountains.

4 North Cascades
NATIONAL PARK
Washington

Summer months draw rafters to the **Skagit** and **Stehekin Rivers** in North Cascades National Park. The emerald green Skagit River usually runs deeper and has nine miles of

In the late 1950s Jackson Lake Lodge in Grand Teton National Park began offering commercial rafting trips using surplus military rafts, ultimately popularizing the sport.

Class II and Class III rapids. Considered ideal for families and those new to the thrill of white water, the Skagit's action happens within a quarter mile, leaving the rest of the journey for easy current coasting. From the superbly remote Harlequin Campground, a boat launch takes rafters to the wild and scenic Stehekin River, known for being crystal clear beyond spring's runoff season. Water temperatures seldom rise above 50 degrees on these Washington waterways, meaning wet suits can be nearly as important as life jackets for preventing injuries.

5 Dinosaur
NATIONAL MONUMENT
Colorado & Utah

River rats know there's more to Dinosaur than old bones. This national monument straddling the Colorado-Utah border also boasts some of the best white water in the West. The park's rivers churn up nasty rapids and whirlpools. The

Yampa and the Gates of Lodore on the **Green** are best done as multiday white-water trips with experienced guides. But a less intense stretch of the Green River through Split Mountain Canyon can be run in a single day. Class III rapids give the Green enough oomph for a mild adrenaline rush, but not enough to preclude kids as young as seven from testing their mettle on the river. The upstream access point is Rainbow Park, not far from the McKee Springs Petroglyphs. Downstream you can exit at several points along Cub Creek Road. Several local outfitters run Split Mountain trips, and the season typically lasts May to September.

6 Grand Teton
NATIONAL PARK
Wyoming

Only a small portion of the 1,000-mile **Snake River** flows through the Tetons, but it's about as gorgeous as a river can get, with ribbons of water set against jagged, snow-covered summits. But that handsome face hides a mean temper, a Wyoming river that should never be taken for granted. Advanced white-water skills are necessary for several sections of the Snake, in particular a stretch of white water in the Rockefeller Parkway between Southgate and Flagg Ranch. Despite a lack of rapids and whirlpools, the river south from Deadmans Bar

Rafters come face to wave with the "bone crusher" rapids of the Middle Fork of the Flathead River in Glacier National Park (p. 165).

can also get pretty nasty, thanks to tricky currents and logjams. Snake River Canyon south of Jackson Hole tenders a much longer stretch of white water. At least half a dozen outfitters in Jackson Hole offer half- and one-day trips along the Snake.

7 | Gauley River
NATIONAL RECREATION AREA
West Virginia

Charging through the mountains of West Virginia, the turbulent **Gauley** is the wicked river of the East. The numbers tell a tale of lurking danger: a drop of 668 vertical feet in 28 miles; a watercourse strewn with more than a hundred rapids, some of them gnarly Class V raft-eaters. That translates into fast, steep, rocky, and wild, a river that demands constant attention and a high degree of technical skill. Rapids like Shipwreck, Suckers Go Right, and Pure Screaming Hell earn their menacing names from reputation rather than whim. The upper Gauley is often ranked as one of the world's top 10 white-water runs. But the river does have a kinder, gentler side: The lower Gauley is less rigorous and open to rafters as young as 12 years of age. Needless to say, it's best to run this river with experienced local outfitters, many of which offer overnight and one-day trips along both the upper and lower sections. Unlike much of the white-water world, summer is not the season for rafting. The Gauley is only open over six weekends in September and October, when water is released from the Summersville reservoir upstream.

8 | Olympic
NATIONAL PARK
Washington

Rivers radiate from Washington State's Mount Olympus like spokes on a wheel, some affording trips that combine white-water adventure and the park's Pacific Northwest woods. The most popular trip is the Elwha River through the heart of the park, but this waterway's extensive restoration project has resulted in vast closures. Instead, the secluded **Hoh River** is a voyage through Oxbow Canyon and the lush rainforest that carpets the park's western fringe. Olympic's best white water is the **Sol Duc River,** where the Class III rapids can get a little daunting as it passes out of the park and into Olympic National Forest. But there's only enough water in the Sol Duc during the winter months, which means rafting and kayaking are limited to November through March—and the water is most definitely cold.

9 Big South Fork

NATIONAL RIVER & RECREATION AREA
Kentucky & Tennessee

This slice of the South seems little changed since the days when this neck of the Kentucky and Tennessee woods was the stomping ground of frontiersmen like Daniel Boone. The **Big South Fork of the Cumberland River** and at least **three of its tributaries** beckon white-water paddlers to a world of thick woods, abundant wildlife, and the occasional Class IV rapid. River conditions vary greatly, with some streams as quiet as a field mouse while others roar like the black bears that still live in these hills. Some waterways are best in summer, others only runnable after winter and spring rains. The toughest run is the 11 miles of the Big South Fork between Burnt Mill Bridge and Leatherwood Ford, where serious rapids are flanked by soaring sandstone cliffs. The section from Leatherwood down to Blue Heron Mine is also risky business, in particular Angel Falls and Devil's Jump.

10 Harpers Ferry

NATIONAL HISTORICAL PARK
Maryland, Virginia & West Virginia

Harpers Ferry might be an icon of American history, but the park at the confluence of the **Potomac** and **Shenandoah Rivers** is also a hub for rafting, kayaking, and tubing. As they tumble down from the Appalachians, both rivers churn up considerable white water upstream from Harpers Ferry. Renowned rapids include the White Horse on the Potomac and the mile-long Staircase on the Shenandoah. No matter which river is run, the trip eventually flows through the middle of the national historical park and the famous water gap at Harpers Ferry. Formed 360 million years ago when the Potomac cut through the Blue Ridge, the dramatic breach is one of the natural wonders of the mid-Atlantic region, a spot where three states (Maryland, Virginia, and West Virginia) collide in a mosaic of water, stone, and forest. Thomas Jefferson called it "one of the most stupendous scenes in nature ... This scene is worth a voyage across the Atlantic." Local companies offer guided trips on both rivers, typically March through November, although river runners in spring and fall should probably invest in a good wet or dry suit.

Gauley Fest

America's premier white-water festival is the annual Gauley Fest, which takes place over the third weekend in September in West Virginia. In addition to rafting and kayaking in nearby Gauley River National Recreation Area and New River Gorge National Park and Preserve, the event includes live Appalachian-flavored music, a white-water gear and equipment market, and close encounters with the nation's top river runners. The fest kicked off in 1983 as a way for paddlers and environmentalists to celebrate their successful battle against a hydroelectric scheme that would have severely impacted white water on the Gauley. Low-cost camping is available on the festival grounds during the entire four-day event.

10 BEST PARKS FOR
FLAT-WATER PADDLING

The nation's oldest boating tradition is the canoe, a craft used by both Native Americans and early Europeans as a means of transport, trade, and exploration. National parks preserve that heritage with canoe and kayak routes little changed from the days when the boats were made from birch bark rather than aluminum or plastic.

1 | Buffalo
NATIONAL RIVER
Arkansas

Known as a canoe stream—yet prime for kayaking to boot—the **Buffalo** is wild, beautiful, and strikingly clear as it freely flows more than 150 miles through the Ozark Mountains. The setting is stunning, too, with bluffs of limestone and sandstone soaring 400 feet from the riverbanks. The river is largely dependent on rainfall—and water levels can change by the day—so check with a ranger to determine the best district for your group's level of adventure. Generally the lower river's Buffalo Point has a quieter current perfect for families and beginners, while the more central Ponca section is most popular for canoeing. And while developed campgrounds and historic cabins dot the length of the river, backcountry camping is also allowed anywhere on the Buffalo.

2 | Big Thicket
NATIONAL PRESERVE
Texas

The name alone is enough to conjure images of paddling through primeval woodlands or swamps. And that's exactly what Big Thicket is: six water corridors and nine land units protecting more than 105,000 acres of east Texas wilderness. Encompassing several ecological zones, among them longleaf pine forest, southern swamp, and midwestern prairie, the park has enough size and scope to accommodate paddles from a couple of hours to a couple of days. The waterways are mostly slow flowing, with plenty of campsites and sandbars, and shorelines rich with indigenous flora and fauna. The **Neches River** on the park's eastern fringe presents the longest, wildest path for kayaks and canoes, winding through tall bluffs and dense vegetation. Paddlers can explore numerous lakes and bayous off the river, or keep following the Neches south of the park all the way to the Gulf of Mexico. The park's other great paddle route is **Village Creek** between the Big Thicket Visitor Center and its confluence with the Neches near Lumberton. Concessionaires rent boats to those who want to explore Big Thicket on their own and offer guided canoe trips of various lengths on both the Neches and Village Creek.

The Buffalo River is predominantly fed via rainfall, not perennial springs, so check the water levels before putting in.

3 | Voyageurs
NATIONAL PARK
Minnesota

Named after the French trappers and traders who roamed the North American interior in the late 17th and 18th centuries, Voyageurs is still the crème de la crème of backwoods canoeing and kayaking. This watery wilderness in northern Minnesota boasts more than 500 islands scattered across a vast labyrinth of **lakes, rivers, channels,** and **ponds.** The international boundary between the United States and Canada closely follows the classic voyageur route along the park's northern fringe, but there are hundreds of other paddling paths that take anywhere from a few hours to more than a week. Further adding to the park's canoe and kayak cachet is the fact that a boat is the only way to reach the 200-plus campsites. Numerous outfitters along the park's southern border rent vessels and camping equipment; check with the park for a list of authorized companies. Additionally, visitors without boats can join short ranger-led paddles that depart from the park's visitor centers during the summer months.

4 Congaree
NATIONAL PARK
South Carolina

Make like a "swamp fox" and paddle the **Congaree River** region of South Carolina. It was amid these watery woods that Continental Army general Francis Marion (nicknamed the "Swamp Fox") hid from the Redcoats during the American Revolution. The riparian scenery hasn't changed much in the years since. Nowadays, Congaree is the nation's largest remaining old-growth floodplain forest and home to some of the tallest, biggest trees in the eastern United States. The wide Congaree River runs roughly 25 miles along the park's southern boundary. But the most popular paddling ground is the **Cedar Creek Canoe Trail** through a gauntlet of huge cypress trees. Rangers lead guided paddles along Cedar Creek on weekends. Otherwise, visitors must provide their own boats. The canoe trail runs seven miles between Bannister's Bridge and Cedar Creek Landing and is easily

▶ *On May 3, 1962, U.S. Supreme Court Justice William O. Douglas completed a three-day trip down the Buffalo River, leading him to advocate for its preservation for "all the people ... as a remnant of the ancient Ozarks."*

paddled in four to six hours. More ambitious paddlers can pick up a backcountry camping permit at the visitor center and venture all the way down Cedar Creek to the Congaree River and a put-out point at the U.S. 601 bridge, a distance of around 20 miles. If water levels are low, paddlers should be prepared for portages and carry-overs.

5 Harriet Tubman Underground Railroad
NATIONAL HISTORICAL PARK
Maryland

On Maryland's Eastern Shore, Harriet Tubman Underground Railroad National Historical Park honors the best known "conductor" among the network of folks who helped enslaved African Americans escape to freedom. The park comprises 25,000 acres of federal,

state, and private land that tells the story of the liberator's life. Part of the expansive park is Blackwater National Wildlife Refuge, near the area where Harriet Tubman was raised. The **Blackwater** and **Little Blackwater** river systems have three water trails that pass through the refuge—the Orange (3.8 miles) and Green (four miles) trails are open year-round, while the nine-mile Purple closes from October until April to protect waterfowl. Here, immersed in the refuge and paddling the marsh, it's easy to let yourself imagine Tubman's time and the many heroic ways a freedom fighter traveled, hid, and survived in the 1800s.

6 Missouri
NATIONAL RECREATIONAL RIVER
Nebraska & South Dakota

The "Big Muddy" **Missouri** may look slow and placid as it flows through South Dakota and Nebraska, but America's longest river is one of its most challenging. Among its dangers and annoyances are submerged sandbars, waterlogged trees, floating debris, swirling eddies, and tricky currents. Careful preparation and caution are imperative. The park is divided into two units on either side of Lewis and Clark Lake. The 39-mile upstream portion threads its way through Karl Mundt National Wildlife Refuge and the Yankton Sioux Indian Reservation before tumbling into the lake. The 59-mile downstream section runs through ranch and farm country between Gavins Point Dam on the Nebraska–South Dakota border and Ponca State Park, Nebraska. There are numerous put-in places along the length of the park, but only a few campgrounds because so much of the shoreline is private property. Primitive camping is allowed on many islands and sandbars. Fully guided kayak expeditions are also available in both portions of the park.

7 | Mississippi
NATIONAL RIVER & RECREATION AREA
Minnesota

Spread along 72 miles of the **Mississippi River** on either side of Minneapolis-St. Paul, this eclectic park protects the upper reaches of America's most celebrated waterway. The Park Service manages the riparian trail in partnership with numerous local entities including counties, towns, nonprofits, and private landowners. The 40 miles of river upstream from Minneapolis is too shallow for large craft, but perfect for canoes and kayaks, an interesting blend of rural and urban landscapes that includes more than four dozen bald eagle nests. Numerous public boat ramps provide access to the river for those who have their own

boats. Those who wish to paddle the entire length of the park must navigate through four locks. Outfitters can set you up with canoes, kayaks, and other equipment.

8 | Niobrara
NATIONAL SCENIC RIVER
Nebraska

A watery wedge between Nebraska's Sand Hills and the wide-open prairie, the **Niobrara River** offers an environment little changed from the days when the Pawnee, Lakota, and bison were masters of the surrounding plains. The park protects a 76-mile stretch of the river as it slices through the Great Plains. While the river's lower stretches (below Meadville) are wide and calm, the upper reaches generate numerous rapids, including the Norden Chute (Class IV) and Rocky Ford (Class III). This western section is bounded by sandstone cliffs riddled with fossil beds and more than 200 waterfalls, including 63-foot

Epic Canoe Voyages

Early voyageurs may have explored the North American interior with birchbark canoes, but these more current record breakers have earned their accolades, too.

In April 1936, Geoffrey Pope and Sheldon Taylor launched a canoe at the foot of 42nd Street in Manhattan and the Hudson River and headed upstream. Sixteen months later they arrived in Nome, Alaska, after paddling thousands of miles through the Great Lakes region and western Canada.

Verlen Kruger and Steve Landick topped that in 1980–83 when they paddled 28,000 miles around North America, including the eastern seaboard, the Pacific coast, the Arctic Ocean, and Alaska, as well as the Missouri, Mississippi, and Colorado Rivers and the Great Lakes.

More recently, British lads Chris Maguire and Neil Armstrong paddled from Canada to the Amazon Basin from 1993 to 1996, a 13,000-mile odyssey that took them all the way down the Mississippi River and along the Texas coast.

Smith Falls, the highest in Nebraska. Paddlers can overnight at any of 10 campgrounds along the river. The usual boating season runs mid-April to mid-October, and outfitters in the nearby towns of Valentine and Sparks rent not only canoes, but camping equipment, too.

9 | Chattahoochee River
NATIONAL RECREATION AREA
Georgia

Country superstar Alan Jackson immortalized the **Chattahoochee River** with his 1993 hit "Way Down Yonder on the Chattahoochee." But the river's repute stretches back to antebellum days when local poets wrote odes to the fluid corridor. The northern Georgia park includes 48 miles of the Chattahoochee between Lake Sidney Lanier (named after one of those bygone poets) to Paces Mill on the outskirts of Atlanta. Conditions range from absolutely flat to Class I and II rapids depending on how much water is released on any given day from Buford Dam. Depending on your speed, it can take anywhere from 17 to 26 hours to paddle the entire length of the river inside the park. Given the fact there are no campgrounds and the park is not open overnight, the paddle will certainly have to stretch over several days.

The unique blue color of the water by Alley Mill, part of Ozark National Scenic Riverways, is due to mineral content.

Morgan Falls Dam presents a formidable barrier 35 miles downstream from Lake Lanier, making it impossible to continue southward without portage between Chattahoochee River Park and Morgan Falls Park. Canoes and kayaks, as well as paddles and life vests, can be rented from several local vendors.

10 | Ozark
NATIONAL SCENIC RIVERWAYS
Missouri

Dedicated in 1971, Ozark was the first U.S. national park intended to protect a wild river system. The name is plural because the park embraces two major waterways: 105 miles of the **Current River** and 41 miles of the tributary **Jacks Fork** in southern Missouri. Float trips are popular on both, especially during the sultry summer months. The lush scenery is enough of a draw on its own, a mosaic of maples, cottonwoods, sycamores, and willows that provides habitat for hundreds of bird, reptile, and mammal species. But the park also contains more than 300 caves and sinkholes, the world's largest assemblage of first-magnitude springs, and historic waterfront buildings like the 1894 Alley Mill. Paddlers can put in at two dozen spots, most of them equipped with campsites and picnic areas. Canoe liveries can be found at Van Buren, Salem, Eminence, and Akers Ferry.

SWIMMING HOLES

As the words of 19th-century naturalist John Muir so often remind us, national parks are places to cleanse our bodies, minds, and souls. This is especially true of the springs, natural pools, and swimming holes found at many parks. The waters—some hot, some cold—refresh and reinvigorate under the guise of recreation.

1 Big Bend
NATIONAL PARK
Texas

Down beside the Rio Grande, the **Hot Springs Historic District** of Big Bend National Park preserves the remnants of an early 20th-century bathhouse that drew health seekers to this remote corner of Texas. J. O. Langford homesteaded the riverside patch in 1909 with visions of transforming the area into a Western version of Hot Springs, Arkansas. Dubbing it the "Fountain of Youth That Ponce de Leon Failed to Find," Langford charged 10 cents a day or $2 for a three-week treatment in waters that could allegedly cure anything that ailed you. The resort endured until the 1930s, when a series of floods destroyed the bathhouse. Modern-day bathers will find that it's a two-mile drive down a gravel road and then a half-mile hike from the parking area to the stone ruins and a 105°F pool on the river's north bank. (Rio Grande flooding can cause sediment runoff to the springs, forcing occasional closures or simply the need to dig a bit before bathing.) It's not a huge pool, but the views across the Rio Grande to the rugged mountains of Mexico are sublime and the water soothing after a day of hiking in the park.

2 Olympic
NATIONAL PARK
Washington

Native American legend says the **Sol Duc Hot Springs** on the Olympic Peninsula are fed by the tears of dragons living in nearby caves. Located in the park's northwest region not far from Crescent Lake, the springs lie amid old-growth evergreen forest alongside a wild river renowned as a coho salmon spawning ground. The rustic hot springs resort offers three mineral pools and a freshwater swimming pool, as well as cabins for overnight stays. The water temperatures vary between 50°F and 104°F depending on the pool and the season. The resort is closed in winter, but during the spring and fall the hot pools are open during twilight hours to expedite a sublime combination of soaking and stargazing. The local Quileute frequented the springs long before European settlers discovered their curative powers. A huge lodge was developed around the pools in 1912, but it was destroyed by fire four years later and never rebuilt.

3 Chickasaw
NATIONAL RECREATION AREA
Oklahoma

Swimming holes are what Chickasaw National Recreation Area is all about, a cluster of **freshwater springs, pools, creeks,** and **lakes** in south-central Oklahoma. The water gushes up from the 500-square-mile Arbuckle-Simpson Aquifer that underlies the region, rising through porous limestone and dolomite to reach the surface. One of the oldest national parks, the spread was originally called Sulphur Springs Reservation, created in 1902 to protect the plentiful mineral springs along Travertine Creek near the town of Sulphur. Within a couple of years it had been renamed Platt National Park, a moniker that lives on in the park's Platt Historic District, where most of the best swimming holes are located. The most popular of these is Little Niagara, but it is also the most crowded. As part of the park's interpretive program, rangers lead creek and spring hikes, explaining Chickasaw's watery wonders. Chickasaw's 1930s Civilian Conservation Corps architecture is another draw. There are plenty of campgrounds for those who want to stretch their swimming over multiple days.

4 | Shenandoah
NATIONAL PARK
Virginia

Virginia's long, leafy national park is rife with swimming holes, most of them located far off the beaten track and accessible only to those who are willing to hike for a cool dip. Many of the trailheads are on Skyline Drive along the crest of the Blue Ridge. One of the easiest

> ▶ *The National Park System includes more than 150,000 miles of rivers and streams, and more than 4 million acres of lakes, reservoirs, and ocean sit within park system boundaries. Check safety information before swimming in any park waters.*

to reach is **Big Rock Falls** near the Byrd Visitor Center, a round-trip hike of roughly three miles. The trail leads downhill from the Milam Gap parking area toward Mill Prong Creek and a wilderness camp where President Herbert Hoover once stayed. The **Moormans River** in the park's deep south is another great place to get wet, reached via trails that start from the Wildcat Ridge parking area on Skyline Drive. Both the South Fork and North Fork of the Moormans have a number of swimming holes, but the best camping is found along the north branch, which also offers a side trail to lovely Big Branch Falls. White Oak Canyon in the park's central section, north of the Byrd Visitor Center, offers natural swimming pools at the bottom of several waterfalls, reached on a steep downhill trail

from either the White Oak or Limberlost parking areas off Skyline. On the other hand, some pools are more easily accessed via roads and parking areas along the park fringe. For instance, Nicholson Hollow and its excellent Hughes River swimming holes are best approached from the Old Rag parking area at the top of Va. 600 out of the town of Nethers.

5 | Sequoia
NATIONAL PARK
California

It takes several days of steady hiking along the High Sierra Trail to reach **Kern Hot Springs** in the backcountry of eastern California's Sequoia. But the effort isn't wasted, as a soaking spot secluded in a deep valley surrounded by Jeffrey pine, incense cedar, and soaring granite faces awaits. The main pool is a scorching 115°F, but the runoff mixes with cold water from the nearby Kern River to create a sublimely warm pool that's just right for dipping. Clothing optional? Of course, this far off the path—although a crude wooden fence around the hottest pool provides some privacy. Those who want to linger can stay at the primitive campsite near the springs. But it's first come, first served and very popular with backpackers trekking between Giant Forest Village and Mount Whitney. For those who enjoy ice-cold water, many of the tarns

May and June are popular months for visitors at Medano Creek in Great Sand Dunes National Park and Preserve.

along the High Sierra Trail are swimmable, including Upper Hamilton Lake and Precipice Lake.

6 | Great Sand Dunes
NATIONAL PARK & PRESERVE
Colorado

The continent's tallest dunes aren't the only point of wonder at Great Sand Dunes National Park and Preserve. Here, Colorado's natural beach draws visitors to wander and wonder as they explore **Medano Creek** and its seemingly mysterious surge flows. Snowfields in the Sangre de Cristo Mountains melt into Medano Lake, which cascades into wide and shallow creeks that flow around the base of the dunes. Medano Creek (*médano* is Spanish for "sand dune") typically hits peak flow in May, and it's at this time (through early June) that surge flow—when underwater sand ridges build up and break, sending down a wave—is most impressive. The phenomenon is so intriguing (and flat-out fun) that weekday visits are recommended to avoid the May and June weekend crowds. And don't delay: Snowpack makes all the difference in the depth and duration of Medano Creek, and by August and September water is typically available only via a one- to three-mile hike up the dry creek bed.

Midnight Hole is a popular swimming spot in Great Smoky Mountains National Park.

7 | Death Valley
NATIONAL PARK
California & Nevada

A former hippie hangout that was later invaded by New Agers, **Saline Valley Warm Springs** is a quintessential California desert escape. Visitors can reach the Saline Valley from Calif. 190 near the park's western entrance, but the easiest access is probably from Big Pine Road in the north. The springs are located seven miles down a dirt track off Saline Valley Road. Three pools are available for soaking, some surrounded by trees, grass, and shrubs in oasis-like settings. Clothing is most definitely optional. Camping is allowed in designated sites around the springs, and some soakers opt to stay for a while. Over the years much counterculture artwork (including many bat motifs) has accumulated in the hot springs area. There's also an impromptu lending library on wooden shelves near one of the pools. Some pretty out-there happenings still go on at Saline (among Park Service rules: Birthing or attempting to give birth in the springs is prohibited). Go with an open mind, and extra sunscreen for body parts that rarely see the light of day.

8 | Great Smoky Mountains
NATIONAL PARK
North Carolina

Once considered a hidden gem for hikers stumbling on the refreshing oasis, the word's now out about **Midnight Hole** in Great Smoky Mountains National Park—and the mountain-fed pool has the amenities to support it. Parking? Restrooms? Easy hike? Check, check, check. To experience the crystal green waters of this North Carolina gem, take the Bent Creek Trail approximately 1.5 miles and listen for the delighted squeals of swimmers. To your left, massive boulders squeeze the river, making a six-foot waterfall with a deep pool beneath. The trail is a former railroad grade that hauled lumber during the logging boom, but today it's mostly paved and just a gradual climb to brave the year-round brisk water and chill out among trees, wildflowers, and other adventure seekers.

9 | Point Reyes
NATIONAL SEASHORE
California

The park is technically a national seashore, but the freshwater swimming is also superb at Point Reyes in Northern California. The peninsula that Sir Francis Drake most likely set foot on 500 years ago is riddled with small ponds and lakes that make for great wading, floating, or even a vigorous backstroke if you are so inclined. **Bass Lake** at the southern end of the park is generally considered the best inland swimming. It's about three miles up the Coast Trail from the Palomarin Trailhead near the town of Bolinas. Bring your own lunch and picnic beneath the shoreline evergreens or browse for thimbleberries and salmonberries in the lakeside vegetation during the summer months. There's a "Tarzan rope" for swinging over the water and many regulars bring blow-up floats to laze away the day on the lake. Bass has long been one of the San Francisco Bay Area's favorite skinny-dipping spots, so bathing suits are optional. The Point Reyes "lake district" includes four other small water bodies, but none of them are especially swimmer friendly.

10 New River Gorge
NATIONAL PARK & PRESERVE
West Virginia

Forced to pick favorites, West Virginians may point to the **Glade Creek Trail** not only for its wonderful waterfalls but also for the swimming sites. In the town of Prince, turn onto the gravel Glade Creek Road and follow it for seven miles to reach the trailhead parking near the convergence of Glade Creek and the New River. Winding the route of an old narrow-gauge rail line, the moderate trail follows the cascading stream for 5.6 miles past rhododendron, dense hardwoods, and hemlock forests. But you don't have to wait very long to beat the heat—a couple swimming holes in the trail's first mile are sure to appease even the youngest hikers. (Consider heading out early for a more private soak.) You can turn around at any point on this out-and-back, or add more activity to the day's excursion: The lower section of the Glade Creek is a catch-and-release trout stream.

Tide-pooling

Exploring seaside tide pools is a popular pastime in several parks. The name derives from the fact that the rocky depressions—and their inhabitants—are only exposed at low tide. Life in a tide pool is fragile, so when it comes to marine creatures, the rule at national park tide pools is "look and don't touch."

Although its primary function is history, Cabrillo National Monument in San Diego boasts excellent tide pools along the western shore of Point Loma. With an underwater kelp forest just offshore, Cabrillo's tide pools are especially rich with diverse animal life, including sea stars, sea urchins, and anemones, limpets and chitons, crabs and young lobsters, and octopuses and small fish trapped by the outgoing tide. Rangers are often on duty to answer questions about the pools and their inhabitants.

Farther north along the West Coast, four of the islands in Channel Islands National Park have great tide-pooling sites. In each case, you need to travel by ferry or private boat from Ventura or Santa Barbara on the mainland, and then hike varying distances to the tide pools.

On the other side of the nation, Acadia National Park in Maine is also well endowed with tide-pooling sites. The most popular include the gravel isthmus that connects Bar Island and Mount Desert Island at low tide, as well as Wonderland and Ship Harbor on the island's southern flank, near Bass Harbor Head Light Station.

Cabins That Cruise

Picture this: After whiling the day away on tranquil waters, you can say goodnight without disembarking. Keep the adventures afloat at these park units known for houseboating.

As much as parks safeguard land for future generations, they also protect water resources—the nation's lakes, seashores, and rivers that sink deep and flow wide. Cruising into the wild blue yonder on a houseboat is one of the more relaxing ways to explore the National Park System.

LAKE LOCALES

Tucked in the high plains of Texas in an area once afflicted by the Dust Bowl, Lake Meredith National Recreation Area is a water-sports paradise. Open year-round, but particularly popular during the Texas Panhandle's triple-digit summers, Lake Meredith was formed by the construction of Sanford Dam across the Canadian River. At the Sanford-Yake boat ramp, houseboats can easily put in and out of the water, while slips rent by the month.

Voyageurs National Park's watery expanse along the U.S.-Canada border in Northern Minnesota is ideal for houseboats. Comprising four large lakes linked by narrow portage channels, this North Woods park has countless places to float. The water might be a little chilly for swimming (even in summer), but houseboaters can spend their days berry picking on land and bird-watching, fishing, and stargazing. For a break from the water, the historic Kettle Falls Hotel offers billiards and a cold brew. There are dozens of designated overnight mooring sites. Skippers with their own craft can use public ramps near the park's three visitor centers, and plenty of nearby outfitters rent fully equipped vessels.

One of the landmark public works projects of Franklin D. Roosevelt's presidency, Grand Coulee Dam created a massive reservoir named after the man who made it all possible. Lake Roosevelt, which stretches more than 150 miles along the Columbia River in eastern Washington State, is one of the most popular houseboating venues in the Pacific Northwest. The national recreation area's attractions include wildlife-watching (moose, bears, deer), angling (30 different species of fish), and a dramatic change of scenery from the arid grasslands around the dam to the forested slopes upstream. Captains can launch their craft from the many public boat ramps, including Crescent Bay right behind the dam to China Bend near Northport. Rentals are also available; the season runs mid-spring to mid-autumn.

TUCKED-AWAY GEMS

Everglades National Park is a welcome respite for houseboaters who delight in bird-watching, wildlife spotting, and recreational fun. At the southernmost tip of the state peninsula, the Flamingo area is especially equipped for houseboats, with plenty of marinas and gear shops. For those who'd rather rent, park concessionaires have all-amenity boats that can fit families of six. You can cruise the River of Grass backcountry, or simply anchor in a quiet bay to cast a line, watch for wildlife, and enjoy the immense night sky above this precious Florida wilderness.

Boats are the only way to reach the barrier islands of Cape Lookout National Seashore off North Carolina's coast. Core Sound, Back Sound, and Barden Inlet on the inside of the barrier islands offer ideal houseboating terrain: calm water, superb scenery, secluded coves, and many small islands. Given the often shallow water, NOAA navigation charts are highly recommended, especially for skippers not familiar with the area. Two of the more

Houseboaters have 186 miles of waterways to play in at Lake Powell in Glen Canyon National Recreation Area.

popular places for houseboat "camping" are Cape Lookout Bight and the inside of Shackleford Banks between Wades Shore and Whale Creek.

ENDANGERED FAVORITES

Climate change and two decades of drought have put lake levels for the country's largest and second largest reservoirs, Lake Powell and Lake Mead, to shocking lows, affecting boating options. Thanks to a snowy 2022–2023 winter, levels have improved, and hopes are that these acclaimed water recreation sites return to sustainable levels in coming seasons.

UNINVITED PASSENGERS

If an ounce of prevention is worth a pound of cure, there's a whole lot of weight going into protecting parks from the harm of invasive species. Zebra mussels are one such culprit. Freshwater shellfish native to the Caspian and Black Seas, these mussels are distinctive in appearance, with a flat-bottomed shell and black wavy stripes. They're also quite small—two inches at their largest—and microscopic in the larva stage. When zebra mussels gather on surfaces, the effect, called "biofouling," can have a vast environmental impact. Easily overwhelming native mussels, and reproducing at a rate of a million eggs per year, zebra mussels can shift local food webs—and wreak havoc on boats. What can be done? The National Park Service has instituted important guidelines for watercraft in efforts to stop aquatic invasive species. Boaters are asked to drain and wash their boats, trailers, and equipment well after they've been in water and wait five days before entering another fresh body of water.

SEA KAYAKING

Long, low, sleek, and lightweight, sea kayaks are ideally suited for water-based touring, especially in wilderness waters. You might see a few canoes in some of the areas described here, but by and large, these are made-to-order kayak trips that ply some of the country's most magical coastal and inland waters.

1 Kenai Fjords
NATIONAL PARK
Alaska

The vast fjords of Alaska's Kenai Peninsula offer superb kayaking in the face of towering tidewater glaciers amid drifting icebergs and harbor seals. If your time is limited, spend a day paddling in **Resurrection Bay** right from park headquarters. You'll see hanging glaciers hundreds of feet high above the water, you'll paddle by the base of cascading waterfalls, and you're likely to see sea otters, porpoises, and seabirds such as auks and puffins. A more ambitious day requires a 42-mile motorboat ride to **Aialik Bay,** which distills all the glacial glories of Kenai Fjords, including the frequently calving, 800-foot face of Aialik Glacier. In the icy waters you might see humpback whales, orcas, fin whales, and even harbor seals catching a rest on the icebergs. With more time, venture

to **Northwestern Lagoon,** a similar environment where you could paddle for hours or even days and not see another soul, then camp in solitary splendor. The park maintains a list of authorized outfitters that organize guided kayak trips.

2 Glacier Bay
NATIONAL PARK & PRESERVE
Alaska

The primordial majesty of Alaska's Glacier Bay must be seen close up. A paddle around the narrow inlets of the **West Arm** in a sea kayak, weaving around just calved icebergs, feels like the beginning of time. The 15,000-foot summits of the Fairweather Range tower overhead. Glaciers seem to flow from every crevice. The sounds are as impressive: water flowing beneath the glaciers and tumbling from mountain streams, the crackling and gurgling ice, and the "white thunder," as the

locals called the calving of tidewater glaciers. For a perfect three- to five-day trip, pick up a rental kayak in Bartlett Cove, hop the sightseeing boat to Queen Inlet, and start paddling west to see massive Grand Pacific Glacier. Once past the glacier and the cruise ships, find solitude in Johns Hopkins Inlet at the mouths of Lamplugh and Reid Glaciers, where you can camp before making the open-water paddle back to Queen Inlet. Glacier Bay's waters are quite protected, but glaciers come with their own weather systems—beware of strong winds coming off them, and, of course, the surge that accompanies calving.

3 Gulf Islands
NATIONAL SEASHORE
Mississippi

The jewel among jewels of the Gulf Coast barrier islands stretching across Mississippi and the Florida

A sea kayaker paddles through the
icy clutch of Bear Cove, off Aialik Bay,
in Kenai Fjords National Park.

Panhandle is **Cat Island,** a gorgeous and fascinating goal at the end of a six-mile paddle from Gulfport, Mississippi. The island rises higher and has more vegetation than the rest of the other park islands, and it's ringed by miles of white-sand shoreline. (Parts of the island are private inholdings; heed posted signs.) Cat Island's "T" shape helps maintain calm conditions, and while en route you might be escorted by jumping mullet or pods of bottlenose dolphins. Ashore, solitary camping (no facilities; bring your own drinking water) is allowed just above the high-tide line. Inland highlights include marshes (watch for gators), and forests of pines, palmettos, and live oaks dripping with garlands of Spanish moss. Fall is the best time to go; the humidity has subsided, thunderstorms are unlikely, and the seawater remains pleasantly warm.

4 | Isle Royale
NATIONAL PARK
Michigan

One of the nation's least-visited national parks, Isle Royale is rugged and wild—and that's a big part of its appeal. **Lake Superior** is known for unpredictable weather, including sudden wave-generating squalls, so paddling these waters isn't for beginners—but it's a true thrill for those with experience. Sea kayaks should be at least 15 feet 8 inches long; recreational kayaks are not encouraged.

▶ *Painted Cave on Santa Cruz Island is one of the world's largest known sea caves. The cave measures 1,215 feet in length (the size of more than four football fields), has a 160-foot entrance, and is almost 100 feet wide.*

Want to self-guide the trip? Remember marine radios are recommended, outer-shore landing sites are scarce, and hypothermia can happen fast. Alternatively, professional guide companies offer all-inclusive, multi-day excursions, many of which can be customized to cater to your group's skills and interests. Beyond paddling along ancient volcanic shorelines, guides lead hiking, fishing, and photography tours, and even help watch for flora and fauna like thimbleberries and moose.

5 | Cape Lookout
NATIONAL SEASHORE
North Carolina

The North Carolina barrier islands of Cape Lookout stretch 112 miles, yet lie just three miles offshore and are without roads or many facilities, making them an ideal place for secluded, protected paddling

in shallow water. The island waters are also a mecca for kayak-based fly-fishing for false albacore. Camping is permitted everywhere; there are no designated spots on the long, unspoiled beaches. You can build a fire, swim, fish, clam, crab, bird-watch, watch for humpback whales and four types of sea turtles—loggerhead turtles nest here in the summer months—and even play in the surf on the windward side. For an easy, short trip, put in at Harkers Island on **Black Sound,** site of the park's visitor center, and paddle over to Shackleford Banks on Onslow Bay, where more than 100 wild ponies roam, as they have for hundreds of years. It's best to paddle early, before the wind picks up, and be sure to pick up a tide chart so you can avoid traveling against a surging current.

6 | Everglades
NATIONAL PARK
Florida

Meet Florida at its wildest and wilderness kayaking at its finest: the 99-mile **Wilderness Waterway** of Everglades National Park. The route courses narrow, mangrove-lined rivers and streams and winds around thousands of mangrove islands from Everglades City to Flamingo. Supreme serenity is certain, and sightings of manatees and dolphins are likely. The route is well marked, but it also invites variations—with the proper charts (available at the Gulf Coast Visitor Center in Everglades City), you can fashion your own itinerary using the waterway as an organizing structure. For example, you can form open loops by choosing a river, paddling it out to the Gulf of Mexico, camping on an island, then following a different waterway back into the mangrove maze. Your course, though, has to be cleared with rangers at the Gulf Coast Visitor Center so that you can be assured of campsites, either on beaches or on the park's camping platforms called chikees. The full waterway takes about 10 days to paddle, but you can also make a shorter loop trip out of Everglades City without retracing your paddle strokes and still have a full-on experience of the Everglades wilderness.

7 Apostle Islands
NATIONAL LAKESHORE
Wisconsin

With 21 islands scattered in the waters of Lake Superior off the tip of Wisconsin's Bayfield Peninsula, plus 12 miles of mainland shoreline, the Apostles are ideally suited for all sorts of kayak touring, from day trips exploring cliffs and sea caves to multiday excursions to the outer islands. For a day trip or a simple overnighter, put in at Meyers Beach and paddle north along the mainland's **Lakeshore Trail,** where red sandstone cliffs tower above

and sea caves and arches make for great paddling. Eighteen of the 21 islands permit camping, so options for longer trips abound. Sand Island, for example, is less than an hour's paddle across open water—skirt the circumference, explore the sea caves at Swallow Point, and tour its lighthouse before setting up camp. With more time, continue on to Oak Island, put ashore, and hike to the overlook for a great view of

all the Apostles. More ambitious paddlers can focus on notching up lighthouses—six in all—or aim to reach Outer Island, the park's most remote spot, for more amazing views and a wonderful sense of privacy. Heed Park Service warnings that Lake Superior is cold and lake conditions can turn rough. Wet suits are recommended, as are guide services. The park's website has a link to a list of authorized outfitters.

War Dogs of Cat Island

If you dig deep into the Gulf Islands National Seashore archives, you might come across this curious tidbit about Cat Island: During World War II, the island was used by the U.S. Army Signal Corps to train service dogs for the military.

The call to press dogs into service came not long after the attack on Pearl Harbor, as concerns about enemy intrusion and sabotage grew. Dogs served as valuable sentries at critical installations. Thus the War Dog Program, more commonly called the K-9 Corps, was born and the Cat Island War Dog Reception and Training Center established.

Quiet and secluded, the island was an ideal location for the hush-hush work, plus it was densely vegetated and humid—similar to the tropical jungle in the Pacific. Sentry dogs were trained in about eight weeks and put to work guarding shorelines where submarine attacks were a concern. As the war progressed, dogs learned to perform reconnaissance, to detect mines, and to deliver messages. As scouts, they warned against enemy ambushes. Nearly 2,000 dogs served nobly in all these regards in both theaters of the war.

8 | Channel Islands
NATIONAL PARK
California

The 300-foot north-facing cliffs that mark the east end of **Santa Cruz Island,** largest of the Channel Islands, have the appearance of Swiss cheese, pocked as they are with some of the world's largest sea caves. Dark, mysterious, and thrilling to explore—ideally in a superbuoyant sit-on-top kayak with a guide—the caves are numerous, long, and subject to seasonal variations, so local knowledge is vital. And always wear a helmet—surges of tides and waves can quickly make a big cave turn small. As one outfitter puts it, it's a low-rigor but high-adventure outing. It can be done in a day—or camp on the island and make a weekend of it. Most tours begin at Potato Harbor, where guides will start off with some easy pass-through caves. One, Surging T Cave, is 354 feet long. Of the island's 100-plus caves, 72 are more than 200 feet long. Some are huge and vaulted, like the immense Painted Cave. Others are best avoided by claustrophobes. Seal Canyon Cave, for instance, winds 600 feet into oblivion and is so narrow that kayakers have to back in, then shoot out—and surge occasionally cuts off that reassuring light at the end of the tunnel. Watch for blowholes in the cliffs that spray out a sudden *ptooey* of foam, and watch for harbor seals everywhere. The park has a list of authorized outfitters.

9 | Pictured Rocks
NATIONAL LAKESHORE
Michigan

The colorful rocks of the park's name are just one highlight of a wild 42-mile **shoreline** that rivals the Apostle Islands as one of the finest sea kayak destinations on chilly Lake Superior. The 50- to 200-foot multihued sandstone cliffs, striped and colored by seeping, mineral-filled water, stretch for 15 miles along the west end of the park. The 500-million-year-old rocks have been carved into arches, spires, and caves that are a delight to explore by kayak. They begin just upcoast from Sand Point; day trippers can put in here or at Miners Castle. Touring kayakers can proceed beyond and camp in permitted backcountry campsites. Beyond the rocks is Twelvemile Beach, a long stretch of white sand crowned by a backdrop of wild North Woods. Beyond the Twelvemile Beach campground is Au Sable Point and its 1874 lighthouse, and the Grand Sable Dunes, perched atop the 300-foot-high Grand Sable Banks. Rangers advise that sea kayaks are the only type of kayak that should be used.

10 | Acadia
NATIONAL PARK
Maine

The Atlantic coastal waters surrounding much of Acadia National Park are not actually in Acadia. But kayaking is a great way to gain perspective on the park's stunning scenery, to poke around its surrounding islands (most are in private hands), and to see wildlife. The best kayaking is on the quiet west side of Mount Desert Island, where it's preferable to go with an outfitter; they can run one-way trips so paddlers don't have to backtrack against the wind or tides, or search for parking. **Blue Hill Bay** and **Western Bay** are well away from cruise ships, ferries, and whale-watching boats, so the experience is quiet and you get to paddle by seal rookeries and see lots of porpoises and birds—loons, terns, black-backed and herring gulls, guillemots, and occasional bald eagles. All the while Acadia's mountains—Bernard, Mansell, even Cadillac—loom above or in the distance. Intrepid paddlers can brave the busier waters of Frenchman Bay out of Bar Harbor to the Porcupine Islands.

SCUBA DIVING & SNORKELING

Whether you're a veteran diver or a novice with mask and snorkel, a cold-water devotee or a tropical lagoon aficionado, the National Park System offers numerous locations where you can get your (fin) kicks in the water. Delights to be found include shipwrecks, coral reefs, colorful fish, and underwater heritage trails. Dive in.

1 | Biscayne
NATIONAL PARK
Florida

Although just about any place in this South Florida park is ripe for scuba, the best underwater terrain is a chain of **six wrecks** on the park's far eastern fringe. Part of the Biscayne Maritime Heritage Trail, the wrecks run the gamut from 19th-century sailing vessels like the triple-masted *Erl King* (sank 1891) to early 20th-century iron-hulled steamships like the *Lugano* (sank 1913). The only way to reach the secluded sites is by boat, tying up at mooring buoys poised above the wrecks. Five of the sites are too deep to preclude anything but diving. However, the remains of the *Mandalay* are in waters shallow enough for snorkeling. Dubbed the "red carpet ship of the Windjammer fleet," the superchic steel-hulled schooner smashed onto Long Reef on New Year's Day of 1966 on an overnight run between the Bahamas and Miami. The National Park Service has produced maps and waterproof information cards on each of the wrecks.

2 | Buck Island Reef
NATIONAL MONUMENT
Virgin Islands

One of the few dedicated marine units in the park system, Buck Island protects an astonishing coral garden and underwater habitat near St. Croix in the U.S. Virgin Islands. Snorkelers flock to the island's east end for a renowned **underwater trail** through warm, translucent water that never reaches more than 12 feet in depth. Submerged signs describe a reef ecosystem rife with brain, star, and elkhorn corals, as well as sponges and soft corals. Sea turtles, rays, and reef sharks are among the larger species that frequent the lagoon; overall the park protects more than 250 fish marine species. Diving is limited to two scuba moorings in about 30 to 40 feet of water above an elkhorn reef. Visitors who desire a break from the water can hike, picnic, sunbathe, or take a siesta on 176-acre Buck Island. Access to the park is by boat only, and concessionaires operate out of Christiansted or Green Cay Marina on St. Croix. To learn more about the protection of the island before you travel, watch the award-winning short film "Caribbean Gem."

3 Amistad
NATIONAL RECREATION AREA
Texas

Diving in the desert may seem like an oxymoron, but scuba diving is a major draw at Lake Amistad in southern Texas. The 65,000-acre reservoir was created in 1969 by the building of the Amistad Dam near the confluence of the Rio Grande and Devils River. The submarine scenery includes sunken boats, underwater cliffs, and a ranch house submerged by the rising lake. One popular site is **Indian Springs,** a spring up the Devils River arm that raises the surrounding water temperature to 72°F. Animal life includes catfish and three types of bass; spearfishing (with a license) is allowed in areas around the lake. Although the lake is colder in winter, the underwater visibility is much better between November and April, often reaching beyond 40 feet. Look for concessionaires to book lake dives and rent small speedboats for private dives.

4 National Park of American Samoa
NATIONAL PARK
American Samoa

Warm, clear tropical Pacific waters make American Samoa ideal for scuba diving and snorkeling. Spread across four different islands, the territory's namesake national park encompasses lagoons

▶ Florida's Coral Reef, the world's third largest coral reef system, stretches 350 miles from Dry Tortugas National Park and includes a significant section of tract in Biscayne National Park.

and coves protected by offshore reefs. More than 200 types of corals and nearly 1,000 fish species have been spotted in waters in and around the park, including moray eels, eagle rays, Moorish idols and mantas, lionfish, stonefish, puffers, and reef sharks. Larger animals like humpback and sperm whales are sometimes seen farther from shore. From poisonous sea snakes to sea turtles, marine reptiles are also present. The park's largest section sprawls across the remote northern shore of Tutuila Island, home of territorial capital Pago Pago. But most of this area is impossible to explore without your own boat. Far better to hop a puddle jumper to the twin islands of **Olosega** and **Ofu** in the Manu'a Islands, where the reefs and lagoons are far easier to reach. Local charter companies offer diving equipment rental as well as crewed boats to scuba diving and snorkeling spots in the park.

5 Channel Islands
NATIONAL PARK
California

Ethereal kelp forests, spooky sea caves, and a multitude of large animals are just three of the factors that make these Southern California islands an incredible place to dive or snorkel. The best conditions are found off **Santa Barbara** and **Anacapa Islands,** as well as the east end of Santa Cruz. Kelp forests are known to harbor as many as a thousand different species, from sea urchins, sea cucumbers, and delicate sea stars to moray eels, sheepshead, and bright orange garibaldi (the California state fish). Santa Cruz is especially rich in sea caves, some of them among Earth's largest, but they should always be approached with caution. Harbor seals, sea lions, and dolphins are Channel Islands regulars, and there is always a chance of spotting gray, humpback, and minke whales—and occasionally orcas—in the open waters around the islands. There are even wrecks, like the remains of the Pacific Mail steamer *Winfield Scott* that went down off Anacapa in 1853.

6 War in the Pacific
NATIONAL HISTORICAL PARK
Guam

Although dedicated to the memory of soldiers and civilians from all nations who lost their lives in the War in the Pacific, the national park

A garibaldi swims through a kelp forest off the shore of Anacapa Island in Channel Islands National Park.

also safeguards a fair amount of Guam nature, including beaches, coral reefs, and lagoons. More than 3,500 marine species have been seen inside the park. U.S. Marines kicked off the Allied invasion of Guam by storming ashore at Agat and Asan Beaches in July 1944. Nowadays, the warm tropical waters off these strands are prime places for snorkeling and scuba diving. The **Asan Beach unit** features two designated scuba diving areas at the west and east ends of the beach, as well as a snorkel area at Adelup Point. The **Agat Beach unit** has four scuba diving areas as well as a snorkel area at both Apaca Point and Ga'an Point. Waters near the national park shelter the remains of many ships and planes that went down during two wars. The artifacts range from a World War I German cruiser called the *Cormoran* to a Japanese Aichi D3A "Val" bomber from World War II. As many as four of these wrecks can be dived in one day. Shore dives are offered at Agat and Asan.

7 | Pictured Rocks
NATIONAL LAKESHORE
Michigan

Diving is a particularly common pastime at the country's first national lakeshore, especially in the **Alger Underwater Preserve,** which was created to protect the cultural resources of this area. Here, the marine world rivals the spectacular

A diver explores a tugboat wreck in Lake Superior.

shore, as vessels related to all types of maritime trade—fur, copper and iron ore, lumber, and passenger freight—can be found on the floor of Lake Superior. An added plus: Many of the major shipwrecks are both well preserved and buoyed. Dozens of wrecks have been documented within the lakeshore boundaries, pummeled by autumn storms with wave heights upwards of 25 feet. Interestingly though, only 15 of the wreck sites are known. The nearby town of Munising is well equipped with a dive shop and an air-fill station, as well as rental boats and charters. Back on land, Pictured Rocks has attractions like lighthouses, waterfalls, and more than 100 miles of hiking trails. And come summertime, eerie remains of the

Sitka (sank 1904) and *Gale Staples* (sank 1918)—smashed and mangled by Superior—can be spotted along the beach at Au Sable Point.

8 | Sleeping Bear Dunes
NATIONAL LAKESHORE
Michigan

Located in the channels that separate the mainland and island portions of this national lakeshore, the **Manitou Passage Underwater Preserve** offers sunken docks, shipwrecks, and natural submarine landscapes. Several wrecks lie in shallow water, which makes them ideal for novice divers and even snorkelers. The remains of the three-masted schooner *Flying Cloud* (sank 1853) and the wooden brig *James McBride* (sank 1857) can be accessed directly from the mainland. Other wrecks lie scattered around North and South Manitou Islands and will require a boat or a combination of ferry passage (from Leland) and hiking to the shoreline wade-in points. Manitou Island Transit makes the 90-minute crossing daily between May and October; divers can choose to return the same day or camp overnight on either island. South Manitou's wrecks include the package freighter *Francisco Morazan* (sank 1960), the wooden steamer *Walter L. Frost* (sank 1905), and the steam barge *Three Brothers* (sank 1911).

9 Isle Royale
NATIONAL PARK
Michigan

As Gordon Lightfoot sang in "The Wreck of the Edmund Fitzgerald," **Lake Superior's** fickle weather has taken more than its fair share of ships to the bottom. A good number went down around Michigan's Isle Royale and are now protected by the National Park Service. Superior's cold water has helped preserve their remains, and the lake's clarity makes them easy to explore. The 532-foot-long *Chester A. Congdon* is the largest of the park's wrecks, a bulk freighter that sank in 1918. Another bulk freighter, the *Emperor* (sank 1947), is the most intact wreck, retaining its engine room and various cabins. All divers must register beforehand at one of the park's visitor centers. With water temperatures between 34°F and 55°F, a full wet suit or dry suit is highly recommended. The scuba diving season runs mid-June to early September, and outfitters are available to run trips in the park waters.

10 Virgin Islands Coral Reef
NATIONAL MONUMENT
Virgin Islands

The only unit of the National Park System that does not include any dry land, this Caribbean preserve protects 12,708 acres of pristine marine habitat around St. John Island in the U.S. Virgin Islands.

Famous Wrecks

Some of the most renowned shipwrecks in American history took place in and around national parks. The U.S.S. *Monitor* lies in 230 feet of water about 16 miles off Cape Hatteras National Seashore. The famed Civil War ironclad went down during an 1862 storm only nine months after her epic fight in Hampton Roads. In April 1554, three Spanish galleons ran aground in a storm along the shore of today's Padre Island National Seashore. Some of the survivors eventually made their way back to the Mexican port of Veracruz, but more than 300 passengers and crew perished in the disaster, one of the worst in Spanish colonial history. The wrecks weren't relocated until 1967 and remains occasionally still wash up on Texas beaches. In terms of sheer numbers, Cape Cod National Seashore is the clear winner—more than 3,000 ships have wrecked on the peninsula's outer side. Among them was the British man-of-war H.M.S. *Somerset III,* which survived combat in the Seven Years' War and the early part of the American Revolution, but not a horrific storm off Cape Cod in November 1778.

Only one of the monument's four segments touches shore, a slice of **Hurricane Hole,** Coral Bay, and Round Bay that includes mangrove swamps and shoreline coral reef. Other than Hurricane Hole—which can be accessed on U.S.V.I. 10 from Coral Bay village at the east end of St. John—the park is difficult to explore without a boat. Further complicating access is the fact that anchoring is forbidden in order to protect the fragile tropical marine ecosystems. Moorings are available for day use in the Hurricane Hole area. The national monument is jointly administered with Virgin Islands National Park and shares the visitor center at Cruz Bay on the western end of St. John. Two of the park's more popular activities are fishing the open ocean south of St. John and snorkeling the Hurricane Hole mangroves.

A DAY AT THE BEACH

From the coconut palm–fringed bays of the Virgin Islands to the chilly wind- and wave-carved Olympic Peninsula coast, some of America's best beaches are protected within the confines of the park system. They tread a delicate balancing act between recreation and conservation of fragile shoreline ecosystems that nourish billions of creatures.

1 | Assateague Island
NATIONAL SEASHORE
Maryland & Virginia

Sprawling across a barrier island shared by Maryland and Virginia, this national seashore boasts **37 miles of white-sand strand** within easy driving distance of several of the eastern seaboard's largest cities. Only a small part of Assateague's beaches are accessible by paved road, which means that sunseekers must either hike to their favorite patch of sand or obtain an over-sand vehicle permit from the Park Service and cruise down the 12 miles of beach on the Maryland side, where driving is allowed. The Maryland side also sports several beachfront campsites, but you may have to share your secluded camp with the island's wild horses. Surf fishing has long been popular along the Assateague coast; crabbing and clamming are possible in the shallow bays on the island's leeward side. The seashore is also one of the best places along the mid-Atlantic to board surf, with consistent surf throughout the year and fairly warm water in summer thanks to the Gulf Stream.

2 | Redwood
NATIONAL & STATE PARKS
California

The trees don't steal the whole show at Redwood. In fact, 40 miles of Pacific coastline have their own allure. While the water seems appealing, the park urges visitors to avoid the hazards of swimming ("sneaker waves" have proven fatal here) and turn instead to tide-pooling. Consult a tide chart and head to **Enderts Beach,** accessed via a half-mile walk along the Coastal Trail, to see how low tide reveals a diverse range of invertebrate animals like California mussels, ochre sea stars, periwinkle snails, and white sea cucumbers. Rangers host seasonal tide pool walks where they explain the concept of ocean productivity (at which the Pacific excels) and how it helps the Rocky Intertidal Zone flourish. **Gold Bluffs Beach**—named for the overlooking hills—reveals a sight of another scale: gray whales. Bring binoculars and keep watch for spouting; migration happens from November to December and March to April, but you're likely to get lucky on a clear day. Plan to stay: On-beach camping promises the idyllic duet of a stunning sunset followed by a Pacific lullaby.

3 | Cumberland Island
NATIONAL SEASHORE
Georgia

The Carnegies and Rockefellers knew a good thing when they saw it, and there's no doubt that what attracted some of the wealthiest

Two herds of wild horses call Assateague home, one on either side of the Maryland-Virginia border.

19th- and 20th-century landowners to Cumberland Island is what still anchors its appeal today. This largest and southernmost of Georgia's barrier islands doesn't wow with amenities. Rather, it's the 17 miles of **pristine, unspoiled beaches** within wilderness maritime forest and lush salt marshes that truly enchant visitors. Beachcombing is popular, and shark's teeth can often be collected from the sandy roads. Watch for feral horses, nesting loggerhead turtles, marshes full of fiddler crabs, and armadillos. The quickest way to the water is by hiking an easy half mile from the Sea Camp dock. There's no lifeguard on duty, but the Sea Camp Ranger Station is mostly staffed throughout the day. Since some of the island is private, it's important to follow signs—public beach crossings are marked with tall striped poles. You can see remnants of the Carnegie lifestyle presented through ruins of the Dungeness and Plum Orchard estates, and even opt to overnight at Cumberland's only commercial establishment, Greyfield Inn. Then again, you can also book a campsite (developed or primitive) to settle in and simply delight in the rare sound of silence.

4 | Indiana Dunes
NATIONAL PARK
Indiana

Anyone's first visit to **Lake Michigan's** Indiana Dunes is a mind-blowing experience: a pristine lakeshore set against a backdrop of smokestacks, cranes, and other bleak industrial architecture. But that is what is so special about this park—the fact that it shares space with steel mills and port facilities. Conservationists kicked off the campaign to save the nearly 200-foot-tall dunes more than a century ago, but it wasn't until the 1960s that their dream became reality. The park extends along roughly 15 miles of lakeshore just east of Chicago, a stretch that also includes Indiana Dunes State Park and a number of private holdings. Swimmers should recognize that a freshwater body as large as Lake Michigan can be just as hazardous as the open ocean, including riptides, large waves, and ice in winter. In addition to locker rooms and showers, West Beach has lifeguards through the summer season. Visitors can overnight at Dunewood Campground or at a site inside the state park.

5 | Cape Cod
NATIONAL SEASHORE
Massachusetts

To naturalist Henry David Thoreau, who made four visits to Cape Cod between 1849 and 1857, the Cape

▶ *In 1903, Italian inventor Guglielmo Marconi sent the first successful transatlantic wireless radio message between the United States and Britain from a spot in what is now Cape Cod National Seashore.*

was a coastal version of Walden Pond, a place to discover and escape back into nature. The cape is still much the same today, a wild thing on the eastern edge of Massachusetts that continues to rebuff civilization. The national seashore portion includes 40 miles of strand, much of it backed by rolling dunes, with 15 different beaches, some town-managed and others under the auspices of the Park Service. They range from the placid bayside sands of **Duck Harbor** to untamed Atlantic strands like **Nauset** to historical shores like **Marconi Beach.** The long stretch of sand between Race Point and Head of the Meadow is open to off-road vehicles with a park permit. The only camping within the national seashore is self-contained vehicle camping at Race Point; tents and camping trailers are not allowed. The nearest other camping is in Nickerson State Park near Orleans.

6 | Virgin Islands
NATIONAL PARK
Virgin Islands

Turquoise water, talcum-powder-fine sand, palm trees swaying in a gentle breeze: What's not to love about the beaches in this Caribbean national park? The more renowned spots are on the north shore of St. John Island, a chain of pearly white strands that includes **Hawksnest Bay, Trunk Bay,** and **Cinnamon Bay.** Trunk Bay boasts a snack bar and an offshore snorkel trail through coral gardens. Cinnamon has a beachfront campground and cottages, and a water-sports shack where various boards and boats are rented. Those who crave seclusion can hoof it to isolated strands on the south shore of St. John and the park's **Annaberg sector.** The beach at **Brown Bay** looks out across Sir Francis Drake Channel to the British Virgin Islands. The one-mile Ram Head Trail leads to an unusual (and normally empty) **blue cobblestone beach.** The park's most isolated beach, located at **Reef Bay,** requires a little more than two-mile hike along a trail of the same name.

7 | Golden Gate
NATIONAL RECREATION AREA
California

There's an argument to be made that Golden Gate has a greater variety of beaches than any other unit of the National Park System. Beaches in the **San Francisco Bay**

Area park range from tiny urban slivers popular with early morning swimmers (Aquatic Park) to massive strands that can accommodate as many as a million people on summer holiday weekends (Ocean Beach) to secluded wilderness beaches that can only be reached by foot (Tennessee Cove). Boulder-strewn Baker Beach offers unsurpassed views of the nearby Golden Gate Bridge. The thin strip of sand at Crissy Field is a popular launching pad for windsurfing and kiteboarding. Muir Beach is home to a Zen meditation center, a popular English pub (the Pelican Inn), and a Monterey pine grove where monarch butterflies often winter. Rodeo Beach shares its magnificent stretch of coast with a wildlife-rich lagoon, and Marin Headlands, on the north side of the Golden Gate, houses vintage World War II Army buildings and a Cold War–era Nike missile site. The park also has three waterfront campgrounds, all of them in the Marin Headlands section.

8 Padre Island
NATIONAL SEASHORE
Texas

Quiet **North Padre Island** remains virtually unchanged from when Spanish explorers first probed its long crescent shore 500 years ago. Eons of waves, tides, and hurricanes created the 70-mile-long landfall and its wilderness beaches, the longest stretch of undeveloped barrier island on the planet, and likely to remain so given its Park Service protection. Five species

Best Beach Camping

Becher's Bay (Channel Islands National Park, California): Steinbeck's California come to life; waterfront camping on uninhabited Santa Rosa Island.

Cinnamon Bay (Virgin Islands National Park, U.S. Virgin Islands): The quintessential Caribbean campground; cook cheeseburgers in paradise.

Garden Key (Dry Tortugas National Park, Florida): Sack out beneath coconut palms on the same island as historic Fort Jefferson.

Holgate Beach (Kenai Fjords National Park, Alaska): Were those grizzly bear tracks outside your tent this morning? You betcha.

Kalaloch (Olympic National Park, Washington): RV and tent campsites with views of the wild Pacific coast; ranger campfire talks at night.

Ocracoke (Cape Hatteras National Seashore, North Carolina): Wild ponies, rolling dunes, and seafood barbecues—the Outer Banks doesn't get any better.

Sea Camp (Cumberland Island National Seashore, Georgia): Splendid isolation in a cool grove of trees; wild horses and armadillos keep you company.

Twelvemile Beach (Pictured Rocks National Lakeshore, Michigan): Secreted in a grove of white birches on a bluff overlooking Lake Superior.

Watch Hill (Fire Island National Seashore, New York): An oceanfront campground with marina, restaurant, and tiki bar? Hey, it's Long Island.

Wildcat Beach (Point Reyes National Seashore, California): Bluff-top campground near beachfront Alamere Falls; site No. 7 has the best ocean views.

York Island (Apostle Islands National Lakeshore, Wisconsin): A favorite among the lakeshore's 18 islands with campsites; beautiful beach and view of anchored sailboats.

Catch Petit Portal Arch (aka "Lover's Leap") at sunset in Pictured Rocks.

Chapel Rock and Miners Castle are named after nearby rock formations. At the far end of the park are the Grand Sable Banks and Dunes, which plunge 300 feet to the lake. Only the hardy go for a dip in chilly Lake Superior (also known for its rip currents), but the beaches in Pictured Rocks make for great walking, camping, and just contemplating.

10 | Cape Lookout
NATIONAL SEASHORE
North Carolina

The main advantage of Cape Lookout over its much more visited northern neighbor, Cape Hatteras, is its remoteness, a solitude born of a lack of bridges to the mainland. A boat is the only way to reach these unsullied North Carolina **barrier island beaches,** and advance planning is necessary, because other than the ferry landings, the islands are completely wild. Swimming, surf fishing, and beachcombing for shells are three ways to while away a lazy Cape Lookout day. The surfing isn't half bad, either, especially along Shackleford Banks and Cape Point. With so little development, wildlife thrives inside the national seashore, in particular shore and migratory birds, and the harems of wild horses that roam the Shackleford Banks. Primitive camping is allowed along the beaches, and there are two sets of rustic cabins (accessible only by boat) at Great Island and Long Point.

of sea turtles lay their eggs along local beaches, and given that it's located on the central flyway between North and South America, the island offers rich pickings for bird-watchers. Recreational opportunities also abound: swimming and surf fishing on the Gulf shore, windsurfing and kayaking on the lagoon side. The national seashore's only vehicular entrance is in the north, near Corpus Christi. Paved road extends only a few miles into the park; south of the Malaquite Visitor Center, the only way to transit the strand is by hiking, biking, or driving on top of the sand. Vehicles are allowed on the beach all the way down to the park's southern tip; four-wheel drive is highly recommended.

9 | Pictured Rocks
NATIONAL SEASHORE
Michigan

It seems counterintuitive that some of the nation's best beaches should be so far away from the sea. However, the Great Lakes aren't exactly ponds and many a coastal resort would love to have the kind of beaches lining Michigan's Pictured Rocks. The park name derives from the sandstone cliffs that etch much of this Upper Peninsula shoreline. Wildly scenic beaches flank the multicolored palisades. **Sand Point's** placid pocket-size beach is protected within the confines of South Bay, whereas **Twelvemile Beach** opens straight onto Lake Superior with views that seem to stretch to Canada. The small beaches at

WIND SPORTS

The breeze blows strong and steady in several national parks that lend themselves perfectly to wind sports like sailing, kiteboarding, and windsurfing. Seaside locations might be expected, but this top 10 also features inland waters where wind sports fans will find recreation amid stunning scenery.

1 | Yellowstone
NATIONAL PARK
Wyoming

America's oldest national park may not seem the most obvious place to set sail, but with its steady breeze, wide-open waters, and dazzling scenery, the park's **Yellowstone Lake** in northwestern Wyoming is an ideal venue for wind sports. In fact, the first European-style craft to grace the lake was a sail-fitted rowboat called *Annie*, part of the 1871 Hayden Survey of the region. Perched at more than 7,000 feet, this is the nation's largest high-altitude lake—as much as 20 miles from north to south and 14 miles from east to west. Bridge Bay Marina on the north shore is the best place to launch sailboats, but all craft should have collapsible masts in order to clear the highway bridge across the harbor mouth. Windsurfers are best launched from Grant Village Marina on the west side, where floating docks and a cement slip provide quick access to the lake. Sailors and surfers must obtain a boating permit from a ranger station before going onto the water. Secluded "boat party" campsites are located along South Arm and Southeast Arm along the lake's roadless southern shore.

2 | Padre Island
NATIONAL SEASHORE
Texas

With miles of flat water and a shallow, sandy bottom, Laguna Madre on the western flank of Padre Island is considered a prime place to learn and practice windsurfing. **Bird Island Basin**—not far from the entrance gate and the Malaquite Visitor Center—is the park's windsurfing hub and the only developed launch area. With an average annual wind speed of 18 miles an hour and some sort of breeze nearly every day, *Windsurfing Magazine* has ranked "BIB" as the best flat-water sailing in the lower 48. The season runs early spring to late fall, although mid-summer can be scorching hot and humid. Private outfitter Worldwinds operates a windsurfing school and rental center right on the beach at Bird Island. Their instruction runs from group beginners classes and kids camps to private lessons in deepwater starts, jibe footwork, and other advanced skills.

3 | Biscayne
NATIONAL PARK
Florida

"Sailing again in Margaritaville" could easily be the theme of this laid-back tropical park at the southern end of Florida. Stretching between South Miami and Key Largo, Biscayne embraces more than 200 square miles of **sultry sea** and **sandy islands.** Needless to say, boating is the only way to explore much of the park. Most people cruise down from the Miami metro area, but boats can also be put in the water via ramps at three county parks (Homestead Bayfront, Black Point, and Matheson Hammock) adjacent to the federal reserve. Broad, shallow Biscayne Bay separates the mainland park from insular portions like Boca Chita and elongated Elliott Key with their picnic areas and primitive campgrounds. A maritime heritage trail along the eastern edge of the park connects a string of six shipwrecks spanning a hundred years of local maritime history. Windsurfing is also popular, especially along the park's mainland coast, where the calm waters are ideal for novices. Owing to numerous reefs and shoals, National Ocean Service (NOS) navigation charts are highly recommended.

4 Cape Hatteras
NATIONAL SEASHORE
North Carolina

"Radical" is the word that veteran windsurfers have used to describe the conditions at Cape Hatteras, known for its big waves and big breeze even when it's not hurricane season. Newbies should stick to calm **Pamlico Sound** along the western edge of the barrier islands. Those who want to ratchet up their heart rates and adrenaline flow can head for the open ocean

▶ *The Wright Brothers consulted the U.S. Weather Bureau when picking a place for their glider experiments, ultimately choosing North Carolina's Outer Banks for the large sand dunes and steady prevailing winds.*

side. Maybe even more than for the Wright brothers, the North Carolina seashore is renowned for its capricious weather. One day it's board shorts and sunscreen, the next day dry suit and hood are de rigueur. Even during summer windsurfers should pack for extreme weather shifts. **Salvo** and **Haulover** on Hatteras Island are two of the more popular spots to put in, especially when there's good wind blowing from the northeast. Booties are recommended to safeguard feet from sharp, broken shells on the shallow seafloor.

5 Golden Gate
NATIONAL RECREATION AREA
California

San Francisco Bay provides one of the world's most dazzling places to sail. The twin arms of Golden Gate National Recreation Area separate the (sometimes) calmer bay waters from the often turbulent open Pacific. **Crissy Field Beach** on the south shore is the most popular place for kiteboarding, a spectacular (and also relatively new) sport that combines surfboard and parachute. The same beach is also a great place to launch Windsurfers for a cruise beneath the Golden Gate Bridge or a quick whip around Alcatraz. **Rodeo Beach** and **Muir Beach** in the Marin Headlands and **Ocean Beach** in San Francisco are other tremendous windsurfing and kiteboarding spots. While most sailors set out from marinas outside the park, Golden Gate does have its own facility—Travis Marina (aka Presidio Yacht Club) on Horseshoe Cove in the Fort Baker section in Marin County. A holdover from when the military occupied Fort Baker, the Air Force–operated marina provides sailing lessons and rentals to both military families and the general public.

6 Lewis & Clark
NATIONAL HISTORIC TRAIL
Oregon

Today's Lewis and Clark National Historic Trail charts the 1803–06 transcontinental expedition led by Meriwether Lewis and William Clark, crossing tribal lands and comprising more than 100 significant sites, interpretive centers, and "sense of place" landscapes that were part of the Corps of Discovery. Of course

Windsurfing enthusiasts flock to Hood River, Oregon, to take advantage of winds channeled through the Columbia River Gorge.

the legendary expedition benefited from Lewis and Clark's prowess at navigating North America's raging rivers. Oregon's Hood River region, known as the Gorge after the confluence with the Columbia River, is a hotbed of sporting, with kiteboarding and windsurfing stealing the show at the **Hood River Waterfront Park and Trail.** This is a great place for people-watching, or dive in yourself, keeping in mind that winds can be wild. The Hook is the park's hub for water sports, with a protected cove specifically for novice windsurfers. There's also a sandbar called The Spit where kiteboarders take on strong currents and unpredictable water levels. Dry off and end the day in the vibrant Hood River community, known for its dining, shopping, and breweries.

7 | Apostle Islands
NATIONAL LAKESHORE
Wisconsin

A labyrinth of sea caves, hiking trails, empty beaches, and pristine wildlife habitats, Lake Superior's **Apostle archipelago** is a rare glimpse of the long-ago Great Lakes. Boating is the only way to explore the islands of this Wisconsin park, especially the smaller and more far-flung landfalls. Thirteen of the islands boast docks where boaters can tie up for the day or overnight, and 18 have campgrounds for those who want to sleep ashore (some campgrounds are on islands without docks). Sailors can launch their craft from the National Park Service ramp at Little Sand Bay or more than half a dozen locations just outside the park boundaries, including the towns of Bayfield, Ashland, and Red Cliff. Overnight docking fees on the islands vary from $15 to $30 depending on the length of boat; alternatively, boats can be anchored offshore for free. While sunny days predominate during the boating season, Lake Superior is famous for dramatic shifts in weather including thick fog and violent squalls. For those who don't have their own vessel, businesses in Bayfield offer day trips and overnight live-aboard cruises.

Go Fly a Kite

While it might seem cool to honor the legacy of Ben Franklin by flying a kite at Independence National Historical Park in Philadelphia, other units of the park system are much more apropos for aerial artistry.

The combination of brisk ocean breeze and rolling green grounds of Castillo San Felipe del Morro make San Juan National Historic Site in Puerto Rico a popular spot for kites.

The annual Salem Merry-time Festival at Salem Maritime National Historic Site in Massachusetts includes kite-flying sessions and lessons, as well as sea chantey sing-alongs and model-boat building.

At the annual Wright Kite Festival, kids can relive the world's most famous flight with kite-flying demonstrations and kitemaking workshops at Wright Brothers National Memorial in North Carolina.

And finally, spring brings kite-themed fun at the Blossom Kite Festival happening each year during the National Cherry Blossom Festival on the Washington Monument grounds in Washington, D.C.

8 | Curecanti
NATIONAL RECREATION AREA
Colorado

Don't be fooled by the fowl name: The **Bay of Chickens** is Colorado's coolest place to windsurf and just one of several where you can "bump and jump" in Curecanti. Strung out along 40 miles of the Gunnison River on the western flank of the Rockies, the recreation area comprises three remote reservoirs. Blue Mesa is the largest of these, home to both the Bay of Chickens and the **Iola Basin,** another popular windsurfing spot. The former is on the north shore between Dry Gulch Campground and the park's Elk Creek Visitor Center. Iola Basin is farther east, a broad stretch of water that is best accessed from the boat ramp at Stevens Creek Campground. Breezes barreling down through the park's steep, narrow semidesert canyons can present quite a challenge, even for experienced windsurfers. Water temperatures in July and August can reach comfortable levels, but the rest of the year a wet or dry suit is highly recommended.

9 | Acadia
NATIONAL PARK
Maine

With a maritime heritage that stretches back more than 400 years, Maine is a cornerstone of American sailing and Acadia one of the few national parks that offers both lake and ocean boating. Small nonmotorized craft can take to the waters of 24 lakes and ponds in the park's **Mount Desert Island** portion. These range from relatively large bodies of water like Long Pond to the cozy confines of The Tarn near the park's Nature Center. Saltwater sailors also

Laying anchor in St. Thomas in the Virgin Islands

have plenty of scope, from the calm waters of the **Somes Sound** fjord and **Frenchman Bay** to the blustery open Atlantic between Mount Desert and Isle au Haut. A number of Acadia's smaller islands are closed during the eagle and seabird nesting seasons; boats can anchor at or circumnavigate the islands then, but stepping ashore is not permitted. The town of Bar Harbor on the park's eastern flank is a huge sailing center, with marinas and sailing outfitters offering family-oriented cruises of Somes Sound and the Cranberry. In Manset Village on the south shore of Mount Desert Island you can also find sailing lessons, personalized sailing tours around Acadia, and 19-foot sailboat rentals.

10 | Virgin Islands
NATIONAL PARK
Virgin Islands

A favorite with both first-time sailors and experienced yachtsmen, the Virgin Islands have everything that makes Caribbean sailing compelling: blue skies, warm weather, and a steady but gentle breeze, as well as great tunes, delicious seafood, and coves that never seem to get too crowded. Virgin Islands National Park sits at the heart of the action, easternmost of the American isles and just across the Sir Francis Drake Channel from the British portion. Many people bring their own boats down from the U.S. eastern seaboard or arrange bareboat charters from adjacent islands like Tortola and St. Thomas. The park provides plenty of safe anchorages and pristine waters in which to swim, snorkel, or scuba dive off the back end of a boat. The town of Cruz Bay is home to both the park visitor center and provision stores where sailors can stock their galleys. Several local outfitters offer crewed charters and catamaran cruises in national park waters. Small sailboats and Windsurfers can also be rented on the beach at **Cinnamon Bay** inside the park. While the height of the local yachting season is December to March, fall and spring are also pleasant just avoid the June to November hurricane season.

SEASONAL ENJOYMENT

The northern lights dance across the sky above Voyageurs National Park.

WILDFLOWER BLOOMS

Mountains may be majestic and waterfalls awe-inspiring, but the wildflowers of our national parks can be unmatched in their display of nature's beauty and variety. Peak seasons vary from year to year depending on rainfall and temperature, so check with rangers before planning a bloom-viewing trip.

1 North Cascades
NATIONAL PARK
Washington

Flowers tend to pick sides in North Cascades National Park, choosing either the more moist, western region of the Cascade ridge or the drier eastern side to lay their claim. Flowering occurs from April until early September, due to the park's variation in elevation and precipitation, but peak season is June through August. The biodiversity of the park's flora is largely unmatched, and visitors can look for **glacier lily, lupine, skunk cabbage, heather, paintbrush, alpine cinquefoil, red columbine, phlox, aster,** and **bear grass,** to name just some. And wildflower enthusiasts aren't the only ones attracted to the park's blooms: Pikas—furry, plant-eating mammals that call North Cascades' rockslide areas home—collect grass and wildflowers in the late summer, storing them under rocks to use as their wintertime food supply. Join a guided hike led by local outfitters for a narrated tour of the colorful displays. The Cascade Pass Trail and Maple Pass Loop have some of the best views, while the ironically named Easy Pass and Fisher Basin area have extensive meadows below glacial peaks.

2 Great Smoky Mountains
NATIONAL PARK
North Carolina & Tennessee

The list of 1,600-plus species of flowering plants in Great Smoky Mountains tops that of any other North American national park—the park's own biologists refer to it as "Wildflower National Park"—in fact, this Appalachian site boasts the greatest biodiversity of any site of similar size on Earth outside the tropics. One major reason is the stability of the landscape: The Great Smoky Mountains are among the planet's oldest ranges and have been relatively untouched by major geological forces for more than a million years, giving life time to diversify. Elevations in the park range from 875 to 6,643 feet, making it possible to (figuratively) travel from the Southeast to Canada within the park's boundaries. The wildflower show at Great Smoky Mountains begins with spring blooms such as **trilliums, lady's slipper orchids,** and **violets,** continues

Methow Valley in the North Cascades bursts with wildflowers like balsamroot and lupines.

with summer species such as **cardinal flower** and **butterfly weed,** and finishes in fall with composites like **goldenrod** and **coneflowers.**

3 | Rocky Mountain
NATIONAL PARK
Colorado

The vast and magnificent peaks of this Colorado park extend the wildflower season for visitors, with displays appearing in spring in the drier montane zone below 6,000 feet, while some species may not bloom on the tundra above tree line until June or even July. In the lower habitat dominated by ponderosa pine, look for **geranium, mariposa lily, miner's candle, lupine,** and tall **penstemon.** Higher up in elevation thrive **calypso orchid, twinflower, pipsissewa,** and gorgeous **blue columbine,** the Colorado state flower. But the truly special wildflowers of Rocky Mountain occur in the alpine zone from around 11,000 feet up to more than 14,000, where the growing season is a brief few weeks and harsh conditions mean plants hug the ground to escape the worst of the bitter winds. Here you find **alpine sunflower, primrose, bistort, avens, mountain dryad, alp lily, moss campion, snow buttercup,** and **sky pilot.** Trail Ridge Road provides easy access to this tundra flower garden, but hikers should take special care not to destroy plants with careless footsteps: A small plant may actually be decades old, and revegetation may take even longer.

4 | Big Bend
NATIONAL PARK
Texas

The words "desert" and "diversity" might seem mutually exclusive to some, but the Chihuahuan Desert of the southwestern United States and Mexico is among the wettest and most biologically varied deserts on Earth. Big Bend serves as the premier showcase of Chihuahuan Desert ecology, with more types of **cacti** (around 60) than any other national park, and has been honored as a World Heritage site for its globally significant biodiversity. As is the case in all deserts, the wildflower bloom at Big Bend is highly dependent on winter and early spring rains, but in most years it's striking, and in wet years it can be truly spectacular. Not only do cacti such as claret cup and prickly pear provide color, but the park lowlands can be a virtual carpet of

Flower Fests

Wildflowers are nothing short of enchanting, and at these vibrant parks, that's something to celebrate, with weekend festivities and events planned from coast to coast.

Each May, botanical workshops, a youth wildflower art contest, guided hikes, and birdwatching make up just a few of the highlights of the Shenandoah National Park Wildflower Weekend, a nearly four-decade tradition.

Since 1951, Great Smoky Mountains National Park has hosted the Spring Wildflower Pilgrimage in May, inviting visitors to guided walks and engaging programming on such topics as journaling, area history, and nature photography.

And every July, Utah's Cedar Breaks National Monument Wildflower Festival packs more than a week's worth of programs and activities to mark the summertime blanketing of wildflowers in the natural amphitheater.

wildflowers such as **bluebonnet, desert baileya, globe mallow, cowpen daisy, desert marigold, Chisos prickly poppy,** and **cenizo,** to name only a few of the scores of species. One of Big Bend's most spectacular blooms is displayed by a plant hardly regarded as a wildflower: the **Harvard agave** or century plant, which features a stalk up to 20 feet high loaded with yellow flowers, blooming just once before it dies.

5 | Mount Rainier
NATIONAL PARK
Washington

When it comes to volcanoes, there's bad news and good news: The bad news is that they sometimes create havoc for humankind and nature alike; the good news is that the ash they scatter acts as rich fertilizer, creating lush growing conditions for plants. After millennia of eruptions, the slopes of Mount Rainier constitute a naturally tended garden of wildflowers. In fact, the park's most popular area was given the name Paradise in 1885 when a settler saw a glimpse of heaven in the panorama of Rainier's peak rising above a field of wildflowers. From **salmonberry** to **paintbrush, lupine** to **penstemon,** and **phlox** to **bunchberry,** Mount Rainier reveals a never ending display from late spring through summer, when the meadows of Paradise are a sea of purples, blues, reds, and yellows in all shades. Note

▶ *Edible plants, fruits, and mushrooms can appear especially appealing in the wild, and our national park units are full of them. Check with park staff before you forage—many allow this limited practice for personal use, while others strictly forbid it.*

that the road to Paradise may not open until late May, depending on weather, and the peak of bloom for subalpine flowers is brief.

6 | Shenandoah
NATIONAL PARK
Virginia

The ancient Appalachian Mountains are renowned for their biodiversity, a circumstance owing in part to the varied terrain and elevations, as well as the abundant moisture. President Theodore Roosevelt, an ardent nature lover and conservationist, once referred to the Appalachians' "marvelous variety and richness of plant growth." Spanning more than 70 miles north to south, with an elevation range of around 3,500 feet, Shenandoah National Park reflects that variety: It's home to

more than 1,400 species of plants, nearly 900 of which are wildflowers. These blooms generally begin in late March and continue through fall; the beginning of May is the park's annual Wildflower Weekend, when guided field trips help beginners learn colorful species such as **bellwort, toothwort, lady's slipper orchid, Solomon's seal, wild ginger, sweet cicely,** and various **violets.**

7 | Tallgrass Prairie
NATIONAL PRESERVE
Kansas

When Europeans first arrived on the shores of North America, tallgrass prairie covered 140 million acres of the continent, supporting spectacular gatherings of wildlife, from bison and elk to wolves and prairie chickens. Today, only about four million acres of this habitat remains, most of it in scattered patches too small to serve as functioning ecosystems. Tallgrass Prairie National Preserve, in the Flint Hills of eastern Kansas, protects a sizable tract of the grassland that once blanketed much of the Great Plains. The 500 species of plants occurring on the nearly 11,000 acres here include **butterfly weed** and other **milkweeds, wild indigo, Indian blanket, liatris, bee balm, prairie clover, coneflower, compass plant, coreopsis, aster,** and **leadplant.** The Southwind Nature Trail and the Bottomland Trail both offer up close looks at prairie wildflowers, as well as big bluestem, Indian grass, and switchgrass growing to heights of eight feet or more.

8 | Indiana Dunes
NATIONAL PARK
Indiana

Perhaps best known as a beach getaway, Indiana Dunes surprisingly ranks among the top 10 most biologically diverse National Park Service sites, with more than **1,100 flowering plants and ferns.** The impetus for the park's creation, in fact, was not recreation but biology—dating back to 1899, when a local biologist wrote a scientific paper describing the site's varied ecosystems. Wildflower diversity is boosted by the presence not only of dunes but of oak savannas, swamps, bogs, marshes, prairies, and forest. Hikers can enjoy blooms such as **moss pink, blue flag iris, Dutchman's breeches, cardinal flower, columbine, lupine, pitcher plant,** and more than two dozen species of **orchids.** Check the park's website for springtime ranger-led wildflower hikes.

9 | Joshua Tree
NATIONAL PARK
California

This vast park (covering 789,745 acres) in Southern California is a favorite destination for travelers who enjoy seeing and photographing wildflowers. Joshua Tree lies at

Joshua Tree is home to 813 vascular plant species, including several species of cholla cactus.

the meeting point of the Colorado and Mojave Deserts, two ecosystems quite different in the composition of their vegetation. The Colorado Desert, in the eastern part of the park, is both hotter and drier than the Mojave Desert in the west. The plant for which the park is named, the Joshua tree (a species of yucca), grows in the latter habitat, blooming profusely in spring. The park also includes part of the Little San Bernardino Mountains, dominated by juniper and pinyon pine, and several oases where fan palms grow. Wildflower season begins in early spring in the southern part of Joshua Tree, progressing through the year to blooms in the highest mountains in

June. Some of the herbs and shrubs growing here include **desert dandelion, desert gold poppy, chuparosa, ocotillo, Arizona lupine, paper-bag bush, mariposa lily, desert senna, indigo bush, brittlebush, chia, fiddleneck,** and several types of **cacti.** How diverse is the vegetation in Joshua Tree National Park? When conservationists proposed saving this area in the 1930s, one suggested name was Desert Plants National Park.

10 | Buffalo
NATIONAL RIVER
Arkansas

America's first national river, the Buffalo is famed for crystal clear water, high bluffs, and canoeing suitable for boaters of all abilities. The Arkansas Ozark Plateau also boasts a splendid diversity of flora. Blooming begins as early as February when the leaves have yet to appear on trees. **Hepatica, bloodroot, toothwort, rue anemone, wild ginger, trout lily,** and various **violets** are among the first to appear, followed by **trilliums, May apple, jack-in-the-pulpit, green dragon, Solomon's seal, columbine, phlox,** and many others. The easy trail in Lost Valley, near Ponca, is a pretty place to start enjoying Buffalo River wildflowers, but visiting a variety of habitats, from streamsides to clearings to rocky ledges, will bring the greatest species total. Trails near Pruitt and the ghost town of Rush are good places to explore.

FALL COLOR

Nature puts on its own display of artistic creation each autumn when hardwood trees from oaks to aspen change color as they prepare for winter, delighting us with blazing reds and yellows, subtle earth tones, and nearly every shade in between. The following national parks offer some of the best fall color to be found in the United States.

1 Acadia
NATIONAL PARK
Maine

Much of the fabulous fall color of this national park on the Maine coast is owed to a disaster of more than a half century ago. The autumn of 1947 brought the driest conditions ever recorded to Mount Desert Island, location of most of Acadia National Park's expanse. A fire broke out in mid-October and raged for 10 days, burning 17,188 acres, including 10,000 acres in the park. Mature spruce and fir forest was destroyed, replaced by fast-growing hardwoods such as **birch, aspen, poplar, and maple.** Instead of the uniform year-round green of conifers, once devastated parts of Acadia now display reds, oranges, and yellows in fall, delighting those who drive its scenic loop or hike or bike along its historic unpaved carriage roads, from which motorized vehicles are barred. A map of Mount Desert Island is often described as looking like a lobster claw; the burned area is located on the central part of the eastern half of the "claw."

2 Guadalupe Mountains
NATIONAL PARK
Texas

The stark Chihuahuan Desert landscape seems too harsh for any but the hardiest plants, and the highest slopes are covered with ponderosa pines and Douglas firs, green year-round. Yet there's a special spot in this national park that attracts thousands of leaf peepers each fall and in fact has been called, with plenty of justification, the "most beautiful place in Texas." It's McKittrick Canyon, in the northeastern part of the park, a chasm known for both its geology and its flora. Here, **bigtooth maples, oaks, walnuts,** and other hardwood trees and shrubs turn shades of yellow, red, and orange each fall, creating a scene so striking that crowds sometimes cause park managers to limit entry to prevent damage to the natural resources. Late October through mid-November is the usual period for peak leaf color in this riparian oasis, where hiking trails of various lengths offer the chance to see the show up close. Be sure to take note of the unusual geology here: These massive mountains actually began as reefs under an ancient warm ocean. The same limestone has been hollowed out underground just north of here to form the network of caves at Carlsbad Caverns National Park. The canyon is designated for day use only, but a drive-in campground is within distance for visitors who'd like to spend more time exploring McKittrick during the cooler days of fall.

3 Cuyahoga Valley
NATIONAL PARK
Ohio

Covering more than 51 square miles along the Cuyahoga River between Cleveland and Akron, this diverse, quasi-urban park protects some of the region's most attractive woodlands. Though perhaps best known for its Towpath Trail and waterfalls, Cuyahoga Valley provides fall foliage in areas of mixed hardwood forest comprising **oak, hickory, maple, beech,** and **sycamore.** Fall color usually peaks in October. The Ohio & Erie Canalway Scenic Byway runs through the center of the park, but for an unusual way to admire the autumn color, buy a ticket on the Cuyahoga Valley Scenic Railroad, sit back, and watch the forest pass outside the train window.

4 New River Gorge
NATIONAL PARK & PRESERVE
West Virginia

One of the country's newest national parks, New River Gorge is famous for having some of the best fall foliage thanks to its proliferation of **sugar maples, red oaks, yellow poplars,** and more. And the superlative is well earned, given that autumn seems to linger longer and more colorfully in West Virginia. For the best leaf-peeping opportunities, begin at the Canyon Rim Visitor Center for paint-palette panoramas, then take a walk in the woods (specifically

▶ *Although we see leaves change their colors in autumn, the red, orange, and yellow pigments are actually present in leaves year-round.*

the Long Point, Endless Wall, or Fern Creek Trails), meander along a scenic drive, or enjoy the last days of tepid float trips through the park to see the trees in vibrant display. Mid- to late-October is prime time for color, and the park plans quite the celebration: The largest single-day festival in the state of West Virginia, Bridge Day, takes place on the third Saturday in October. This celebration of the 1977 completion of the New River Gorge Bridge, the third highest in the United States, includes BASE jumpers, artists, live music, and plenty of food, with the colorful trees framing it all.

5 Blue Ridge Parkway
NATIONAL PARKWAY
North Carolina & Virginia

Timing is everything in viewing fall foliage, as environmental factors can cause peak color to vary from year to year at any one spot. Stretching 469 miles mostly north to south through Virginia and North Carolina, the Blue Ridge Parkway gives travelers assurance that they'll run

into the best colors somewhere along the way, as autumn moves south (figuratively speaking) across the eastern United States, and specifically across the Blue Ridge Range of the Appalachian Mountains. In addition, the parkway links two of our best national parks for fall foliage: Shenandoah and Great Smoky Mountains, each worth exploration by road and trail. **Maples** are the stars of the scenic show along the drive, with a supporting cast of **oak, poplar, birch, black gum, sassafras, tulip poplar** (a species of magnolia that grows very tall in these mountains), **sumac,** and other hardwoods. The great elevation differences along the Blue Ridge Parkway and in the two national parks means that different areas reach top color at different times, adding to the likelihood that a visitor will find eye-popping hues. Peak color usually arrives at Shenandoah National Park in mid-October, with Great Smoky Mountains National Park typically a little later.

6 Delaware Water Gap
NATIONAL RECREATION AREA
New Jersey & Pennsylvania

Set in the Delaware River Valley of Pennsylvania and New Jersey, this park includes historic villages, farmland, and grassland, but it's mostly covered by a maturing hardwood forest, slowly growing back from

Autumn brings an array of colors to foliage along the Blue Ridge Parkway.

Worth the Wait

Predicting peak autumn color for a particular location is a tricky business. Though some general assumptions can be made, the exact period of brightest shades varies annually depending on factors such as rainfall, temperature, and amount of sunshine. So what's a would-be leaf peeper to do? Increasingly state and local tourism agencies, as well as meteorologists, are providing up-to-the-moment reports on fall foliage, allowing travelers to make last-minute plans to see those flaming maples. Check tourism agency websites and social media accounts for the latest on when and where to hit the road.

periods of logging in previous centuries. Extensive tracts of various **oaks** give a range of fall colors from dull yellow to red-orange. Other autumn hardwoods include **maple, birch, hickory, dogwood, beech,** and **sycamore.** More than 200 miles of roads wind through Delaware Water Gap; some good ones to explore include Old Mine Road in New Jersey and U.S. 209 and River Road in Pennsylvania. Three scenic overlooks along Route 611 provide great views of the Delaware River as it passes through the Water Gap, an eroded low point in Kittatinny Ridge. For a closer view of the park's forest, hike some of the 27 miles of the Appalachian National Scenic Trail running through the area.

7 | Zion
NATIONAL PARK
Utah

Best known for its spectacular geological formations and slot canyons, this southwestern Utah park is also well regarded regionally for the fall foliage of its hardwoods. In riparian areas along the Virgin River and other streams grow **Fremont cottonwoods,** which turn a beautiful shade of pale gold in autumn, while **bigtooth maple, box elder** (another type of maple), **birch, oak,** and **hackberry** add their hues, as well. In both streamside locations and higher on slopes, **quaking aspens** turn a more brilliant gold. Because Zion National Park encompasses a large elevation range, the period of fall color extends from September (in the high mountains) well into November (in the lowlands). When sunrise or sunset light turns Zion's rocks a gold-red color and the leaves add their spectacle, few parks can match it for sheer visual impact.

8 | Effigy Mounds
NATIONAL MONUMENT
Iowa

Iowa is not all flat—as evidenced by this park in the extreme northeastern corner of the state, where 400-foot-high bluffs loom over the Mississippi River. Neither is the state all farmland and prairie; for proof, just walk some of the 14 miles of trails in Effigy Mounds National Monument through upland forests of **oak, sugar, maple, hickory, aspen, walnut, Kentucky coffee tree,** and **basswood,** all of which show beautiful colors in a brief blaze of glory late September through early October. The national monument was set aside to protect more than 200 Native American ceremonial mounds (including 31 in the shape of animals such as birds, bears, and deer), but the park designation had the side effect of preserving woodlands, wetlands, and 81 acres of native tallgrass prairie. In all, the natural features of this historical park make it one of the most diverse sites in the upper Midwest for everyone from bird-watchers to hikers to weekend scenery oglers.

Zion National Park's Virgin River winds past red-rock walls and brilliant fall foliage.

In the fall, the white bark and bright yellow leaves of quaking aspen make for a spectacular display.

9 | Rocky Mountain
NATIONAL PARK
Colorado

One species of tree dominates the fall foliage at this central Colorado park: **quaking aspen** (*Populus tremuloides*). This deciduous tree's stunning colors of greenish yellow to glittering gold make it worthy of a visit all by itself. Beginning in mid-September, aspens start to change their hues, continuing into October before the harsh Rocky Mountain winter strips the leaves from the branches. Aspens grow where fire or past logging has removed the dominant conifer forest and are found up to 10,000 feet on mountainsides. They often grow in clumps or groves and can reach heights of up to 100 feet. These trees "quake" because the leaf petiole (stem) is flat instead of round as in most trees, causing the leaves to quiver in the slightest breeze and adding to the attractiveness of the golden masses; in addition, when the aspens quake, they make a delightful soft rustling sound unlike any other tree ruffled by the wind. Large areas of aspen grow throughout Rocky Mountain National Park, easily seen from Trail Ridge Road, Bear Lake Road, and many other park locations.

10 | Little River Canyon
NATIONAL PRESERVE
Alabama

This little-known park in northeastern Alabama combines tree species of the Appalachians with those of Deep South hardwood forests for a diversity that adds to the color palette in fall. Take the Canyon Rim scenic drive off Highway 35 along Little River Canyon for a fine overview of the environment. Several overlooks include one at charming Little River Falls, where the river cascades over rock ledges beneath tree-covered hillsides. **Maple, oak, hickory, sweet gum, black gum,** and **tulip poplar** trees span the color range from pale yellow to vivid orange. The gorge, one of the deepest in the country east of the Mississippi River, looks especially stunning at the peak of color, usually from mid-October through early November.

10 BEST PARKS FOR
NIGHT SKIES

Just as the parks are dedicated to preserving landscapes, cultural sites, and habitat for wildlife, so are they concerned about overhead resources—particularly the kind of skies that reveal the beauty of the nighttime firmament. A major component of the parks' night-sky programs is to guide visitors toward an appreciation of darkness.

1 | Bryce Canyon
NATIONAL PARK
Utah

Bryce has long staged one of the most enjoyed night-sky shows in the National Park System. Beginning in the park visitor center, everyone moves outdoors to view the sky through dozens of telescopes provided by the park and volunteers. And what a sky—so dark and clear that the **Andromeda galaxy** can be spotted with the naked eye, a claim some astronomers don't believe until they come to Bryce and see it for themselves. While program participants wait for a scope, they can see the Milky Way arcing across the sky from horizon to horizon. So dark is the sky here that Jupiter and Venus throw shadows, and if the moon is out, a third shadow might be cast by the heavens. If the moon is full the star show is diminished, but the park then leads full-moon hikes to see what happens when shadows befall the park's mysterious hoodoos. Stay tuned in June for the annual astronomy festival, which features both daytime and evening programming. Plan ahead and make reservations—all experiences, like Bryce, are quite popular.

2 | Natural Bridges
NATIONAL MONUMENT
Utah

Remote Natural Bridges is about as close as you can come to experiencing the night sky as it was before human lights trespassed upon it. The monument's dark sky is rated a Bortle Class 2, Class 1 being the sky as it would have been before the invention of the light bulb. In fact, a look out over the horizon from the parking lot reveals no artificial lights. The monument's arid climate and elevation (6,500 feet) contribute as well. Little wonder, then, that Natural Bridges was named the world's first International Dark Sky Park by DarkSky International. The park runs a regular program through spring and summer that starts with a lecture and a naked-eye guided tour of the **Milky Way**'s intricately interwoven stars and clouds of dust and gas, vivid without the aid of optics, followed by telescope viewing sessions.

3 | Chaco Culture
NATIONAL HISTORICAL PARK
New Mexico

Very few units in the National Park System have their own observatory. Thanks to a private donation, this International Dark Sky Park has a 25-inch Dobsonian telescope at its disposal, a powerful scope whose specialty is deep-space viewing. In tandem with a digital imaging system, the scope can track and photograph changes in the heavens, such as **supernovas** and

asteroids. Many of its thousands of images are incorporated in the park's frequent night-sky programs. Those programs, held from April through October in an outdoor amphitheater, include the opportunity to view the heavens directly through the big Dobsonian, as well as a number of smaller but still very powerful telescopes provided by volunteers. Of course, this being a historical park, cultural connections are a significant part of the presentation, which illustrates how certain doorways and structures built by the Chaco people receive the rays of the sun at the time of the solstices. Full moon? No problem. That's when the park conducts special full-moon tours of some of the ancient Chacoan structures.

4 | Hawai'i Volcanoes
NATIONAL PARK
Hawaii

Hawai'i Volcanoes shares some of the conditions that make nearby Mauna Kea one of the world's great observatories—namely, distance from cities and an islandwide lighting ordinance that ensures dark skies. The park adds the bonus of programs that pay cultural homage to the Polynesians, who were skilled celestial navigators. On select Tuesday evenings visitors can gather at the Kīlauea Visitor Center for special After Dark in the Park events. Afterward, consider

▶ *Astronomer and artist Dr. Tyler Nordgren coined the campaign catchphrase "half the park is after dark" in 2010 to promote the nighttime allure of national parks.*

claiming a pullout along Chain of Craters Road for an astounding, open-horizon view out over the lava fields beneath the **Southern Cross constellation.**

5 | Voyageurs
NATIONAL PARK
Minnesota

The stars must figuratively align to see the **aurora borealis:** The aurora must be active and the sky must be clear. Wintertime's longer nights, while certainly cold, increase your chances. Voyageurs National Park is one of the country's best bets for spotting the seemingly elusive northern lights. And while there are many open horizons at the International Dark Sky Park, two viewing areas stand out: the parking lot at the Rainy Lake Visitor Center and the Meadowood Road Day Use Area. Even if you don't see the dazzling dancing green lights, there's plenty of star-studded fun at this North Woods park, such as

celebrations featuring free astronomy events like special speakers, meteor shower viewings, films, campfire s'mores, telescope sessions, and more. Enjoy the chase!

6 | Capitol Reef
NATIONAL PARK
Utah

Distant from cities and a mile high, Capitol Reef has what astronomers call magnitude 7 skies, meaning very, very distant objects can be seen with the naked eye. Which means the unpolluted sky is very, very dark and the star show is amazing. After the regular evening ranger presentations, park staff brings out the Meade telescope and the star party starts around 10 p.m. with dazzlers like **Saturn** and **Mars** just before they fall below the horizon. Then come more distant showpieces and their accompanying stories, like the **Dumbbell Nebula,** revealing amazing detail—its spiral arms, its tidal interactions with neighbors, the **Lagoon Nebula** and the **Triffid Nebula**—9,000 light-years away in Sagittarius. Naked-eye viewing is also dazzling—**Mizar** and **Alcor,** binary stars in the Big Dipper, and **Albireo,** which appears to be one bright star until the Meade reveals its companion. Unless a midsummer monsoonal flow clouds the skies, viewing is superb throughout the park, but nowhere more dramatic

The Milky Way lights up the sky in Voyageurs National Park.

Great Basin National Park is home to three groves of rare Great Basin bristlecone pine trees.

globular star cluster and the binary star Albireo, and naked-eye views of the dramatic galactic center of the Milky Way. An hour away in the town of Ely, you can hop aboard the Great Basin Star Train, an out-of-this-world experience so popular it typically sells out a year in advance.

8 | Craters of the Moon
NATIONAL MONUMENT & PRESERVE
Idaho

In August 1969, four Apollo 14 astronauts explored a lava landscape, meticulously selecting scientifically valuable specimens to represent the **moonlike volcanic geology** they traversed. They weren't on the moon, though. The group was in Idaho, training at the only national park unit named for a celestial body. The successful Apollo 14 mission didn't happen until 1971, but it was this time at Craters of the Moon National Monument and Preserve that the astronauts credit for their successful space preparation. And the park isn't known only for what's underfoot. Located in a sparsely populated region of interior Idaho, the monument is ideal for night-sky viewing, too, and earned its designation as an International Dark Sky Park in 2017. **Star parties** are held in the summer and fall, and volunteers from the local astronomical society provide telescopes and plenty of pointers for enthusiasts.

than from Panorama Point. The annual Heritage StarFest is a fun time to celebrate the dark skies of Capitol Reef and surrounding Wayne County. Check the park's website for dates and details.

7 | Great Basin
NATIONAL PARK
Nevada

Nevada may be synonymous with bright lights for some, but Great Basin, with its isolated setting near the Utah border, is hardly on the same planet as Las Vegas or Reno. The International Dark Sky Park is blessed with exceptionally dark skies and, outside of its high forested sections, a nearly 360-degree view of the horizon, where the Milky Way takes center stage on summer nights. The Great Basin Observatory opened here in 2016, and weekly programs at the visitor center's Astronomy Amphitheater stress the importance of dark skies and the impact of light pollution (check with the park for schedule and for details on the annual Astronomy Festival). Following the talk, everyone steps outside for a telescope viewing session of such distant objects as the **Hercules**

9 Big Bend
NATIONAL PARK
Texas

It's no surprise that arid, remote Big Bend, 225 miles from the nearest city, has night skies among the darkest in the national parks. At least 2,000 stars are visible to the naked eye on a clear moonless night, as well as planets and shooting stars. A broad, 250-mile-distant horizon helps, too, as does night-sky-friendly lighting around the main visitor centers at headquarters and Chisos Basin. Park rangers and volunteers host regular night-sky interpretive programs, including star parties, moonlight walks, and telescopes. On a clear night, the telescopes can see as far as two million light-years away to the **Andromeda galaxy.** Check with the park for the current schedule. Winter brings the darkest and cleanest skies—in part due to low humidity and infrequent cloud cover—and there's no particular "best" place in the park to view them. Just look up.

10 Acadia
NATIONAL PARK
Maine

Because Acadia is largely surrounded by ocean and its nearby communities honor dark-sky ordinances, the park has one of the darkest skies in the eastern United States. Sand Beach is a particularly perfect place to feel infinitely small: Imagine lying in this protected cove surrounded by ancient rocks with a multitude of stars (plus the **Milky Way, Jupiter, Saturn,** and more) twinkling above. Ranger programs and nighttime hikes illustrate how humans can easily adapt to the dark, and these efforts are celebrated each fall during Acadia Night Sky Week. Plan for lectures, star parties, and many opportunities for gazing at the Milky Way. On your own, Cadillac Mountain is a great place for a 360-degree dark-sky view (check the park website for seasonal vehicle closures), but be prepared for a cold and windy session.

Animals After Dark

Nighttime can be prime time in our national parks—at least for our feathered, fuzzy, and four-legged friends. By protecting the night skies, the Park Service preserves the chance for many species to hunt, mate, escape heat, and avoid predators. It's common for birds to use the stars to navigate, while sea turtle hatchlings depend on the reflection of the moon and stars to guide them to the safety of the ocean.

Parks After Dark programming at units like Pennsylvania's Hopewell Furnace National Historic Site and Valley Forge National Historical Park highlight the value of dark skies with sessions on fireflies, moths, and owls. Across the country in California's Death Valley National Park, critters like coyotes, large-eared jackrabbits, kit foxes, and kangaroo rats choose the much cooler nights to scavenge.

Of course, the king of the evening remains the bat, with the Carlsbad Caverns National Park Bat Flight Program In New Mexico ranking among the top-attended ranger events in the entire park system. More than 50 unique species of bats live in our national parks, where darkness isn't just helpful—it's essential to their well-being.

Are We There Yet?

Even on the best days, family travel can feel like you're juggling logistics and luck. It's worth the work, say these practiced parents who have made national park venturing a priority.

"There was this time in Grand Teton where we were all sitting by the river watching the kids throw rocks in the water. Kids want to touch, feel, hear, and truly immerse in things. Sitting by the water, it felt like nature was soaking into us," recalls Preethi Harbuck of a recent trip she took with her husband, Daniel, and their six children.

For the Harbucks, who run the travel blog Local Passport Family, national parks were a natural extension of their on-the-go lifestyle. In fact, five of the Harbucks have been to all 63 national parks. But it took a cross-country move to officially start the count. "We really caught the bug after moving to California, where there is not only the most national parks of any state, but some really spectacular ones." The family checked many, many more parks off their list when they moved into an RV for a season during the COVID-19 pandemic. "That's when we started feeling more at home in the parks," Harbuck says.

For Sarah Stewart Holland and her husband, Nicholas Holland, national park trips have been a more recent addition to the family's prolific travels. Sarah, author of two books and co-host of the Pantsuit Politics podcast, remembers it being the second year of the pandemic when her family ordered national park passports and a scratch-off poster. "That's when Nicholas and I set a goal," she says. "We decided we wanted to visit all of the national parks before we retire—rather than waiting *until* we retire because, you know, life is to be lived."

The more parks the Hollands visit (they're at 29 as of this writing), the more they love them—and their three sons do, too. "We've told our kids what an incredible gift the national parks are, but it's so much fun to actually get to see diverse geography and some really interesting parts of the country," Sarah says. "Parks encourage you to slow down, enjoy each other, enjoy nature, and enjoy history. They help kids see that they're small, but still part of this big, beautiful whole."

THE BEST-LAID PLANS

A lot of thought goes into Sarah's itineraries—such as school schedules, weather, proximity of parks, plus predicted crowd levels—before settling on dates and locations. From there, the Hollands work to secure lodging—typically opting to stay inside the parks to reduce the daily driving time for the kids, "which has turned out to be a great hack for us," Sarah says.

The daily activity planning falls to Nicholas, who aims to strike the right balance of outdoor adventure and must-sees for their kids' ages and abilities. (The Hollands often travel with three generations, while the Harbucks have smaller campers to consider.)

When they arrive at each park, the Hollands start at the visitor center for passport stamps, Junior Ranger books, and to chat with a ranger. "Rangers are like librarians—they love to help," says Nicholas. "They'll give you the skinny on whatever it is you need to know, and if something is closed, they'll have helpful alternatives."

The Junior Ranger programs are a top priority for Harbuck's family, too, who adds that she appreciates how the books recognize Indigenous lands and history. "The whole point of the program is to get kids loving the parks and interacting with the land that we're fortunate enough to be able to recreate on."

A tactile map helps visitors experience history at Wright Brothers National Memorial in Kill Devil Hills, North Carolina.

REMEMBER THE REAL VALUE

While it hasn't all been rosy, both families say park travel is more than worthwhile. Plan for the nuance of your particular crew, Nicholas advises, which means for longer trips, consider building in slower-going days with museums or driving tours rather than hikes. Ultimately, Sarah encourages folks to push aside their intimidation: "People sometimes act like we've taken our children to the moon, but we haven't. Pick the closest park to you and go for a day. Yes, the parks are big and nature is powerful, but start small and see what the experience is like."

Harbuck reminds us that there's no award for doing a 15-mile hike versus a half mile. "It's easy to get caught up in trying to get the 'full experience,' but it's the interaction with nature, with the land, and with the people that makes the experience authentic and meaningful." She and her husband recently conquered Half Dome with their two oldest kids. "We'd been to Yosemite a dozen times before, and we finally did it. But hiking Half Dome was not what made us feel most connected to Yosemite. The experiences that allow us to continue to appreciate nature as a family are more meaningful and long-lasting than any of the individual accomplishments."

10 BEST PARKS FOR
SUNRISE & SUNSET

Although it happens twice a day each and every day, more often than not the natural phenomenon of the rising and setting sun is overlooked. Where this event does command a lot of attention is at national parks, where perhaps an extraordinary landscape or a prominent feature accentuates this astonishingly beautiful event.

1 | Capitol Reef
NATIONAL PARK
Utah

The soaring monoliths **Temple of the Sun** and **Temple of the Moon** mark millions of years of geological activity. They got their modern name from Charles Kelly, the first superintendent of Capitol Reef National Park, but it's easy to imagine that Indigenous settlers had similarly noble monikers for these imposing formations. Locating them in Capitol Reef's Cathedral Valley requires a vehicle with high ground clearance since rain in the rugged, remote region can cause impassable mud. The monoliths—made of fine-grain sandstone—rise 400 and 200 feet, respectively, above the valley floor, making for a spectacular golden view at sunrise. (Check with a ranger the evening before to confirm the route will be clear for a sunrise viewing.) Capitol Reef also has a number of end-of-day spots for admiring the sun's course. **Sunset Point** has an astonishing view of the sun illuminating the walls of the Waterpocket Fold. Its parking lot (located just off Highway 24) fills quickly, so come early and settle in. Another option is aptly named **Panorama Point,** where sweeping sights show the glow of the descending sun on the west-facing red cliffs.

2 | Arches
NATIONAL PARK
Utah

At sunset in Arches National Park, **Delicate Arch** seems to ignite with the flare and fire of the desert sun, its iconic image symbolizing the American Southwest. It's roughly 1.5 miles from Wolfe Ranch to the arch via the Delicate Arch Trail, so time your hike to arrive at least 30 minutes before sunset and simply follow the cairns that mark the route. The trail pitches up and around the final corner where, pierced by wind and sand, the center of the "sandstone fin" has created a 46-foot arch that, when the sun falls, changes like a desert chameleon, filtering sunset through a color wheel of red and orange, crimson and gold.

3 | Joshua Tree
NATIONAL PARK
California

As the afternoon fades into evening over Joshua Tree's cactus and pinyon, cool clouds fingerpaint the sky above this Southern California desert park. Adding an aural layer to the vivid spectacle of sunset is the distinct bay of howling coyotes. Roads and ridges that run north and south persuade travelers to reach peaks that provide ever changing vistas as the world turns. Should a vehicle be able to negotiate off-road

A hidden gem in the National Park System, Capitol Reef boasts canyons, cliffs, and spellbinding monoliths.

The Bass Harbor Head Light Station provides a perfect setting for watching the sun set at Acadia National Park.

trails, views can improve through access to little-visited areas populated by cacti, junipers, yuccas, and, silhouetted against the horizon, the most special part of this park's sunset, the eponymous **Joshua trees.** Framed by bands of color, darker above and brighter below, these otherworldly plants appear as inky black splotches against the sky.

4 | Acadia
NATIONAL PARK
Maine

Long before the rest of the country is up for coffee, early risers awake at Acadia National Park in time to catch **America's earliest sunrise.** Between October and March, the first light of day to fall upon the United States shines at 1,528-foot **Cadillac**

Mountain in the heart of Acadia, on Maine's coast. It's a wonderful sensation to feel the warmth of the sun without having to share it with 300 million others, and that experience is heightened by spectacular panoramic views from an overlook at the peak. The sun rises above the Atlantic Ocean's horizon and casts streaks of color—oranges, reds, pinks, depending on atmospheric conditions—across the water, while in the foreground the exposed rock of the mountains glows warmly in the sun. From the parking lot walk to the Summit Trail and find a spot (out of the wind) facing east. Plan this ascent. Gates are open 24 hours at the park, but weather conditions and seasonal vehicle reservations can close or limit the road to the top of Cadillac Mountain. Sunrise occurs around 6:30 a.m. in October, about a half hour later in November. Dress for the weather and bring along a camera, snacks, and something warm to drink while waiting for the start of the daily show.

5 Yosemite
NATIONAL PARK
California

Nature photographers will never lack for an amazing image as long as there are **sunsets at Half Dome.** Near the end of each clear day, the usually harsh sun softens to cast an even and gentle glow throughout the eastern end of Yosemite Valley, deep within this California national park. Within minutes, the exposed northwest face of Half Dome begins to change hues with the setting sun and, depending on the season, the scene may vary between a brilliant reddish orange and a soft wintry gray. Other mountains are also illuminated across America, certainly, but Half Dome's broad wall of granite seems to scoop

up every ray of the setting sun, creating an inspiring glow across what could be the world's largest sundial.

6 Canaveral
NATIONAL SEASHORE
Florida

The government determined in the 1950s that the scientists, engineers, and astronauts working on America's space program on the southern end of Cape Canaveral needed some privacy. To protect the cape from further development while ensuring a buffer, in 1975 Congress preserved 58,000 acres of seashore,

land, and lagoons along with 24 miles of protected coastline to create the longest undeveloped beach on Florida's Atlantic coast. Arrive here for sunrise and infinity lies to the east. The barrier island beaches—Apollo, Klondike, and Playalinda—are largely absent of people, so this will reveal a Florida **sunrise in its natural state,** an experience that is a pure pleasure. This is especially true of **Klondike Beach.** Sandwiched between the two others and accessible only by foot, Klondike is designated a backcountry beach, and the park restricts the number of visitors to its

Native American Sky-watchers

With instant access to information about weather and tides, sunrises and sunsets, phases of the moon and seasonal changes, it's easy to take the cycles of nature for granted. This wasn't so a few hundred— or a few thousand—years ago when Native Americans, primarily in the Southwest, observed celestial bodies and tracked the seasons to predict with incredible accuracy the cycles of the sun. In New Mexico, in the remote Chaco Canyon at Chaco Culture National Historical Park, shafts of light called "sun daggers" pierce spiral petroglyphs on Fajada Butte. At Aztec Ruins National Monument, the Aztec West great house aligns with the sun's location on the horizon both at the summer solstice sunrise and the winter solstice sunset.

wave-lapped sands. Bring a beach chair—at daybreak the music of nature begins. Here comes the sun.

7 | Petrified Forest
NATIONAL PARK
Arizona

With Petrified Forest National Park an already magical setting for a great American sunrise, one day in particular may influence anyone's travel schedule. With the sun's rays tracking a slightly different path throughout the year, for 10 days before and after June 21 (with the highlight being on the **summer solstice**), the Earth's alignment with the sun impacts more than a dozen **"solar calendars"** created throughout the park by prehistoric peoples, with the spiral and circular petroglyphs being intersected by or interacting with the sun's rising rays. Ancient tribes took time to place them here. Take time to marvel at the confluence of ancient science and nature.

8 | Gateway Arch
NATIONAL PARK
Missouri

Sun salutations have even more meaning at Gateway Arch National Park, where **sunrise yoga**—with the famed arch in the foreground—has become a local weekly favorite from May through October. Sponsored by the Gateway Arch Park Foundation, the class is free, but pack your

▶ *After sunset in Arches National Park, the ground rapidly loses heat to the night sky and ambient air temperatures may drop significantly. Temperature fluctuations of more than 40 degrees in a 24-hour period are not uncommon here.*

own mat, towel, and water bottle and meet at Kiener Plaza next to the Old Courthouse. Deep breaths and gentle stretching will leave participants feeling centered both mentally and physically within the iconic sites of St. Louis. The all-levels sessions are weather pending, so check for updates on social media if in doubt. Namaste.

9 | National Mall
NATIONAL MALL & MEMORIAL PARKS
District of Columbia

With national icons framing each end of the east–west National Mall in Washington, D.C., dawn and dusk softly and gradually illuminate silent sentinels of American history. When the crisp blue and orange sky breaks in the east, the gleaming white **Capitol dome** topped by the Statue of

Freedom is bathed in the refreshing rays of daybreak, and the effect elicits a natural sense of hope. At dusk, try to find a spot near the west front of the Capitol building. The Mall is adorned in the comforting rays of sunset that first descend behind the **Washington Monument** before backlighting the **Lincoln Memorial.** Once again, that sense of optimism returns, knowing that now and for the next several hours it will shine from here to the Pacific as it falls across 3,000 miles of America.

10 | Badlands
NATIONAL PARK
South Dakota

The cliché image of a cowboy riding into a beautiful sunset magnifies the cowboy's independence as well as nature's power. That feeling still exists in Badlands National Park, where a lack of development leads to a refreshing sense of solitude. In the eastern reaches of the park, a series of overlooks have been carved out from the colorful buttes for a perfect vantage point and sanctuary, where the lonely wide-open prairie, protected in adjoining Buffalo Gap National Grassland, leads to a feeling of oneness with nature. From atop a north–south ridge are **commanding views at dawn and dusk,** and after the sun disappears in a swirl of pink and orange clouds, the night sky is soon aglow with a shimmering sheath of stars.

10 BEST PARKS FOR
PICNICKING

A picnic shouldn't be just a meal quickly consumed on the way to something else—it should also be an opportunity to savor time and place. The following picnic spots in the national parks make enjoying even a simple sandwich outside a real occasion, with scenery, atmosphere, and recreation all important parts of the experience.

1 Shenandoah
NATIONAL PARK
Virginia

Shenandoah National Park, which runs in a long, narrow swath along the Blue Ridge from Front Royal, Virginia, to Rockfish Gap and the northern entrance of the Blue Ridge Parkway, has several large, scenic picnic areas that are most stunning in the fall when the leaves take on vibrant autumn hues, creating a canopy of brilliant reds, yellows, and browns. Perfect spots for a large outdoor gathering like a family reunion or cookout with friends, all of these picnic grounds have tables, grills, and seasonal restrooms in the vicinity, and several of them are also within easy reach of hiking trails that lead to lovely mountain scenery and overlooks of the broad Shenandoah Valley. The picnic sites are accessible along Shenandoah's most famous attraction, **Skyline Drive,** which runs along the crest of

the mountains through the entire length of the park. The **Dickey Ridge picnic area,** at milepost 4.6, provides access to three short, easy trails—Dickey Ridge, Fox Hollow, and Snead Farm—that meander past streams, old orchards, a cemetery, and fields. From the **Big Meadows picnic area** (milepost 51.2), the Lewis Falls Trail—a steeper

hike—leads down to an observation point of the 81-foot-tall cascade. A moderately strenuous 2.6-mile round-trip hike from the **South River picnic area** (milepost 62.8) leads to a view of the 83-foot-high South River Falls.

2 White Sands
NATIONAL PARK
New Mexico

Brilliantly white under a strong sun, 275 square miles of sand dunes form an expanse that's starkly beautiful and constantly shifting as wind blows the malleable terrain of White Sands National Park. In the very heart of the dunes, off Dunes Drive, are **picnic areas with individual shelters** scattered through the sands. Considered both retro and futuristic, these shelters were crafted by architect Lyle Bennett to reflect the architectural style of Frank Lloyd Wright and are designed to shade visitors from the harsh New Mexico sun. Those picnicking under one of these shelters are guaranteed a truly memorable experience: Looking out over the dazzling expanse of gypsum dunes while enjoying a meal, alfresco diners feel alone, very small, and set adrift in the vast sea of sand. It's an interesting and unusual place to enjoy a picnic lunch, with only one hitch: Be prepared to do whatever possible to keep sand out of the food. And don't forget sunglasses!

3 | Whiskeytown
NATIONAL RECREATION AREA
California

For a picnicking experience that's a bit farther off the beaten track than California's larger national parks, but that still offers lovely views and chances to spot wildlife, head to Whiskeytown, which has several nice picnic grounds, all beautifully maintained by the Park Service. There are impressive views of the 50-foot cascade of **Crystal Creek Falls** from two of the picnic areas, and at the **Brandy Creek** and **Oak Bottom** areas of the park, other picnic grounds are invitingly shaded by Douglas firs, canyon live oak, and ponderosa pines. These latter two areas are wheelchair accessible, and there's access to the beach, fishing piers, and hiking trails as well. All of the national recreation area picnic sites have tables, grills, and restrooms, as

▶ *It's one thing to raise a glass to the national parks, but before you raise one in a national park, confirm that alcohol is permitted in the place you're planning to picnic.*

well as offer the occasional wildlife sighting. Whiskeytown—at the ecologically rich junction of the Klamath, Cascade, and Coast Mountains—is home to, among other creatures, bald eagles and black-tailed deer. But, of course, picnics are not to be shared with the animals and diners are reminded never to leave food unattended.

4 | Pu'uhonua o Hōnaunau
NATIONAL HISTORICAL PARK
Hawaii

It's hard to improve upon Hawaii's gorgeous coastlines and halcyon weather, but you can add a flourish to the outing by packing a picnic to enjoy under the shade of coconut trees near the shore. The **picnic area** at Pu'uhonua o Hōnaunau National Historical Park on the Big Island has gorgeous views of the ocean, beaches, and palm tree groves, all backed by clear blue skies. In the tide pools near the picnic grounds,

look for colorful native fish swimming amid corals and seaweed. The picnic grounds are at the back of the park, right on the water, and are shaded enough to be comfortable during the day. Or come in the evening for a stunning view of the sunset; the park closes shortly thereafter, however, so plan accordingly. Expect to find tables and barbecue pits, but there's no food to purchase in the park, so you'll need to arrive fully provisioned. For convenience, stock up at a store along Hawaii 11 before heading into the park.

5 | Mount Rainier
NATIONAL PARK
Washington

At 14,410 feet, Mount Rainier is the highest point in Washington State, and the **picnic area at the Sunrise Visitor Center,** at 6,400 feet, is the highest place in the national park (and actually in the state) that is accessible by car. The picnic area is located in a unique and beautiful setting—in addition to lovely views of glacier-draped Mount Rainier across a large scenic valley, the site is surrounded by the intriguing spectacle of gnarled, stunted trees, whose growth has been impeded by frigid winter gales. A picnic here affords not only the enjoyment of eating outside in a stunning setting—grills are available if a barbecue is on the menu—but also easy access to the restrooms and

Sheltered picnic tables appear otherworldly against the sugar white backdrop of White Sands National Park.

exhibitions of the visitor center. The center's displays on the volcano's history and the area's flora and fauna will make an invigorating post-lunch hike on one of the many trails that originate here all the more meaningful.

6 John Day Fossil Beds
NATIONAL MONUMENT
Oregon

John Day Fossil Beds National Monument, which protects one of the most comprehensive fossil records

of ancient life in North America, is a fascinating place for a lesson in natural history and geology, but it's also a great place for an alfresco meal. The **Painted Hills unit** of the park is a particularly striking setting for a basket lunch—the landscape is dominated

Wolf Trap's sloped lawn offers visitors a unique performance venue.

by bluffs with colorful ash layers in shades of yellow, red, gold, and black that chronicle ancient volcanic activity. A landscaped picnic area at Painted Hills has shaded picnic tables, restrooms, and outdoor exhibits on the region's history. The vividly colored striations of the hills are best seen in the afternoon, so plan for a late lunch if possible, or stop by for a mid-afternoon snack. The area is also celebrated for its wildflower blooms in the spring, so picnicking here in late April or early May is all the better for the added profusion of color. After lunching, be sure to take advantage of one of the short scenic trails that begin in the vicinity of the picnic area and lead to beautiful overlooks of the Painted Hills.

7 | Wolf Trap
NATIONAL PARK FOR THE PERFORMING ARTS
Virginia

There aren't many places where you can combine a world-class performance with open-air seating, a picnic, and a glass of red wine, but Wolf Trap National Park gives you the chance to do just that. The Filene Center's elegant **outdoor amphitheater** hosts more than 80 performances during the season, ranging in genre from musical acts to dance and theater productions. Audience members who watch from the sloping lawn are encouraged to pack their own basket of food and bring a bottle of wine or six-pack to share during the show. Picnicking isn't allowed in the fixed, covered seating areas along the stage, so many seat holders come early to picnic on the lawn before the show begins. For those attendees who'd rather not pack a basket, Wolf Trap has an official caterer that prepares fantastic sandwiches, salads, and charcuterie when you preorder before the show date. Try to arrive early as the lawn can soon become a sea of blankets. You'll agree it's worth it: Watching a wonderful performance under the stars is a pretty incredible way to picnic.

8 | Indiana Dunes
NATIONAL PARK
Indiana

Fifteen miles of diverse Lake Michigan shoreline draw visitors to Indiana Dunes National Park, and it's quite common for them to stay all day. Picnic tables are provided throughout much of the national park—many are sheltered and a few are even reservable. Parking is almost always at a premium though, so it's wise to plan to arrive early and work up a hunger exploring the vast park. While the picnic shelters at **Lake View Beach** are first come, first served, this is the best setting overlooking the lake. After you dine, stroll down Lake Front Drive to spot five historic homes. These curious structures (closed to the public except for a tour day each fall) were part of the 1933 Century of Progress World's Fair and are all in differing states of restoration. The area around **Mount Baldy,** a dune rising 126 feet above shoreline, is another highlighted spot on the eastern side of the park. Its Beach Trail is short but quite steep, and since the only way out is back up, up, up, coolers aren't ideal. Rather, dine at the shelter near the parking lot before taking a hike. (The summit is closed except via ranger-led programming.)

9 | Pictured Rocks
NATIONAL LAKESHORE
Michigan

The colorful variegated sandstone cliffs of Pictured Rocks National Lakeshore are a stunning setting for picnicking. Water and wind chiseled sandstone into striking, irregular cliff formations that tower above Lake Superior on Michigan's Upper Peninsula, and nowhere is this phenomenon better witnessed than at Miners Castle. There's a beautifully constructed picnic site at the head of a trail that leads down to an overlook of **Miners Castle,** and it's a lovely setting to bring a basket lunch and fuel up before ambling down the path, taking in sweeping views of Lake Superior and Grand Island on the way. **Miners Beach,** about six miles north of Miners Castle, is another idyllic—and more secluded—spot for a picnic. Here gorgeous beaches stretch along the lakeshore, where picnickers can enjoy a meal to the sound of gentle

waves rolling in off the lake. To speed digestion afterward, take a lazy walk up the shore.

10 | Gateway Arch
NATIONAL PARK
Missouri

You can enjoy a picnic lunch just about anywhere on the grounds of Gateway Arch National Park, home of the iconic urban monument set aside the Mississippi River. When Gateway Arch changed its status to a national park in 2018, an extensive

redesign of the grounds and museum followed. Now, meticulous landscaping features walkways lined with London plane trees, circles of bald cypress, a native grass meadow, ponds, and even Zen gardens. While the park doesn't have picnic tables or designated picnicking sites, visitors are encouraged to settle in and enjoy these **tranquil public spaces.** Sound appealing? At the monument, pick up premade take-out items from Arch Café & Restaurant, or order local favorites like St. Louis ribs or toasted ravioli to-go to enjoy outdoors.

Picnics Are for People

Want to make sure your much anticipated grizzly sighting doesn't occur at close range as you're getting ready to bite into your turkey sandwich? You're not likely to be interrupted by a grizzly bear, but the National Park Service has done a lot of research over the years and discovered that, when it comes to food, grizzlies engage in reward-reinforced behavior. In layman's terms, that means that if a grizzly realizes it can come by an easy bite to eat in a certain place, that incentive will begin to override its innate tendency to avoid people. This applies not only to bears, but to deer, raccoons, skunks, squirrels, and many other animals. That means picnickers need to clean up and dispose of food intentionally when they leave the grounds. The parks usually have specific instructions for food storage and disposal in places where animals are likely to look for leftovers, and it's always illegal to feed wildlife. So follow the rules and be tidy—the reward is the full benefit of an outdoor meal in a beautiful, unspoiled setting.

10 BEST PARKS FOR
NATURE SOUNDS

A park getaway is the chance to trade honking horns and ringing phones for the vast variety of natural sounds. The wind in the pines soothes campers to sleep, birdsong serves as a gentle alarm clock, and a rushing creek creates background music for a streamside picnic. The hills are alive, for certain, and they're letting you know.

1 | Kings Canyon
NATIONAL PARK
California

The Roaring River didn't earn its name for passivity. Having traveled through Deadman and Cloud Canyons, the river explodes through a narrow granite chute and cascades into its pool below, releasing a **roar** that surpasses its appearance. The sound of the waterfall, located in the Cedar Grove section of Kings Canyon National Park, changes throughout the year but is at its loudest in the late spring and early summer. Typically accessible from late April to mid-November via Highway 180, it's just a short, shady, and paved walk from the parking lot to the Roaring River Falls. You'll hear it loud and proud before you arrive. Note: As appealing as the clear pool seems, it's not a safe swimming hole; currents below the surface can easily drag visitors under.

2 | Yellowstone
NATIONAL PARK
Idaho, Montana & Wyoming

No sound is arguably more thrilling for a nature lover than the **howling of wolves.** An audio icon of wilderness, a wolf's call symbolizes nature's endurance against habitat loss and persecution, as well as the resilience of ecosystems that are allowed to function naturally. Though gray wolves are most abundant north of the lower 48 states, for most travelers they're easiest to see and hear in Yellowstone National Park, where they were reintroduced in 1995 after being extirpated in the 1920s. Thousands of people see them annually in Yellowstone, primarily in Lamar Valley in the northeastern part of the park. But exciting as seeing a wolf may be, there's nothing quite like hearing a wolf's howl, and hopefully an answering call, to thrill the senses. Hearing the howl is most likely to occur at dusk or early in the morning, and fall through spring is usually the best time to experience the sounds of this predator. Wolves howl for many reasons, including communication within the pack and as a way to find mates. They may also howl during a hunt. On a still night, the long, shrill tones of a wolf pack can send a chill down the spine, as we feel a mixture of apprehension, respect, admiration, and gratitude for sharing such a wondrous moment in nature.

3 | Mojave
NATIONAL PRESERVE
California

Travelers in desert country have long observed that sand dunes sometimes make a long, low **booming or "singing" noise** when some disturbance (like a person sliding on them) creates a mini-avalanche of sand down a slope.

Roaring River Falls in Kings Canyon National Park is at its loudest in late spring and early summer.

The sounds of traffic and aircraft (such as flightseeing rides over the Grand Canyon) can intrude on natural sounds in parks, lessening the wilderness experience for visitors and impacting communication for animals. One organization addressing this issue, Quiet Parks International, works to raise awareness of the natural soundscape. In doing so, members hope to influence land managers to consider soundscape management and prevent noise pollution just as they do water pollution, invasive species, and other threats to parks. The process is strict, but such park sites as Haleakalā, Glacier, Great Sand Dunes, and the Niobrara National Scenic River have been evaluated by the nonprofit.

The phenomenon has baffled scientists, with the best guess on the source being friction between individual grains of sand. Lately, though, a theory speculates that the sound really originates in a sort of echo between dry sand on the surface and a layer of wet, hard-packed sand beneath. Whatever the reason, it's fun to play on sand dunes and, when conditions are right, cause them to "sing." A good place to try this is the Kelso Dunes area within Southern California's Mojave National Preserve. Located far off the beaten path, about 42 miles southeast of Baker, the dunes rise to 700 feet high and cover more than 45 square miles. Check with a park ranger about conditions before setting out along the backcountry roads for the Kelso Dunes—and once there, try running down a dune to set the sand in motion.

4 | Point Reyes
NATIONAL SEASHORE
California

Despite being nearly wiped out by hunting in the early 20th century, elephant seals have made a good comeback and now number more than 2,400 at Point Reyes on the central California coast—with the population continuing to grow. They can be observed easily from December through March from the Elephant Seal Overlook at Chimney Rock above Drakes Bay, where they come ashore to breed. Another option for viewing is on the beach adjacent to the park visitor center. And while the sight of these marine mammals is impressive (males can weigh more than two tons), their sound is just as astounding: The males' **powerful trumpeting** can travel for more than a mile, revealing the seals' presence long before they come into view. In addition to the raucous trumpeting, you'll experience a veritable symphony—grunts, snorts, belches, whimpers, squeaks, and squeals. Seeing the teeming mass of animals and their interactions—mothers raising pups, males challenging each other for dominance—and hearing the commotion evokes the wonder of ocean life.

5 | Rocky Mountain
NATIONAL PARK
Colorado

In many parts of the country, one of the most telling signs of fall's approach is the **"bugling" of elk** during the rut, or mating season, as the males attempt to gather or increase their harem of females.

Though the call is commonly known as bugling, it's really more of a screech or whistle—one that starts off deep and resonant and then fades to a high-pitched squeal, ending with a series of grunts. However it is perceived, the bugling is as much a part of autumn as the quaking aspen leaves turning gold. Older males have stronger calls than do young animals, which helps their mating success. Elk are common in several national parks, but there's perhaps no better place to enjoy their bugling than Rocky Mountain National Park, with a backdrop of the rugged peaks of the Continental Divide.

6 | Everglades
NATIONAL PARK
Florida

Alligators are always impressive—even more so when a big male raises its head and tail above the water, inflates its throat, and gives forth with the deep, rumbling growl that's often called **bellowing.** Head to Everglades National Park in the early spring mating season, find a spot where gators have congregated, and listen in as they try to attract mates while scaring away rivals. It's probably the closest humans will come to experiencing a dinosaur in action, or at least it's easy to think so. A gator's size influences the tone and intensity of its bellow. At times several individuals roar in a chorus,

▶ *From birdcall to thunder crack to Indigenous song, PARKTRACKS (nps.gov/subjects/sound) is 12 transportive minutes of tranquility. Tune in, turn it up, and find yourself immersed in the sounds of our national parks.*

seeming to encourage each other in a wild concert that can last several minutes and travel great distances across the water.

7 | Glacier Bay
NATIONAL PARK & PRESERVE
Alaska

This Alaska Panhandle park is known for having one of the finest guided boat tours within the National Park System, an all-day, 130-mile summertime cruise around Glacier Bay to see wildlife and, of course, glaciers. Often passengers get to witness the sight and sound spectacle of a glacier "calving" an iceberg, losing a portion of the glacier to the water. When stress where the river of ice meets the water causes a massive chunk of ice to break off, it does so with a **loud crack** like a rifle shot amplified many times, crashing into the water

with a huge splash that sends spray and waves in all directions. As the tour boat approaches the park's tidewater glaciers (those that reach water rather than ending on land), icebergs drift through the bay, and the boat pauses for up to a half hour to increase the chances that passengers witness the calving. For those who experience the birth of a new iceberg, it's a sight—and sound—long remembered.

8 | Voyageurs
NATIONAL PARK
Minnesota

One of the essential delights of the North Woods, the call of the male common loon is probably best heard at dawn, echoing over a tree-fringed lake when the air is still enough that the water reflects like a mirror. (The smell of bacon frying in a skillet helps, too.) Sometimes called **"laughter,"** the sound is something like a falsetto yodel—and the inspiration for the expression "crazy as a loon." Breeding loons are common in Voyageurs National Park, where they build their nests along the shores of lakes both large and small. These black-and-white birds superficially resemble ducks, but they are not related to those waterfowl. Loons are designed for efficient diving, with legs set so far back on their bodies that they are unable to walk on land and must push themselves

A female wolf raises its head and howls to its pack or mate, making the quintessential sound of the wild.

make a **pleasant rattling noise** when the waves go in and out. All in all, the sounds here enhance a visit to one of the most picturesque coastlines in North America.

10 | Mammoth Cave
NATIONAL PARK
Kentucky

You may think you've experienced silence—in a remote forest, camping in the desert, in an empty basement—but have you, really? Wasn't there a slight breeze, a distant bird chirping, the subtle hum of a manmade device? To experience **true, absolute silence,** visit Mammoth Cave National Park, renowned for its quiet even in the context of caves. Mammoth differs from many caves in that portions of it are extremely dry. A layer of sandstone and shale above the limestone of the cavern keeps water from seeping down, which means it has none of the slow dripping sounds of water splashing into underground pools or striking stalagmites. When it's quiet in Mammoth Cave, it's truly quiet—so much so that it can be startling for people experiencing it for the first time. Famed naturalist John Burroughs wrote of Mammoth: "When no word is spoken, the silence is of a kind never experienced on the surface of the earth, it is so profound and abysmal ... [T]he sense of hearing is inverted, and reports only the murmurs from within."

along on their chests. Listen for them when they return from their wintering grounds in May, and throughout the summer nesting season.

9 | Acadia
NATIONAL PARK
Maine

The rolling of ocean waves is one of nature's most viscerally stirring sounds. The strikingly rugged coast of Maine ranks among the best places to enjoy **surf sounds,** with the added bonus of a special location in Acadia National Park that provides a variation on the theme. At Thunder Hole, located along the park's scenic Loop Road just beyond Sand Beach, a crack in the pinkish granite rock has been widened over millennia to a long rectangular opening; heavy waves compress air within this chamber, causing a low, resonating boom like distant thunder. The thunderous *whoomp* doesn't occur on calm days, and even at times of high waves you sometimes have to wait a few minutes for the right conditions. For a more soothing sound, head to the lovely pebbled beach nearby, where multicolored cobblestones

SCENIC DRIVES

There is no substitute for getting out and seeing a national park up close, whether on foot, on horseback, on a bicycle, or in a canoe. But it's also entertaining, and certainly less exhausting, to be presented with one superlative view after another in a relatively short time, as is the case with our most spectacular scenic drives.

1 Glacier
NATIONAL PARK
Montana

"One of the world's most spectacular highways" is Glacier National Park's own description of **Going-to-the-Sun Road,** and no one who has traveled this 50-mile mountain route between the park's east and west entrances would argue. Stretching over the Continental Divide in northwestern Montana, the road's glacier-carved Rocky Mountain locale provides countless eye-filling vistas, scenic overlooks, waterfalls, and access to short trails. Dedicated in 1933, the drive was named for nearby Going-to-the-Sun Mountain. How the mountain got its name is uncertain—some say from an ancient Native American legend, while others say the "legend" was created by an explorer in the 1880s. The route, open seasonally, reaches its literal high point at 6,646-foot Logan Pass, where, in addition to knife-edged ridges, sheer cliffs, and sharp-pointed peaks, there may be mountain goats and bighorn sheep within sight. East of Logan Pass there's a view to the south of Jackson Glacier, one of the few glaciers in the park visible from a road. Lesser snowfall on the eastern side means Going-to-the-Sun Road stays open longer and higher in fall here than it does on the western slope. Plan ahead: Based on the season and hours you plan to embark, reservations for the road are most likely required.

2 Shenandoah
NATIONAL PARK
North Carolina & Virginia

The **Blue Ridge Parkway** in western North Carolina and **Skyline Drive** in western Virginia are administered by two different National Park Service units. It makes sense to treat them as one route, though: The northern terminus of the 469-mile Blue Ridge Parkway links with the southern end of 105-mile Skyline Drive (part of Shenandoah National Park) at Rockfish Gap, Virginia, creating a 574-mile drive that takes in some of the most beautiful scenery of the central Appalachian Mountains. In addition to dozens of overlooks providing views of forested ridges and valleys, attractions along the combined drives (many open only seasonally) include visitor centers, lodges, campgrounds, restaurants, the Folk Art Center, the Museum of North Carolina Minerals, the historic Skyland resort, and hiking trails leading to mountain summits, waterfalls, and other natural features. Both drives are well marked with mileposts, making it easy for travelers to keep up with their location and find special sites. With limited entrances and exits, no billboards, and speed limits of 45 miles an hour on the Blue Ridge

As Yosemite's Glacier Point Road winds to a high point above Yosemite Valley, it offers glimpses of iconic Half Dome.

Parkway and 35 on Skyline Drive, this combined route offers an experience as relaxing as it is beautiful.

3 Rocky Mountain
NATIONAL PARK
Colorado

Just before **Trail Ridge Road** was completed in 1932, the director of the Park Service described its appeal: "You will have the whole sweep of the Rockies before you in all directions." Winding for 48 miles to cross the Continental Divide between Estes Park and Grand Lake, Colorado, this scenic route 60 miles northwest of Denver rises to a high point of 12,183 feet, making it the highest continuous road in the United States. It takes drivers from ponderosa pine forest through the zone of spruce and fir to the open tundra, with magnificent views of valleys, lakes, rivers, and mountains, including 14,259-foot Longs Peak, the park's highest summit. Overlooks such as Many Parks Curve, Rainbow Curve, and Rock Cut provide panoramas that truly give you the impression of being on top of the world. People with difficulty breathing should beware the thin air. At Fall River Pass, a visitor center has exhibits interpreting the harsh environment above tree line, where only the hardiest plants and animals can survive. Heavy snow may close Trail Ridge from mid-October until late May. Rocky Mountain is one

▶ *Park visitor centers hold a wealth of information, including free scenic road guidebooks highlighting suggested stops, interesting history, and traveler tips. Check park websites, too, as some have engaging audio tours you can download and stream as you drive.*

of the country's busiest national parks, and timed entry permits are currently required.

4 Yosemite
NATIONAL PARK
California

Three separate roads in Yosemite are close enough together that all can be traversed in five hours or so, if one was so inclined. **Yosemite Valley,** 195 miles southeast of San Francisco, boasts far more than its share of natural icons visible from the road: Highlights include 620-foot Bridalveil Fall, the massive granite monolith of El Capitan, 1,000-foot Horsetail Falls, Yosemite Falls (three falls totaling 2,425 feet), and Half Dome. The famed overlook Tunnel View, on the road out of Yosemite

Valley toward Wawona, offers what is perhaps the park's single most famous vista. To the south, **Glacier Point Road** climbs up to Glacier Point, 3,200 feet above Yosemite Valley, another awesome panorama of the valley, peaks, and waterfalls. Bisecting the park to the north of the valley, **Tioga Road** runs west–east to cross the Sierra Nevada via the 9,945-foot Tioga Pass. Originally a wagon road built in 1883, Tioga Road ascends into an alpine world of snowy peaks, crystal lakes, and meadows bright with wildflowers.

5 Crater Lake
NATIONAL PARK
Oregon

Rangers at southwestern Oregon's Crater Lake are occasionally asked what they put in the water to make it so blue. The truth is that the striking deep azure color is completely natural, a result of the purity of the water and its stupendous depth. At 1,943 feet deep, Crater Lake is the deepest lake in the United States and one of the 10 deepest in the world. The lake is actually a caldera, a crater formed after a massive volcanic explosion 7,700 years ago. The park's **Rim Drive** (closed November through May, depending on snowfall) winds for 33 miles around Crater Lake, with more than 30 scenic overlooks offering an ever changing perspective. Must-see stops include Watchman Overlook, with a great view of Wizard Island, a cinder cone rising in the lake; Phantom Ship Overlook, where a small island resembling a ship is visible; the spur road to the Pinnacles, a collection of 100-foot-tall spires; and Vidae Falls, where a creek drops 100 feet over a series of ledges.

6 | Organ Pipe Cactus
NATIONAL MONUMENT
Arizona

Located 120 miles southwest of Tucson, Organ Pipe Cactus rewards travelers with a beautifully rugged volcanic landscape and an environment so biodiverse that the area was designated an international biosphere reserve in 1976. The 21-mile **Ajo Mountain Drive** loops east of Ariz. 85, circling the Diablo Mountains to skirt the Ajo Range, providing a good overview of the park. The gravel and asphalt road is suitable for all vehicles and takes about one to two hours to drive. Much of the breathtaking scenery is composed of rhyolite, rock formed by lava cooling on the earth's surface. In many places, cliffs display bands of dark rhyolite paralleling lighter bands of tuff, volcanic ash compressed and "welded" into rock. Among the plants visible in the desert are organ pipe cactus, elephant tree, and the saguaro cactus, symbol of the Sonoran Desert. Nine miles into the loop, a natural rock arch 90 feet wide appears on a cliff above the

A bison crosses the road in Theodore Roosevelt National Park.

road. In higher elevations the desert gives way to a landscape of oaks, junipers, and jojoba, a shrub favored by bighorn sheep, which might be spotted here.

7 | Theodore Roosevelt
NATIONAL PARK
North Dakota

It's worth the time to stop and set the scene before embarking on the **South Unit's scenic drive.** At the South Unit Visitor Center in Medora, a brief film and a self-guided tour of the 26th president's Maltese Cross Cabin give insight into the rigorous North Dakota landscape that became the backdrop of Roosevelt's formative

20s. "I never would have been President if it had not been for my experience in North Dakota," he once wrote. With this insight as your guide, the views of beautiful badlands, prairie dog towns, and abundant wildlife become all the more evocative. Although traffic is nearly always light in this park, plan at least two hours to take in the drive, which is speckled with overlooks, interpretive signage, and trailheads leading to hiking and biking paths. Erosion is quite common in this region, so check with rangers at the visitor center regarding vehicle closures.

8 | Capitol Reef
NATIONAL PARK
Utah

Located in southern Utah, Capitol Reef encompasses striking rock formations typical of the Colorado Plateau, intriguing historic sites, and a wonderland of canyons, buttes, washes, and back roads to explore. The **Cathedral Valley Loop Drive** covers 58 miles of rough, unpaved road that can be impassable at times. Check with the park visitor center on Utah 24 before setting out. The route, which can take upwards of eight hours, begins with a ford of the Fremont River that is passable most of the time, except during high water after storms and for more extended periods during spring runoff. The road continues to pass alongside multicolored mesas, spires, and other rock formations with names such as Walls of Jericho and the Temple of the Moon. Quite visible in many places are black boulders that look different from the rest of the landscape. Remnants of lava flows that capped nearby mountains about 20 million years ago, they were eroded and later moved to lower locations by various processes including glacial melting. After heading northwest, the loop turns back southeast on Caineville Wash Road. There are several panoramic vistas along the route, which returns to Utah 24 at the community of Caineville.

Building the Roads

The stories behind the construction of some of the parks' scenic drives are almost as fascinating as the natural wonders surrounding them. Shenandoah's Skyline Drive was begun in 1931, during the Great Depression. The original construction crews included out-of-work farmers and apple pickers. In 1933, workers of the newly formed Civilian Conservation Corps arrived; eventually, 10 camps of "CCC boys" helped complete Skyline. Glacier's Going-to-the-Sun Road was finished in 1932; it took 11 years to build the 50-mile-long road. The initial three-month survey work for the road was so challenging that the crew had a 300 percent turnover. Still considered a major engineering feat, the road is listed as a national historic landmark.

9 | Colorado
NATIONAL MONUMENT
Colorado

At Colorado National Monument, taking the scenic route is the most popular pastime. The 23-mile **Rim Rock Drive** winds up wild switchbacks to the tip-tops of mesas and along steep canyon walls before descending into the valley, featuring overlooks, trails, and interpretive panels along the way. Consider starting in Fruita, Colorado: The drive entrance is just four miles from the visitor center, and you'll want to pick up geology pamphlets to inform your route. If heights give you pause, Rim Rock's tight drop-off sections may prove difficult to tolerate. But those that take the challenge can take in red-rock canyons, bighorn sheep, soaring eagles, and layers of rock dating back 1.7 billion years. Just remember to share the road: Bicyclists find the scenery equally stunning, and the narrow navigation requires alert drivers.

10 | Acadia
NATIONAL PARK
Maine

Maine is justly famous for its ruggedly beautiful rocky shoreline, where waves continuously crash into reddish granite ledges while the boats of lobstermen cruise by. An especially striking section of this coast is a highlight of Acadia National Park's 27-mile **Park Loop Road,** which winds around the eastern section of Mount Desert Island, south of the town of Bar Harbor. Open from about April 15 through November, the road provides access not only to the coast but also to lush forests, lakes, the park's famed carriage roads—broad paths suitable for hiking or biking—and an extension leading to the top of 1,530-foot Cadillac Mountain, the tallest summit on the East Coast. Coastal sites reached by the scenic drive include Sand Beach, where those who can tolerate 55-degree water swim; Thunder Hole, where heavy surf compresses air in a hole in the rocks, causing a low booming sound; and Otter Cliffs, one of the most picturesque sections of rocky coast.

WILDLIFE

A parliament of burrowing owls meet at their burrow's entrance in Wind Cave National Park.

BIRD-WATCHING

The diverse range of habitats within our national parks, from deserts to seashores to alpine peaks, means a correspondingly high variety of birds. Some species are confined to small areas, while others can be found across large swaths of land. These parks offer a superb selection of environments to tempt traveling birders.

1 Big Bend
NATIONAL PARK
Texas

When it comes to the number of bird species spotted in the National Park System, Big Bend and Point Reyes National Seashore are neck and neck. That said, Big Bend's number of specialty birds tends to give this out-of-the-way site in western Texas the top spot on the bird-watching ranking. The park's list of more than 450 bird species owes its length in part to the fact that Big Bend is often described as being three parks in one: lush riparian vegetation along the Rio Grande, a vast area of Chihuahuan Desert, and the Chisos Mountains, rising to more than 7,800 feet in the center of the park. The star species is the little **Colima warbler,** which nests nowhere else in the United States. It takes some effort (and a little bit of luck) to find such species as **flammulated owl** and **black-capped vireo;** more common and conspicuous are the colorful **acorn woodpecker, Mexican jay, pyrrhuloxia,** and **Scott's oriole.** The secret to finding a lot of bird species is visiting the park's many different habitats, spending time in places such as Hot Springs Village on the Rio Grande, lower Green Gulch in the desert, and Boot Canyon in the mountains.

2 Chiricahua
NATIONAL MONUMENT
Arizona

Southeastern Arizona undoubtedly ranks as one of the country's most popular birding destinations, and within that region the Chiricahua Mountains are among the very best sites. One of several mountain ranges popularly called "sky islands," the Chiricahuas rise up from the surrounding arid lowlands, cloaked in dense forest like an oasis in the desert. Chiricahua National Monument is best known for its spectacularly varied rock formations, eroded from 27-million-year-old volcanic material called rhyolite. It's also home to many southeastern Arizona specialty birds, from the aptly named **Arizona woodpecker** to the beautiful and graceful **painted redstart.** The eight-mile scenic Bonita Canyon Drive leads up to 6,870-foot Massai Point, passing through forests of pine, spruce, sycamore, Douglas fir, Arizona cypress, and oak. Bird-watchers search along the road and on hiking trails for **zone-tailed hawks, Anna's hummingbirds, white-throated swifts, dusky-capped flycatchers, Mexican chickadees, bridled titmice, Mexican jays, black-throated gray warblers,** and **Scott's orioles.** Most birders combine a trip to the national monument with a drive up Pinery Canyon Road to visit adjacent Coronado National Forest, home to even more regional specialties.

Ocotillo shrubs in Big Bend's Chisos Basin provide food for birds like this Scott's oriole.

3 | Dry Tortugas
NATIONAL PARK
Florida

For a few weeks each spring, this tiny island park—a few specks of land in the Gulf of Mexico, 70 miles west of Key West—provides an unparalleled spectacle. Thousands of birds flying northward from their wintering grounds in Latin America to nesting areas in the United States and Canada, tired and hungry after crossing the Gulf of Mexico, drop out of the sky at Garden Key (the main Dry Tortugas island), covering shrubs, buildings, and even the lawn, so unwary that they can be approached closely. There's no predicting what might show up at any given moment. In addition, thousands of **sooty terns** and **brown noddies** (also a type of tern) nest on Bush Key, just 200 or so yards from Garden Key. **Magnificent frigatebirds, brown boobies, masked boobies,** and **white-tailed tropicbirds** are among other seabirds that might be spotted here, and for the lucky, a **black noddy** might be found perched on the old coaling docks. For birders there's nothing like Dry Tortugas in spring.

4 Blue Ridge Parkway
NATIONAL PARKWAY
North Carolina & Virginia

The late Ludlow Griscom, one of the original gurus of American bird-watching, once said, "Be near Asheville, North Carolina, the third week in April and you will see warblers pour across the mountains." Certainly spring migration is a wonderful time to be in the southern Appalachians, but these mountains are also known for a diverse collection of breeding birds more reminiscent of Canada than of the American Southeast. With elevations up to 6,684-foot Mount Mitchell (the highest U.S. point east of the Mississippi River), the southern Appalachians are home to coniferous forests that mimic habitats far to the north. So, too, does the bird fauna seem more like Quebec than North Carolina or Virginia. Birds found on Appalachian ridges include the **ruffed grouse, saw-whet owl, yellow-bellied sapsucker, common raven, red-breasted nuthatch, brown creeper, golden-crowned kinglet, veery, black-throated blue warbler, Blackburnian warbler, Canada warbler, dark-eyed junco,** and **red crossbill.** Many excellent sites for seeing these and other birds can be accessed along the 469-mile Blue Ridge Parkway. Overlooks, picnic sites, and trails offer chances to leave vehicles and enjoy the varied birdlife.

▶ *Although the spotted species list tops 300, only seven bird types nest in Dry Tortugas National Park regularly.*

5 Point Reyes
NATIONAL SEASHORE
California

Around 490 species of birds have been spotted at Point Reyes, which represents about 50 percent of the total for North America. Many of those species are extreme rarities—birds that lost their way and ended up at this Pacific Coast peninsula, an hour north of San Francisco. Combining seabirds with birds of shore, grassland, scrubland, and forest, Point Reyes has year-round interesting birding and is home to the threatened **northern spotted owl** and **snowy plover,** which nests on park beaches. **Peregrine falcons** sometimes cruise along beaches and mudflats, looking for prey. Visitors to Bear Valley, Limantour Beach, Abbotts Lagoon, and the cliffs around the lighthouse will possibly see such species as **sooty shearwater, brown pelican, Brandt's cormorant, pelagic cormorant, osprey, California quail, black oystercatcher, common murre, rhinoceros auklet, tufted puffin, Allen's hummingbird, Nuttall's woodpecker, Pacific-slope**

flycatcher, Hutton's vireo, chestnut-backed chickadee, wren-tit,** and **California towhee.** But the real magic of Point Reyes is simply the fact that anything can show up here, anytime, making every birding visit an adventure.

6 Pinnacles
NATIONAL PARK
California

The severely endangered **California condor**—the largest land bird in North America—has a wingspan close to 10 feet and can weigh about 20 pounds. Pinnacles National Park is a release and management site for this spectacular and rare species. Since condors do not migrate, they may be found here year-round, with an emphasis on "may." The five-mile, quite strenuous hike to High Peaks in early morning is by far your best bet for attempting to spot the small population of these majestic, albeit rather ugly, giants. (Don't be deceived by their more common, yet much smaller, relative the turkey vulture.) While condors are king at Pinnacles, the park's dramatic cliffs and spires are exceptional nesting habitats for **prairie and peregrine falcons, golden eagles,** and **canyon wrens,** while the river regions around Bear Gulch and Chalone Creek have **black phoebes, yellow warblers,** and **house wrens,** and the chaparral shrub ground cover attracts **spotted towhees** and **California thrashers.**

7 | Everglades
NATIONAL PARK
Florida

Geographically speaking, it's no surprise Everglades National Park hosts a distinctive set of birds: It's located at the southern tip of peninsular Florida, just 120 miles from the tropics. Bird-watchers come here in search of the **short-tailed hawk, limpkin, white-crowned pigeon, mangrove cuckoo, gray kingbird,** and **black-whiskered vireo.** Much more obvious, though, are flocks of **herons, egrets, ibises, roseate spoonbills,** and **wood storks.** These species may appear abundant at times, but in fact the population of wading birds in South Florida has dropped 90 percent or more over the past century because of development and disruption of water flow through the River of Grass. The Everglades is the only place in the United States where **greater flamingos** can be found with regularity (check at the end of Snake Bight Trail). Shark Valley is a good spot to look for **snail kites,** a raptor that feeds almost entirely on large snails. Along the Anhinga Trail, at the Royal Palm Visitor Center, birders can get excellent close-up views of numerous species—including **anhingas.**

With a wingspan of more than 50 inches, the great white egret is one of the largest wading birds in the Everglades.

8 | Rocky Mountain
NATIONAL PARK
Colorado

Spanning the Continental Divide in northern Colorado, this park is home to a delightful variety of Rocky Mountain foothill and high-elevation birds. Driving Trail Ridge Road across the park means ascending to more than 12,000 feet, with amazing vistas of mountain peaks in all directions. Simply stopping along the way can bring sightings of birds such as **Steller's jay, gray jay, Clark's nutcracker, pygmy nuthatch, Townsend's solitaire,** and **western tanager.** Above tree line, lucky birders might find **white-tailed ptarmigan** or **brown-capped rosy-finch.** It usually takes a bit of exploring to find species such as **dusky grouse, northern pygmy-owl, three-toed woodpecker, Williamson's sapsucker, Cassin's finch,** and **pine grosbeak.** Look along streams for **American dippers** and at blooming flowers for **broad-tailed hummingbirds.** The beautiful **mountain bluebird** frequents open areas at lower elevations. There may be no park in North America that combines such rewarding birding with such spectacularly accessible scenery.

Bird Blast

Bird is the word when it comes to festivals and programming in our national park units. Here are a few top avian events.

Acadia National Park (May–June): For more than 25 years the Acadia Birding Festival has celebrated Mount Desert Island's bird community via keynote lectures, walks, guided kayak tours, and catamaran cruises.

Indiana Dunes National Park (May): The Indiana Dunes Birding Festival is an annual showcase of the region's biodiversity and abundance of migratory birds. A highlight is the Dunes Big Day, a guided birding blitz with the goal to log more than 100 species in one day.

New River Gorge National Park and Preserve (May): Savor spring in the Appalachian Mountains at the New River Birding & Nature Festival. Open to all skill levels, the week-long event features the area's crucial stopover habitat through guided excursions and world-class speakers.

9 | Cape Hatteras
NATIONAL SEASHORE
North Carolina

America's first national seashore (established 1953), Cape Hatteras protects 70 miles of barrier islands on the North Carolina coast. Bird-watchers know it as a place of **wintering waterfowls** and **migrant raptors** and **shorebirds,** often in significant numbers and variety. Contained within the national seashore is Pea Island National Wildlife Refuge, where marshes and impoundments can host swans, geese, ducks, herons, ibises, rails, and shorebirds. Groves of trees, such as Buxton Woods, provide shelter for migrant songbirds in fall, when **hawks** and **falcons** appear, heading south along beaches. October is a good time to see the magnificent **peregrine falcon.** Nesting birds include the threatened **piping plover, American oystercatcher, least tern,** and **black skimmer.** The famed Cape Hatteras lighthouse is a good place, too, from which to scan the Atlantic Ocean for **northern gannets, shearwaters, jaegers,** and **gulls.**

10 | Channel Islands
NATIONAL PARK
California

Called the Galápagos of North America, Channel Islands National Park consists of five largely untouched islands off the coast

The largest breeding colony of western gulls in the world is found on Anacapa Island in the Channel Islands.

of California cradled within the six-mile buffer of Channel Islands National Marine Sanctuary. As expected, this pocket of protection draws incomparable diversity—in the sea, on land, and in the sky. Bird-watchers flock to Santa Cruz Island, California's largest, for a chance to eye the **island scrub-jay,** the only island endemic bird species in North America. Dark blue, known for its intelligence, and larger than its mainland relatives by almost a third, this monogamous jay lives only on Santa Cruz Island, giving it the continent's smallest range. Scorpion Ranch and Prisoners Harbor are the most common locations of sightings. Many other birds take refuge in the Channel Islands and rely on their marine resources, including nine raptor species, 30 species of shorebirds, and a dozen species of seabirds. Once highly endangered, **peregrine falcons** and **bald eagles** have been successfully reintroduced. Anacapa has large colonies of **California brown pelicans, western gulls,** and **Scripps's murrelets.** The Channel Islands are also the breeding ground for half of the world's population of **ashy storm-petrels.** Learn about the park's continuous—and critical—conservation efforts on ranger-guided walks offered throughout the year.

10 BEST PARKS FOR
QUIRKY CRITTERS

When it comes to wildlife, the big guys get most of the attention. Visitors generally want to see bears, elk, moose, alligators, and other types of photo-friendly large animals. Yet such an attitude ignores a whole range of wildlife that, though smaller in size, can be just as fascinating in appearance, lifestyle, oddity, and charm.

1 Carlsbad Caverns
NATIONAL PARK
New Mexico

From spring through October, there's more inside this famous southern New Mexico cave than spectacular formations: Hundreds of thousands of **Brazilian free-tailed bats** roost in cave passages by day, leaving each evening shortly after sunset, with their exit flight continuing for up to three hours. For reasons that are not fully understood, the bats always spiral out of the cave in a counterclockwise direction. Visitors gather at a natural amphitheater at the cave's entrance to watch this spectacle, with park rangers providing commentary from Memorial Day weekend through October. The best bat flights occur in August and September, when newborn bats join their parents to swell the numbers leaving the cave, spending the night searching the sky for miles around to feed on insects. So large are the bat flights that visitors below can hear the sound of the mammals' wings (around 11 inches from tip to tip) and even smell them (not all species of bats have an odor). The bats' return to the cave at dawn is also an exciting sight, with the bats performing zooming dives from hundreds of feet in the air. Like many species, the Carlsbad Caverns bats migrate south for the winter, spending the season in caves in Mexico.

2 Great Smoky Mountains
NATIONAL PARK
North Carolina & Tennessee

The cool yellow flashing of fireflies is one of the joys of a summer evening, and this park on the Tennessee–North Carolina border offers a special twist on the theme: One species of "lightning bug" here is famed for flashing synchronously, with hundreds of individuals blinking together to create an amazing effect on wooded hillsides. Though there are more than a dozen types of fireflies in Great Smoky Mountains National Park, the **synchronous firefly** (*Photinus carolinus*) is the only species in North America that can synchronize its flashing patterns. The peak period for this display occurs during a two-week period, usually in mid-June. Viewing the synchronized flashing has become so popular that the park has instituted a lottery for vehicle passes, trying to assure that traffic, headlights, and noise won't detract from the visual experience. And the critters don't stop with fireflies: These mountains have also been called the "salamander capital of the world," with 30 varieties found in the park. Not easy to see, **salamanders** are amphibians showing a spectrum of beautiful glossy colors.

Brazilian free-tailed bats swarm out of Carlsbad Caverns for nightly feedings spring through fall.

Many of those here are lungless salamanders that breathe through the walls of blood vessels in their skin and in the linings of their mouth and throat. A patient visitor can observe salamanders by going out at night with a flashlight after rain or by carefully watching for ground movement in moist areas.

3 Rocky Mountain
NATIONAL PARK
Colorado

Though some varieties can be a nuisance in suburban yards, the active little animals of the **squirrel family** have undeniable appeal. Their sleek forms and frenetic, inquisitive behavior often make them seem as though they have some urgent business to attend to. They're fun to watch—and easily so, too, which can't be said for some mammals such as elusive weasels and shy foxes. Rocky Mountain National Park in north-central Colorado is one of the best places in the country to enjoy a variety of small mammals, from the diminutive least chipmunk to the noisy chickaree (or red squirrel) to the chubby yellow-bellied marmot, which often greets hikers on mountain summits, sunning itself and occasionally giving a loud whistle. Most appealing of all is the tassel-eared or Abert's squirrel, which with its long tufts of ear fur seems like a cross between a rabbit and a squirrel. Many types of squirrels will

Horseshoe crabs date back at least 100 million years before the dinosaurs. Spot one while enjoying a visit to the Atlantic? Snap a photo and log its location and condition at fws.gov/crabtag. These citizen scientist findings help researchers study the ancient species.

approach visitors to beg for food, but of course feeding animals in a national park is prohibited.

4 Saguaro
NATIONAL PARK
Arizona

Not nearly as spoken of as bird-watchers or leaf peepers, reptile enthusiasts have learned to love the amazing variety and beauty of turtles, lizards, and—yes, it's true—snakes. Hot places host many more reptiles than do temperate areas, so it stands to reason that Saguaro National Park in southern Arizona would be one of the best places to observe them. With around 50 species in its two geographic divisions east and west of Tucson, Saguaro provides a home for scaly critters

such as the **threatened desert tortoise; desert iguana; regal horned lizard** (actually a variety of horned toad); **Gila monster,** one of only two poisonous lizards in the world; **Sonoran mountain kingsnake;** and six species of **rattlesnakes,** including sidewinder and the very dangerous Mojave rattlesnake. The best time to see reptiles is at dusk or dawn during the summer monsoon-type rains of July and August.

5 Channel Islands
NATIONAL PARK
California

The word "endemic" indicates a species found in only one place in the world. That place might be a country, a mountain range, or even a small island. The Channel Islands, off the coast of Southern California, have been called the Galápagos of North America in part because they host more than 150 endemic species. Some are very rare and hard to see, while others might be spotted by travelers who make the boat or seaplane trip to this relatively little-visited national park. One such animal is the **island fox,** a smaller relative of the gray fox found across North and South America. After a population decline of more than 90 percent in the 1990s, the island fox has seen a good recovery on the six islands where it lives, and lucky park visitors may spot the cat-size mammal hunting for insects, mice, or crabs; it's more active in the daytime than most foxes. The **island night lizard** is found only on Santa Barbara Island in the park and two small islands owned by the U.S. Navy. These reptiles can live more than 20 years in a range of less than 200 square feet. Any park visitor seeing one of these animals can feel privileged, for each is unique to this single spot.

Wind Cave
NATIONAL PARK
South Dakota

Above one of the world's longest cave systems is a rolling prairie landscape with its own system of tunnels. Excavated by a bustling population of prairie dogs, the burrows in Wind Cave National Park are also used by other species, including the **burrowing owl.** Rather than the expected nesting zones in the tip-top of trees, these bright-eyed owls instead find shelter and raise their chicks in the shallow underground tunnels, earning their name. The burrowing owl, unlike other owls, is active both day and night, but traditionally prefers dawn and dusk for hunting. When threatened, the small bird will make rattling or hissing noises within the burrow, and can even be mistaken (in sound only, of course) for a rattlesnake. Wind Cave has 30 miles of hiking trails throughout its 33,000 acres, and prairie dog towns—and thus, burrowing owls—can be spotted along park roads like Highway 385.

The environment at White Sands offers natural camouflage to an earless lizard.

White Sands
NATIONAL PARK
New Mexico

In most places the world of nocturnal creatures remains virtually unknown to wildlife-watchers. The comings and goings of night-active mammals, reptiles, and invertebrates might as well be happening on the moon, so seldom are they noted. At White Sands National Park in southern New Mexico, however, where the fine material of the gypsum sand dunes records the surprising variety of nocturnal residents, their tracks are like an open book that can be read by interested visitors. If there has been little wind to disturb the dunes overnight, an early

morning observer might see the X marks of **roadrunner** feet, the large prints of a **kit fox,** the neat claw marks of **lizards,** the treadlike evidence of a **centipede,** the paired dots of **pocket mice,** the parallel depressions of a **darkling beetle,** or dozens of other types of prints. The Dune Life Nature Trail, a one-mile loop, is a good place to test your footprint identifications skills.

Fire Island
NATIONAL SEASHORE
New York

One of the oddest creatures found on the shores of North America, the **horseshoe crab** looks like the offspring of a stingray and a

Butterflies aren't the only creature you can spot at Big Bend. Its gray foxes are one of only two members of the canine family that climb trees.

motorcycle helmet, its fearsomely spiky appearance causing unease among some people who encounter one on a beach. Nonetheless, this animal—not really a crab but more closely related to spiders and scorpions—is harmless. Its vast numbers of eggs provide essential food for fish and migratory birds, and horseshoe crabs have been used in important medical research (a protein in their blood has been critical in pharmaceutical and medical device testing). Fire Island National Seashore on Long Island is a good place to see and learn about horseshoe crabs when they come ashore to mate and lay eggs during evening high tides in May and June. Check with the park for times to attend a ranger walk to observe this bizarrely intriguing animal.

9 | Big Bend
NATIONAL PARK
Texas

In recent years ever growing numbers of people have taken up butterfly-watching, using close-focus binoculars to enjoy the brilliant colors and intricate patterns of these "flying jewels." Big Bend, in western Texas, has long been known for its birds, but its **butterflies** are equally diverse. In fact, the park list of more than 170 species is higher than the totals of most states. Such striking species

as variegated fritillary, southern dogface, Texan crescent, California sister, and queen are common, and a lucky observer might spot such uncommon and beautiful butterflies as orange-barred sulphur, Sandia hairstreak, Theona checkerspot, and Chisos metalmark.

10 | Theodore Roosevelt
NATIONAL PARK
North Dakota

Theodore Roosevelt described **prairie dogs** as the "most noisy and inquisitive animals imaginable," and Lewis and Clark wrote that they found the animals "in infinite numbers" on their 1804–06 western expedition. Prairie dogs are still noisy and inquisitive, but

their numbers have been greatly decreased by persecution over the past century or so. One good place to observe these animals—which were called "dogs" by early explorers for their barking call, but are in fact squirrels—is Theodore Roosevelt National Park in western North Dakota. Their "towns" of interconnected burrows can be seen along the South Unit's Scenic Drive, as well as other places in the park. Highly social, prairie dogs live in family units called coteries, greeting each other with what looks like a quick kiss and hopping up to give a comic *jump-yip* call that tells others that no predators are within sight. A visitor who sits watching quietly and inconspicuously can be entertained for quite a while by the endless antics of these engaging little mammals.

Banana Slug

In the running for quirkiest critter in our national parks, the banana slug is a mollusk (like a snail) with no shell: an elongated, bright yellow, slimy blob whose oddness has made it an animal celebrity in the Pacific Northwest. It's the official school mascot of the University of California at Santa Cruz and the designated California state mollusk, and its form decorates countless T-shirts taken home as souvenirs by visitors to the region. Especially in the rainy season, banana slugs are easy to find (and admire?) at several parks, from Muir Woods National Monument in California north through Olympic National Park and onward toward the Alaska Panhandle.

Run With the Big Dogs

At some national parks, dogs don't just play, they work. Denali's sled teams do what machines oftentimes cannot, and in doing so, carry on a tradition thousands of years in the making.

As the only working sled dog kennels in the National Park System, Denali's teams have been tasked for more than 100 years with patrolling for poachers, hauling gear to and from remote locations, and supporting many scientific studies. Their role is time-honored; their work, necessary; their faces, fuzzy wuzzy.

When Denali National Park and Preserve established its first kennels in 1922, it continued a significant aspect of Alaska's cultural and natural history. Mushing—using freight-hauling dogs for travel—was once a common part of every-day life in Alaska as families went by sled to trap and trade. While modern technology has largely diminished their use, at Denali, canine rangers remain a critical part of the park.

Jason Reppert, Denali's assistant kennels manager, explains that when the 1980 Wilderness Act designated two million acres of Alaska wilderness to Denali, park staff was mandated to use the minimum tools necessary to do their work. "In the age of snowmobiles and helicopters and planes, these dogs truly fit the definition of minimum tools, and it's one of the best natural history lessons for the public to see that the dogs are doing real work here in Denali."

MEET AND GREETS

The work of the kennels happens in two distinct seasons. In summer's high tourist time the kennels are open to the public daily, with three engaging programs highlighting the dogs and their role in the park. Every bit as friendly as they are fluffy, the huskies are a popular visitor draw.

"On our biggest days we may have a thousand people visit with the dogs, and we always tell people that it's their off-season, since they run really hard in the wintertime," Reppert explains. While the Alaskan husky is traditionally a social breed, Denali's dogs have to be especially accustomed to all kinds of visitors: "Our dogs are very easygoing and mellow, and able to form connections with everyone who comes to see them." Visitors ask questions (What do they eat? How much do they sleep? How far and how fast can they run?), interact with the dog team, and get to see firsthand just how well the huskies are treated. "It's a dynamic day full of demonstrations as the dogs serve as amazing ambassadors for the park."

TRAIN FOR SUCCESS

When temperatures begin to fall, the dogs' days become much more regimented as the kennels move into full training season. Human and canine rangers work together, exercising skills and strength, and building camaraderie along the way. "The simple consistency of the day helps these dogs perform at their best, so we feed them at the same times, run the team anywhere from 10 to 30 miles each day—depending on our projects—and then sometimes we'll take on longer, multiday trips that involve staying in our remote cabins," he explains. "But essentially the dog's day is eat, run, eat, and sleep." While each season's goals vary, teams are typically run four or five days in a row before everyone is given a few days off.

KEEPING HISTORY ALIVE

While the work is significant, Reppert stresses that there's plenty of warmth, too. "The relationship that we share with these dogs is really, really important to us. Yes, they're working dogs, but they are also essentially our pets." To

Denali's happy canines race through the powdery snow as part of a long tradition of sled dogs at the park.

that end, routine health checks happen three times a week and kennel facilities receive regular attention.

"We're not only continuing the tradition that started over a century ago in Denali, we're continuing to keep sled dogs moving and working in the state of Alaska," Reppert says. "Mushing as a sport, as a hobby, and as a lifestyle is really dwindling. So on top of all the park missions that we're working to fulfill, we're also always trying to contribute to the greater dog-mushing world by being an exemplary example of how to care for these animals."

SO WHAT ABOUT YOUR FUZZY FRIENDS?

We may not all be ready to mush across the Alaska tundra, but what if you could hike with Hank, sail with Scout, bird-watch with Bailey, and camp with Cooper? Most national parks allow pets in developed areas, and some units, such as Acadia and Petrified Forest National Parks, are known for being especially dog friendly. Call ahead or check online to confirm pet guidelines.

And while you're there, mind your manners. The Park Service applauds the hard work of good doggos (and their owners) who abide by the B.A.R.K. Rangers pet policies:

Bag pet waste and dispose of it properly.
Always use a leash with a six-foot maximum lead.
Respect wildlife by giving other animals plenty of space.
Know where you can go—and, moreover, where you can't.

10 BEST PARKS FOR
CHARISMATIC MEGAFAUNA

Park managers call them "charismatic megafauna": large animals that all visitors want to see, from bears to bison. Rangers are quick to point out that national parks aren't zoos—animals go where they want to go and sightings aren't guaranteed—but these park units offer some of the best opportunities to observe North America's most remarkable wildlife.

1 | Yellowstone
NATIONAL PARK
Idaho, Montana & Wyoming

Sometimes referred to as the American Serengeti—a comparison with eastern Africa's famed wildlife-rich plains—Yellowstone hosts the largest concentration of mammals in the lower 48 states. A very lucky visitor might see **grizzly** and **black bears, bison, elk, moose, mule deer, bighorn sheep, pronghorns, gray wolves,** and **coyotes,** to list only the largest of Yellowstone's 67 mammal species. The best places to look for the park's wildlife are Hayden Valley, along the Yellowstone River between Canyon Village and Fishing Bridge, and Lamar Valley, in the northeastern part of the park. Hayden Valley is also a good place to see **big birds,** including sandhill

cranes, white pelicans, trumpeter swans, and bald eagles. Lamar Valley, on the other hand, is probably the best bet for gray wolves. Two other top wildlife-watching areas are the park's northern entrance in

Gardiner, Montana, where you might see bighorn sheep and pronghorns, and along the Blacktail Plateau Drive.

2 | Katmai
NATIONAL PARK & PRESERVE
Alaska

This huge park is on our list essentially for one animal—but what an animal it is, and what a spectacle it provides. At Katmai, you get up-close looks at **brown bears** as they feed on salmon, sometimes with dozens of bears visible simultaneously. The park has built special bear-viewing platforms at falls on the Brooks River near its mouth at Naknek Lake (the largest lake within any unit in the National Park System), where bear numbers peak in July and September. No

A brown bear patrols the shores of Nonvianuk Lake in Katmai National Park and Preserve.

Caribou graze on a hillside at Denali National Park and Preserve.

roads lead from the outside world into Katmai. Most park visitors fly to the community of King Salmon, then travel by floatplane or boat to Brooks Camp. Arranging a visit isn't hard, though, since more than a hundred authorized outfitters operate trips in the Katmai area. Backpackers and canoeists who venture into the park's backcountry may find **moose, caribou, gray wolves, wolverines,** and **lynx** in addition to brown bears.

Wind Cave
NATIONAL PARK
South Dakota

Located in southwestern South Dakota, this national park in 1903 became the first in the world created specifically to protect a cave. But it's what's above ground that attracts wildlife-watchers to this underappreciated park: easily visible herds of **bison** and **prong-horns,** as well as **elk, mule deer, white-tailed deer,** and **coyote.** Fourteen bison were reintroduced to Wind Cave in 1913, and six more in 1916. All were believed to be descended from true wild bison, with no cattle genes intermingled from crossbreeding. Disease free, and with good genetic makeup, Wind Cave's bison population is an important factor in the continued survival of this iconic American mammal. Pronghorn (often called antelope, although this North American mammal is not related to true Old World antelopes) are

regularly seen at Wind Cave as well. Most visitors also take time to visit adjacent Custer State Park while they're in the area—the nearly 1,500-strong bison herd here is one of the largest in the world.

Rocky Mountain
NATIONAL PARK
Colorado

The scenic showplace park of the central Rockies is a great place to enjoy wildlife characteristic of the region. In fall, **elk** come down from the high mountains to open, grassy areas for breeding season, and the "bugling" calls of males echo around valleys. Herds can be seen often in places such as Moraine Park, Kawuneeche Valley, and Horseshoe Park. **Bighorn sheep** have made a comeback in the park and now number around 400. They're most easily seen in early summer at Horseshoe Park, where they come to obtain minerals. In late summer, a hike to The Crater site may bring sightings of big-horns. **Moose** are seen most often on the west side of Rocky Mountain National Park, in wetlands of the Colorado River Valley where they browse on willows. **Mule deer** and **coyotes** are common in the park. Black bears and mountain lions are seldom seen, but hikers need to be aware of their presence and know what to do if they encounter one along a trail.

5 | Denali
NATIONAL PARK & PRESERVE
Alaska

This national park protects the tallest mountain in North America, which at 20,320 feet earns its Athabaskan name of Denali, or "high one." Only one road leads into the park's vast heart, a 91-mile, mostly unpaved route from which private vehicles are banned beyond the first 15 miles. Most people take one of the several types of summertime shuttle buses to enjoy the scenery and the park's legendary wildlife: **grizzly** and **black bears, moose, gray wolves, caribou, Dall sheep, lynx, wolverines, coyotes, porcupines,** and **hoary marmots.** Protected from gawking tourists who leave their vehicles and approach too closely to get that "perfect" photograph, Denali's animals often stay put when buses pass or stop for observation, allowing excellent viewing through binoculars and photography using telephoto lenses. Bus drivers on interpretive tours are experienced in identifying wildlife and providing descriptions and commentary.

6 | Glacier
NATIONAL PARK
Montana

Glacier National Park in northwest Montana is home to 71 mammal species. Even more impressive, of the large mammals present hundreds of years ago, only two species are no longer found in the park: bison and caribou. You have a chance to see **lynx, mountain lions, black** and **grizzly bears, gray wolves, wolverines, white-tailed** and **mule deer, elk, moose, bighorn sheep,** and **mountain goats,** to name only the largest species. Glacier's mammal list is unique in the lower 48 states. The park provides a sanctuary for the largest remaining population of grizzly bears south of the Canadian border, considered to number slightly more than 300 individuals. One of the best ways to see Glacier wildlife is simply to find a spot in the alpine habitat, such as along a trail near 6,646-foot Logan Pass, and scan the landscape with binoculars. Mountain goats and bighorn sheep are often seen this way, and if lucky, you might also spot a wandering bear.

Keep Your Distance

No matter how many lectures park rangers give, or how many warning signs are posted in parks, some people continue to do questionable things around big animals—and pay the ultimate price. People have been killed in both Glacier and Yellowstone National Parks when they got too close to grizzly bears while trying to take photos; the same happened in Yellowstone when would-be nature photographers approached bison. In the Everglades, a woman lost her arm as she was feeding a ham sandwich to an alligator. In Yellowstone, regulations require that visitors remain at least 100 yards away from bears and wolves and 25 yards away from other wildlife; many parks have similar rules. Feeding any animal, from chipmunks to birds to bears, is prohibited in all units. Use binoculars and telephoto lenses to maintain a safe distance, and remember: If an animal gives a sign that it notices your presence, you're too close.

Great Smoky Mountains National Park's Cades Cove is home to both domesticated and wild animals.

7 | Great Smoky Mountains
NATIONAL PARK
North Carolina & Tennessee

The advance of settlement across the eastern United States in the 18th and 19th centuries led to the extirpation of bison, elk, mountain lions, black bears, and even white-tailed deer over large areas east of the Mississippi River. As a result, travelers who want to enjoy such creatures usually head to parks out West. But at Great Smoky Mountains National Park, on the Tennessee–North Carolina border, you can experience a little of what eastern North America was like before European settlement. The park's most famous wild inhabitant is the **black bear,** which thrives here in one of the most lush and diverse environments on the continent. Around 1,900 bears live in the park, which means that the odds of seeing one are good. **Elk** disappeared from this region in the mid-1800s, but beginning in 2001 several dozen have been reintroduced into the park. The best place to look for these animals is the Cataloochee area in the southeastern section of the park, with dawn and dusk the ideal times. **White-tailed deer** are common in Great Smoky Mountains National Park. While visitors sometimes report seeing mountain lions, no hard evidence has been found that they have returned to the area.

8 | Everglades
NATIONAL PARK
Florida

People have different definitions of charismatic, but there's no doubt that nearly everyone who visits this expansive national park in southern Florida wants to see **alligators.** These reptiles and their crocodilian kin are the closest we can come in modern times to seeing a dinosaur, and their sheer size makes sighting one a memorable event. Gators are easily seen at freshwater locations including Shark Valley, the Anhinga Trail at Royal Palm, and Eco Pond near Flamingo—although they're also likely to be seen in any roadside pond, creek, or ditch. In the dry season (November–May) alligators may be seen in great concentrations at shrinking ponds. It's not so easy to see an **American crocodile,** an endangered species that has made

▶ *South Florida is the only region where the American alligator and American crocodile species coexist in the wild.*

an encouraging comeback in Florida in recent years. Crocs inhabit salt water and are sometimes seen around Flamingo.

9 | Theodore Roosevelt
NATIONAL PARK
North Dakota

The badlands of southwestern North Dakota played a big part in Theodore Roosevelt's life, not least because his experiences while hunting and ranching here turned the future president into an ardent conservationist. Today, the park named in his honor, made up of two units, ranks among the best places in the Great Plains to observe wildlife. Reintroduced in 1956, **bison** now number several hundred in the park and often are observed at very close range along roads. The speedy **pronghorn** wanders the park, as well, though it's seldom seen in the North Unit. **Mule** and **white-tailed deer** occupy both units, while **elk** are found only in the South Unit (look for them in the Buck Hill area).

Horses are maintained as part of the historical setting in the South Unit only; Theodore Roosevelt National Park is one of the few areas in the West where free-roaming horses can be observed. Also in the quasi-domestic category are the park's **longhorn cattle,** residing in the North Unit.

10 | Grand Teton
NATIONAL PARK
Wyoming

This near neighbor to Yellowstone hosts much of the same megafauna, but the abundance and variety are different enough to earn Grand Teton its own spot on this list. As is the case with Yellowstone, there are good wildlife-viewing opportunities from roads at Grand Teton. Simply driving along Teton Park Road will probably bring sightings of **pronghorns, elk, bison, coyotes,** and **mule deer.** Other good spots to check from the side of a road include Blacktail Ponds, just north of Moose Junction, for **moose** and **elk;** the meandering Snake River for bison, moose, and **bald eagles;** and Oxbow Bend, for moose, elk, **white pelicans,** and **river otters.** In winter, thousands of elk gather in the Jackson Hole area, many in National Elk Refuge adjacent to the national park. Viewing opportunities here include winter sleigh rides to see elk and the drive to Miller Butte to see wintering **bighorn sheep.**

10 BEST PARKS FOR

OCEAN LIFE

Some of the country's finest natural areas are located around our coasts and islands, where habitats range from rocky shores and salt marshes to sandy beaches and coral reefs. Viewing wildlife here isn't always as easy as watching elk graze in a meadow, but the diversity of species found in the sea makes the effort worthwhile.

1 Kenai Fjords
NATIONAL PARK
Alaska

Though this park 125 miles south of Anchorage covers more than 1,000 square miles, most of it is inaccessible to the average visitor: A massive ice field makes up much of the area, and only one short road enters the park. For many the highlight of a trip to Kenai Fjords is a boat trip into Resurrection Bay and the Gulf of Alaska to explore the fjords, experience tidewater glaciers (ones that meet the sea), and see wildlife. Tours leave from the town of Seward; some ships feature Park Service rangers to provide narration and identify wildlife. All-day trips make a cruise of more than 100 miles to Aialik Bay, and some travel past the Chiswell Islands, part of Alaska Maritime National Wildlife Refuge. Marine mammals often seen on cruises include **humpback whales, orcas** (also called killer whales, though they are dolphins), **harbor seals, Steller sea lions, Dall's porpoises,** and **sea otters.** Depending on the season, a cruise might also bring sightings of **minke** or **fin whales.**

2 Virgin Islands
NATIONAL PARK
Virgin Islands

Many naturalists have compared coral reefs to rainforests for the diversity and beauty of life they host. With just a snorkel mask you can make the ocean your proverbial oyster at Virgin Islands National Park. Just offshore from the island of St. John, Trunk Bay has a 300-foot Underwater Snorkel Trail featuring signs that identify elements of undersea life and the animals that depend on the corals for survival. Elkhorn, staghorn, brain, and star are just a few of the **40 species of corals** that grow in these warm waters, becoming the perfect playground for other reef animals. More than **400 species of fish** have been identified in the park, plus **thousands of invertebrates** such as crabs, lobsters, squid, sea urchins, and sea stars. **Eels, sharks,** and **barracudas** may be spotted sporadically, but an even more exciting find is one of the islands' three species of **sea turtles:** hawksbill, green, or the enormous leatherback.

3 Channel Islands
NATIONAL PARK
California

The five national park islands off the coast of Southern California boast a fascinating natural history, including more than 150 endemic species and the largest breeding colonies of seabirds in Southern California. In addition, the ocean within six nautical miles of the islands is protected as Channel Islands National

Sea lions frolic in the kelp forests off the coast of Anacapa Island in Channel Islands National Park.

A staggering number of fish, coral, and other sea life call the waters of American Samoa home.

4 National Park of American Samoa

NATIONAL PARK
American Samoa

About 2,000 visitors a year explore American Samoa, which, at 2,600 miles southwest of Hawaii, is the only Park Service unit south of the Equator. Those that do travel here find some of the world's most beautiful beaches and coral reefs that are home to nearly **1,000 native species of tropical fish,** such as moray eels, groupers, cardinal fish, butterfly fish, angelfish, wrasses, parrotfish, stargazers, puffers, and gobies. With **giant clams, hawksbill** and **green sea turtles,** and more than **250 species of corals,** the reefs are home to the greatest marine biodiversity of any site in the United States and its possessions. The island of Ofu has excellent coral reefs and offers the best snorkeling to boot.

5 Glacier Bay

NATIONAL PARK & PRESERVE
Alaska

The spectacular scene at Glacier Bay National Park and Preserve appeared very different 250 years ago, when a massive glacier filled the entire bay. Since then, glaciers have steadily retreated, exposing "new" land that has slowly revegetated. For many the most rewarding activity is the daily boat tour. A park ranger is aboard to help see

Marine Sanctuary, including a vast undersea kelp forest. The sanctuary is home to **seals, sea lions, sea otters,** more than two dozen species of **whales** and **dolphins,** and hundreds of species of **fish** and **invertebrates.** Park rangers and volunteers conduct a variety of guided programs, including tide-pool walks to view creatures such as **anemones, sea stars, sea urchins, limpets, periwinkles, chitons, barnacles,** and **mussels.** A summer program at Anacapa Island features park rangers diving into the Pacific Ocean with video cameras, allowing those on dry land to see undersea life, including sea stars, colorful fish, and the occasional mammal. You can ask questions of the rangers via a communications system.

and identify wildlife and to provide information on natural history and geology. Among the marine mammals often sighted are **humpback** and **minke whales, orcas, harbor** and **Dall's porpoises, Steller sea lions, harbor seals,** and **sea otters**—not to mention seabirds such as tufted and horned puffins, common murres, double-crested cormorants, and black-legged kittiwakes (a small gull).

6 | Biscayne
NATIONAL PARK
Florida

Ninety-five percent of Biscayne National Park acreage is made up of the waters of Biscayne Bay and the nearby Atlantic Ocean. A quarter-mile trail on the mainland at Convoy Point offers a glimpse of the above-water world of Biscayne, but to truly experience the park requires a boat. One popular activity is a three-hour trip on a glass-bottom vessel, during which you can see some of the **512 species of fish** that frequent the park, and possibly **sea turtles, moray eels,** and **dolphins** as well. This park encompasses the northernmost islands of the Florida Keys and the northernmost section of one of the world's largest coral reefs. Among the creatures you might see around the coral reefs are **spiny lobsters, sea cucumbers, Christmas tree worms, sponges,** and **fish,** including

such colorful species as butterfly fish, parrotfish, damselfish, wrasses, angelfish, gobies, and barracuda.

7 | Olympic
NATIONAL PARK
Washington

Best known for its glacier-capped mountains, rushing rivers, and lush temperate rainforests, Olympic National Park also has 73 linear miles of Pacific coastline. In recognition of the rich marine life here, 3,310 square miles of ocean—extending 25 to 50 miles from shore—have been designated Olympic Coast National Marine Sanctuary. Some 29 species of marine mammals have been recorded in this region. **Gray whales** migrate northward to their summer range from March into May, while **sea otters, harbor seals, northern fur seals, Steller sea lions,** and **California sea lions** are also seen with some regularity. For sheer diversity, though, nothing beats a trip to a tide pool, where dozens of species can live in just a few square yards. Park rangers lead

Sea Otters

The undeniably charming sea otter, the largest member of the weasel family (males can weigh up to 100 pounds), once numbered in the hundreds of thousands in the northern Pacific Ocean. Unlike blubber-insulated whales, sea otters possess extremely dense fur to protect themselves from frigid ocean temperatures. That fur made them targets of relentless hunting, and by the early 1900s the world's population is thought to have totaled fewer than 2,000. Thanks to various conservation measures, some sea otter populations have made a comeback, and national parks have provided important refuges. These mammals can be seen at Kenai Fjords and Glacier Bay National Parks in Alaska and Olympic National Park in Washington.

regular beach walks and tide-pool explorations at sites such as Mora and Kalaloch.

8 | Point Reyes
NATIONAL SEASHORE
California

It's a measure of the ecological importance of Point Reyes National Seashore that it has been recognized by the United Nations as the Central California Coast Biosphere Reserve. Those interested in marine life will find vast biodiversity here, thanks in part to the upwelling of cold, oxygen-rich ocean water offshore, supporting animals from tiny invertebrates to whales. In addition, the park's marine habitats include lagoons, estuaries, tide pools, marshes, and beaches. A good place to begin learning is the Kenneth C. Patrick Visitor Center, featuring exhibits on maritime exploration, marine fossils, and ocean environments. Cliffs near the 1870 Point Reyes Lighthouse are popular for spotting migrating **gray whales.** Around 20 species of whales and porpoises have been seen around Point Reyes. At nearby Chimney Rock, a breeding colony of **northern elephant seals** can be observed. **California sea lions** and **harbor seals** also frequent the park. Less obvious, but no less interesting, are shoreline invertebrates such as **geoduck** and **gaper clams, hermit crabs, mole crabs, basket whelk**

▶ *The National Park Service has 10 units with coral reefs, and the designation helps protect threatened species like the elkhorn, staghorn, and pillar corals.*

snails, and the jellyfish known as **by-the-wind sailor.** The best way to learn about these unique creatures is to take a ranger-led beach walk.

9 | Padre Island
NATIONAL SEASHORE
Texas

Padre Island National Seashore preserves the longest tract of undeveloped barrier island in the world. Its 70 miles of sandy beach, the great majority of which remains pristine, offer you the chance to find solitude and enjoy the crashing surf, the salty sea breeze, and the raucous sounds of birds. But it's **sea turtles** that make Padre Island such a special place for nature lovers. All five species of sea turtles found in the Gulf of Mexico are officially classified as either threatened or endangered, and all five have been seen at Padre Island. The Kemp's ridley, the smallest and world's most endangered sea turtle, has been the focus of recovery efforts

in Texas, and breeding in the national seashore has increased. In June and July, park visitors can attend releases of hatchling sea turtles from an incubation facility. The western (non-Gulf) side of Padre Island is also fascinating. Laguna Madre is a rare hypersaline lagoon, with a salt content much higher than normal ocean water. Its complex ecosystem provides vital nursery areas for **shrimp** and other **marine invertebrates** and **fish.**

10 | Assateague Island
NATIONAL SEASHORE
Maryland & Virginia

Intriguing animals such as **horseshoe crabs, ghost crabs,** and **blue crabs** frequent this 37-mile-long Atlantic Ocean barrier island at the Maryland–Virginia state line, while **bottlenose dolphins** swim just offshore. Assateague offers many of the pleasures people associate with seashores, from swimming to fishing to beachcombing. Yet there's also plenty here for those who'd like to explore nature, especially marshes and estuaries teeming with life. The half-mile Life of the Marsh Trail wanders through cordgrass, salt grass, and glasswort, some of the components that make salt marshes the most productive of all wetlands. To explore Chincoteague Bay further, you can hike or take canoes to backcountry campsites.

Assateague Island's marshes and estuaries teem with wading birds, such as egrets and herons.

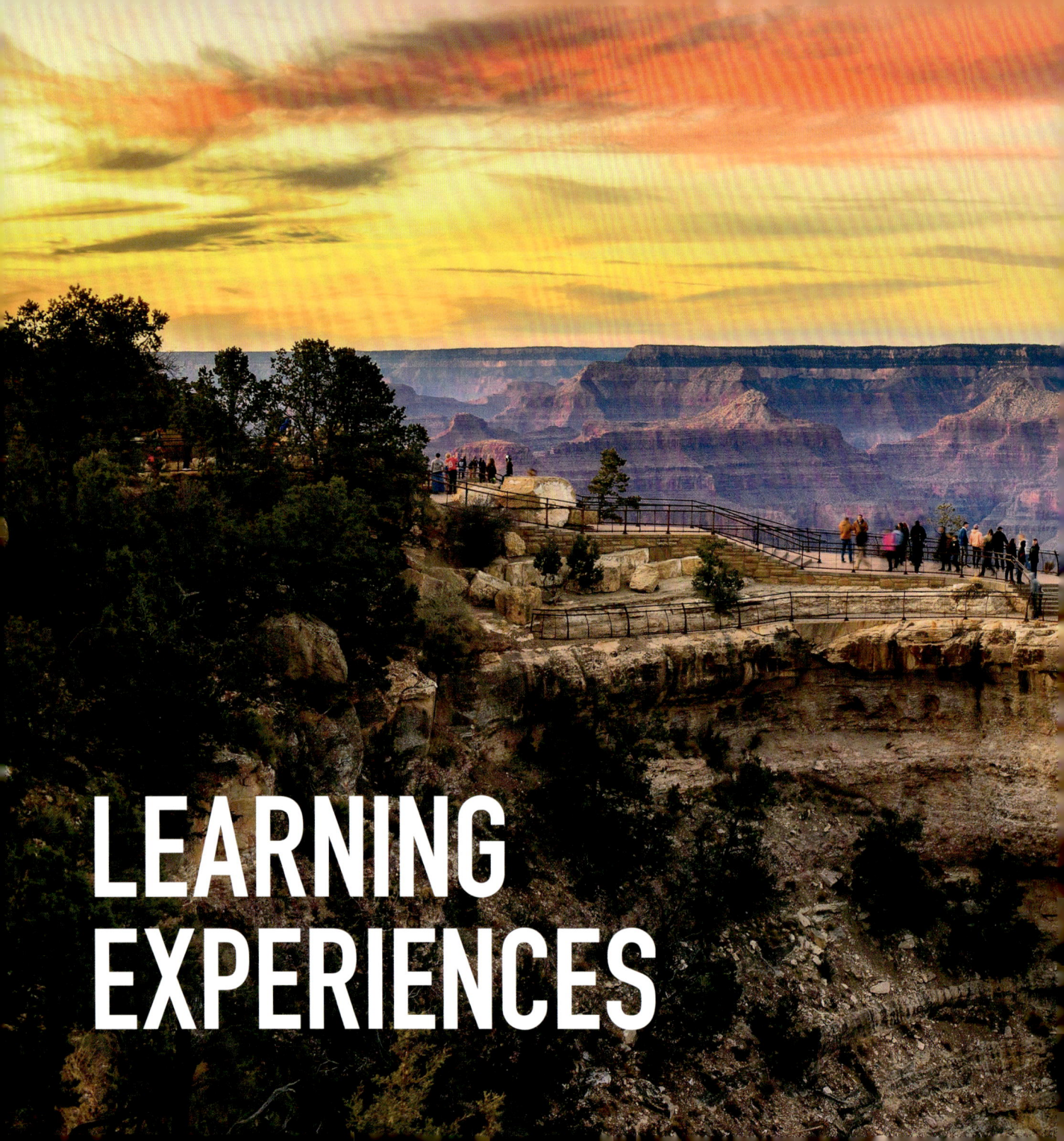

LEARNING
EXPERIENCES

A group congregates at Grand Canyon National Park's Mather Point to take in the sunset.

GUIDED BOAT TOURS

Put away the hiking boots and give tired feet a rest at these national park units where scenic beauty can be enjoyed from a boat, barge, canoe, or kayak. Some of these tours are the sit-back-and-relax type, while a few require paddling. All feature guides who will help visitors see and appreciate the surrounding landscape and wildlife.

1 Apostle Islands
NATIONAL LAKESHORE
Wisconsin

Apostle Islands National Lakeshore in northern Wisconsin encompasses 12 miles of Lake Superior shoreline on the mainland, but the attraction for most is the park's 21 islands, many of them officially designated wilderness. Historic lighthouses are located on six of the islands and some offer guided tours in summer led by park rangers or volunteers. The most popular way to experience an overview of the Apostle Islands, though, is the 55-mile **Grand Tour scenic cruise** (800-323-7619, apostleisland.com), offered at a variety of daily times by an authorized park concessionaire from mid-May to September. Featuring narration on local history from knowledgeable guides, the three-hour trip travels through the heart of the archipelago, passing lighthouses on close-in Raspberry

and northernmost Devils Islands, as well as coastal caves and rock formations. Other cruise options include visits to Raspberry Island and Michigan Island to explore their historic beacons with a Park Service guide.

2 Voyageurs
NATIONAL PARK
Minnesota

During its summer season this national park offers several guided boat and canoe tours at all three of its visitor centers: Rainy Lake, Kabetogama, and Ash River. One, the **North Canoe Voyage** (877-444-6777, nps.gov/voya), lets participants paddle a 26-foot replica of the type of craft used by the original voyageurs, legendary fur traders of the late 18th and early 19th centuries who paddled canoes between northwestern Canada and Montreal. Various ranger-guided

excursions aboard the motorized *Voyageur* journey to an abandoned gold mine, stop in at the historic 1913 Kettle Falls Hotel, grant passengers the opportunity to spot bald eagles and other wildlife, visit a scenic rock garden, and cruise the waters at sunset. The late-night **Starwatch Cruise** showcases the impact of light pollution, and park rangers explain exactly what it means to be a certified International Dark Sky Park. Get the timing right, and you may be lucky enough to catch a glimpse of the aurora borealis as you float on the water.

3 Glacier
NATIONAL PARK
Montana

Although best known for its spectacular glacier-carved mountains and famed Going-to-the-Sun Road, Glacier National Park is also home to some fabulously scenic lakes. Glacier

Get up close to natural wonders—sea caves, sandstone formations, cliff-hugging forests—on the waters of Apostle Islands National Lakeshore.

Park Boat Company (*406-257-2426, glacierparkboats.com*), an authorized concessionaire, offers **narrated boat cruises** in four locations: Many Glacier (cruising on both Swiftcurrent Lake and Lake Josephine), Two Medicine Lake, Rising Sun (St. Mary Lake), and Lake McDonald. Knowledgeable guides provide commentary on the park, expounding on the cultural and natural history of the area and pointing out geological features and wildlife. In addition, at Two Medicine Lake and Many Glacier you have the option of disembarking for a guided hike or exploring on foot (and with a buddy) for a half day before returning on a later cruise back to the departure point. At Many Glacier the guided hike reaches Grinnell Lake; the Two Medicine hike visits Twin Falls. The tours are conducted on six wooden boats, some dating back to as early as 1926 and four of which appear on the National Register of Historic Places.

4 Everglades
NATIONAL PARK
Florida

This South Florida national park is a very big and very wild place, with relatively few ways for the average person to access areas away from roads and visitor centers. Canoeing and kayaking open up new opportunities but can be daunting for inexperienced paddlers. For alternatives, head to the Gulf Coast Visitor Center in Everglades City, in the western part of the park, and sign up for a **guided boat or canoe trip** (305-242-7700, nps.gov/ever). Park concessionaires offer boat tours of the Ten Thousand Islands area daily. Narrated by a naturalist, the tours last 1.5 hours and you'll likely see wading birds, alligators, and other wildlife on the relaxed cruise. You can also bring your canoe or kayak—or rent one—and join a private guide on a strenuous, four-hour **paddling trip** for a more intimate look at the Gulf Coast and its mangrove forests. If weather permits, you'll have the chance to step onto dry land and explore an island. Another option is in the Flamingo area, in the southern part of the park, where boat tours explore the Whitewater Bay backcountry north of Florida Bay. Contact the Guy Bradley Visitor Center in Flamingo (239-695-2945, nps.gov/ever/plan yourvisit/gbvc.htm) for information.

5 Chesapeake & Ohio Canal
NATIONAL HISTORICAL PARK
District of Columbia & Maryland

In what was for its time a massive undertaking, investors in 1828 spearheaded the construction of a commercial canal from the District of Columbia along the Potomac River into Maryland. Twenty-two years and more than 180 miles of infrastructure later, the Chesapeake & Ohio Canal began carrying barges of coal, lumber, and produce from the interior of the United States to the East Coast. Flood damage and competition from railroads put it out of business within a few decades. Now partly restored, the Chesapeake & Ohio Canal takes you back to the 19th century, as park rangers in period dress narrate tours aboard replica 1870s-era canalboats. As part of the **Great Falls Canal Boat Program** (301-739-4200, nps.gov/choh), the replica Charles F. Mercer canal excursion boat—located at the Great Falls Tavern in Potomac, Maryland—travels at a leisurely pace while rangers describe what life was like for the families who lived and worked on the canal. Along the way, passengers get to experience going through one of the canal's locks.

Cruise Crater Lake

One of the most unusual national park boat trips is the summertime two-hour, ranger-guided cruise around Oregon's Crater Lake, the

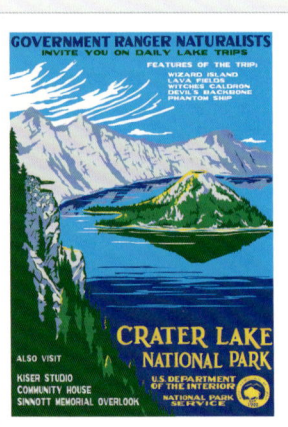

deepest lake in the United States. Occupying the caldera (collapsed magma chamber) of Mount Mazama, a volcano that exploded 7,700 years ago, Crater Lake is known for water of such deep, rich blue that it seems almost unnatural. Some tours allow you to disembark and explore Wizard Island, a cinder cone that grew from an eruption after the caldera formed. Wizard Island hikers, then, are climbing a volcano inside another volcano.

(Although an extensive rehabilitation has these mule-drawn canal tours on temporary hold, expanded operations will be available upon the project's completion.) In historic Georgetown along the start of the canal's route, Georgetown Heritage *(202-480-9540, georgetownheritage.org)* conducts **canal tours** Wednesday through Sunday from spring through October. Swapping mules for engines, these boat rides highlight the history and technology that make the C&O Canal so iconic.

6 | Gateway Arch
NATIONAL PARK
Missouri

Steamboats reigned along the St. Louis riverfront in the 1800s, providing not only industry but also entertainment for visitors and locals alike. Today the tradition continues aboard **paddle-wheel riverboats** at Gateway Arch National Park *(877-982-1410, gatewayarch.com/experience/riverboat-cruises)*. A ticket takes you on an hour-long tour along the legendary Mississippi River while you hear about the heyday of steam travel, learn about the river's role in St. Louis history, and discover its lasting significance. Narrated by a captain or Park Service ranger, tours take place daily from March through November. Adding an element of fun, **themed rides,** such as Oktoberfest, brunch and dinner cruises, and

> ▶ *Known for their strength and intelligence, mules—a cross between a female horse and a male donkey—were once the engines of the Chesapeake & Ohio Canal, pulling boats along their routes.*

PJs and Pancakes With Santa, fill the calendar. In October, the can't-miss **Lock-N-Dam Cruise** takes guests on a daylong journey to the confluence of the Missouri and Mississippi Rivers, complete with a passage through the lock system while a guide explains the history of this intricate transportation technology. A lunch buffet and cash bar round out the experience. Even kids can get in on the fun with the new Riverboat Explorer Junior Ranger program.

7 | Pictured Rocks
NATIONAL LAKESHORE
Michigan

The handsomely colored Pictured Rocks of Michigan's Upper Peninsula formed as sandstone in an ancient shallow sea. Along 15 miles of the Lake Superior shore these rocks create scenes of great beauty, in some places rising from the water as cliffs 180 feet high. Bands of color

reveal minerals in the sandstone: red and orange for copper, green and blue for iron, black for manganese, and white for calcium carbonate. Although the Pictured Rocks can be viewed from some land lookout sites, the best way to see them is from the water. From mid-May to mid-October, Pictured Rocks Cruises, a park concessionaire *(906-387-2379, picturedrocks.com),* operates a tour boat that departs the town of Munising and **cruises along the Lake Superior shoreline** for about three hours. Noteworthy sites along the way include the historic East Channel Lighthouse, Miners Castle (one "turret" of the castle fell in 2006, but it remains a striking rock formation), Rainbow Cave, Indian Head, and Spray Falls.

8 | Channel Islands
NATIONAL PARK
California

The five main islands of Channel Islands National Park feature abundant wildlife and striking scenery. And while some visitors arrive on private crafts or by seaplane, most people experience the park via **commercial boat tours.** Park-authorized concessionaire Island Packers *(805-642-1393, islandpackers.com)* operates boats from harbors in Ventura and Oxnard, with tours promising sights like seals, sea lions, dolphins, whales, sea otters, and the largest breeding colonies of seabirds in Southern California. Sea kayaking allows closer exploration, including Santa Cruz Island's Painted Cave, one of the largest and deepest sea caves in the world. Several concessionaires have **guided sea kayak trips,** giving inexperienced paddlers a greater degree of confidence and safety in dealing with weather and sea conditions that can change quickly.

9 Lake Chelan
NATIONAL RECREATION AREA
Washington

Most activities at this remote and beautiful area in north-central Washington center on the small community of Stehekin, which can be reached only by boat, seaplane, or trail. Historic sites, waterfalls, and varied outdoor activities attract visitors, and trails lead into the wilderness of the adjacent South Unit

of North Cascades National Park. The National Park Service–operated Golden West Visitor Center in Stehekin, open daily from mid-March through mid-October, offers advice, backcountry camping permits, and ranger-led programs. The **boat trip to Stehekin** is a big part of the adventure, as the concessionaire-operated *Lady of the Lake (509-682-4584, ladyofthelake.com)* ferries passengers across spectacular Lake Chelan from its south end to Stehekin in the north, making stops along the way at such sites as Moore Point, Lucerne, Prince Creek, and Fields Point. Set in a valley gouged out by glaciers, the 55-mile-long lake is the third deepest in North America. The scenic trip passes areas such as the Narrows, where the lake is constricted to a quarter mile wide, and Domke and Bridal Veil Falls. While the boat serves as

a ferry service, the captain offers light narration, making this trip (four hours one way) a hidden guided experience you might otherwise overlook. A speedier option is the *Lady Express* (run by the same company), which does the trip in two and a half hours, or the seasonal 70-minute *Lady Liberty.*

10 Acadia
NATIONAL PARK
Maine

The rocky coast of Maine ranks among America's most beautiful landscapes, and Mount Desert Island has its share of the rugged scenery: crashing waves, tall lighthouses, conifer-draped mountains, and hardwood forests that blaze red and gold in autumn. Viewing the shore from the water is an exhilarating experience, and from mid-May into fall Acadia National Park offers ranger-guided boat trips that focus on diverse topics. The five-hour **Baker Island Tour** *(207-288-2386, barharborwhales.com)* takes you to a small island, with a hike to explore nature and history at a homestead and the Baker Island Light Station. The **Acadia National Park Morning Cruise** *(207-276-5352, cruiseacadia.com)* travels to Little Cranberry Island for a visit to the fishing community of Islesford and explores glacier-carved Somes Sound, the only fjord on the Atlantic coast of the United States.

Guided boat tours are a great way to take in Acadia's shoreline.

FIELD EXCURSIONS

Learning is a lifelong process, but learning outside of a classroom's walls, well, that may be the ultimate educational experience. Partnering with some national parks are field institutes that offer courses led by historians, scientists, authors, and other experts who can turn an ordinary vacation into an educational adventure.

1 Yosemite
NATIONAL PARK
California

For more than a century a nonprofit partner has worked to support the projects and programs that preserve Yosemite National Park's visitor experience and ensure its lasting impact on future generations. The **Yosemite Conservancy** was founded as the Yosemite Museum Association in 1923 with the mission to manage funds to construct a museum at the park. Today its portfolio of programs includes not just guided day hikes, backpacking trips, sunset strolls, stargazing, and cultural and natural history lessons, but also classes on writing, photography, birding, and art. Seeking something else for a solo trip, a family, or a group? Fully customized adventures led by Yosemite naturalists can also be designed (starting at $400 for a half day or $700 for a full day). As an added benefit, programming fees booked through the conservancy go toward funding important work inside the national park. *yosemite.org/experience*

2 Arches & Canyonlands
NATIONAL PARKS
Utah

The **Canyonlands Field Institute** uses the combined 400,000-plus acres of Arches and Canyonlands—and several thousand more outside the parks' boundaries—as its campus. Lessons and courses take place at national monuments, in national forests, and on Native American reservations. Seminars can be personalized, so a one-day hike could become a multiday hike that tacks on river running and camping. Attend a lecture, enroll in a literary symposium, learn how to conduct oral histories, embark on a river study course with a group, or get away from everyone by reserving a fleet of rafts for a private excursion. A natural history course may be as simple as a nature hike or one that requires the use of a cooking yurt followed by an overnight in a tipi. Exploring this land on water can cover numerous topics: While polishing skills at white-water rafting and kayaking, students also absorb information on geology, water plants, wildlife, and human and cultural history. *cfimoab.org*

3 Zion
NATIONAL PARK
Utah

The soft swirl of mountains and the odd colorations found in Zion's rocks and canyons are only part of this southern Utah park's personality. It has far more to offer. The **Zion National Park Forever Project** runs customized field experiences for dedicated adventurers, whether individuals, families, or school and faith groups. Some experiences can last a few hours, others take days, and overall the range is as wide as Zion itself with studies of insects, lizards, and snakes and of botany through native-plant seed propagation, low-desert wildflowers, and explorations of the hanging gardens of Zion. Geology is a natural topic, with lessons that explain the mysteries and evolution of the land, from canyon floor to rim. Art courses in poetry, painting, journaling, and photography, meanwhile, draw inspiration from the land, flowers, and sky. *zionpark.org/discover*

4 | North Cascades
NATIONAL PARK
Washington

Classes in the fields of art, literature, and science at the **North Cascades Institute** take students past the implied barriers of the Washington park and deeper into its forests, higher into its mountains, and farther down its rivers so the resonance of nature helps change thinking and lead to preservation and conservation. Geared for all ages and personalities, courses range from relatively quiet programs such as silk painting and wildflower photography to more active endeavors that involve multiday backcountry treks to explore the upper Skagit River,

▶ *In its 100 years of programming, the Yosemite Conservancy has provided more than $140 million to Yosemite National Park, allowing for 700-plus conservation-focused projects.*

hone map- and compass-reading skills, and learn how to track bears and cougars. An acclaimed family camp even offers a memory-making adventure for multiple generations. A variety of lessons help students embrace the outdoors through combination canoe trek/yoga courses, improve their journaling and poetry skills, and interpret the land through watercolors. *ncascades.org*

5 | Denali
NATIONAL PARK & PRESERVE
Alaska

The **Murie Science and Learning Center,** a partnership between Alaska Geographic and the National Park Service, is Denali's offering to the ever inquisitive. Naturalists, biologists, botanists, authors, and artists present hands-on programs, with a focus on adventure, creative arts, and science. The landscape—a frontier expanse of forest,

mountains, tundra, slopes, bogs, and rainforest—is home to prolific wildlife and sometimes rarely seen wildflowers and rivers filled with trout. Field excursions may involve wildlife tracking and studying the habitat of sheep, caribou, moose, wolves, and grizzly bears. Rafting expeditions disappear into the wilderness in search of some of Denali's 165-some species of birds. Attending classes here won't be a night at the Ritz. Some courses involve hiking, kayaking, and no-frills rustic camping, which may be offset by sessions in which you literally draw inspiration from your surroundings during drawing, painting, and photography classes. Some field courses are even programmed specifically for teacher accreditations. As broad as the course selection is, however, everything boils down to a simple lesson: preservation and conservation. *akgeo.org/field-institute*

6 | Grand Canyon
NATIONAL PARK
Arizona

The vast Grand Canyon has the dimensions to conceal countless marvels and mysteries—many of which can be deciphered in more than 300 courses offered by the **Grand Canyon Conservancy Field Institute.** Many programs are simple—yoga classes, a casual walk along the rim, and even writing

Hone your photography and backcountry skills amid the snow-capped peaks and wildflowers of the North Cascades.

retreats—but others are designed to lead students into a more natural (and demanding) environment. From floating down the Colorado River to snaking down canyon trails and trekking into the deepest recesses, courses range from beautiful to brutal, so beware of physical limitations before embarking. Art instructors know where to find the most

perfect plants and flowers and how to sketch, paint, and photograph them. Loveliness abounds in and around the canyon and, in a number of à la carte classes, can be found in the sky as well. Far from the city lights, astronomy sessions will reveal the sort of heavens our ancestors enjoyed. Now we can enjoy it, too. *grandcanyon.org/programs-tours*

7 Great Smoky Mountains
NATIONAL PARK
North Carolina & Tennessee

Lessons at the **Great Smoky Mountains Institute at Tremont** emphasize sense of place, diversity, and stewardship. Considering this park's heritage, it's a smart combination. Clear-cutting led to a host

John Oliver Place is the oldest log cabin on the Cades Cove loop road in Great Smoky Mountains National Park.

of environmental tragedies, and the story of the ecosystem's rescue forms a logical foundation for the institute's lessons. Relocated when this area became a park, the families of Cades Cove left behind a unique mountain culture whose story is shared in courses on cultural diversity. Logging inspires day hikes that celebrate the largest old-growth forest east of the Mississippi, just as streams, rivers, and wetlands help explain the hydrology, structure, and habitats of the southern Appalachian Mountains. The park's countless habitats make it a photographer's paradise. Courses focused on light, composition, and equipment take place in fields of wildflowers and at vantage points perfect for sunrise and sunset

scenes. Family Camps host weekend or weeklong lodging and bonding opportunities for families of all sizes. And don't miss the autumn Appalachian Celebration, a festive exploration of the region's culture in music, storytelling, food, and dance. *gsmit.org*

8 Point Reyes
NATIONAL SEASHORE
California

The **Point Reyes National Seashore Association** features courses nearly as varied as the seashore's dramatic coastline landscape, with field classes whose duration (as little as a couple hours) and price (some free; others starting at $50) make them readily accessible and equally

popular among long-distance travelers as well as day-trippers heading in from nearby San Francisco. Glimpses of the natural world are captured in classes that interpret the environment through paints, pastels, and pencils, while scientists and naturalists help focus observational skills on insects, examine the medical and nutritional value of plants and mushrooms, lead kayaking expeditions on Tomales Bay and other waterways, and conduct evening hikes that skirt past mudflats to a vantage point to see the setting sun. Given the seashore's unique biodiversity, popular day- and weekend-long courses study songbirds, pelagic seabirds, shorebirds, raptors, ducks, and other migratory fowl, while photography courses focus on these birds as well as landscapes, flowers, waves, and sunsets. *ptreyes.org*

9 Yellowstone
NATIONAL PARK
Wyoming

The fascinating human and natural history of northwestern Wyoming provides the content for classes offered by the **Yellowstone Forever Institute.** Budding naturalists can learn about the park's wolves, raptors, and bison as well as geological marvels, including hot springs, mud pots, and geysers. Curious historians can study the days when tourists arrived via horse and buggy and U.S. Army soldiers were on patrol against poachers, vandals, and robbers. Yellowstone's abundance of natural beauty inspires art courses with plein air painting sessions, sketch classes, and instruction in photographing the park's thermal features and brilliant autumn leaves. Some courses reach back in time to turn city slickers into mountain men and women capable of identifying footprints and scat to track animals, studying the habitat and behavior of birds, and learning how beavers change our environment by building their own. And if your vision of Yellowstone includes mountains, forests, and a gently flowing stream, sign up for fly-fishing lessons. *yellowstone.org*

10 | Joshua Tree
NATIONAL PARK
California

Desert Institute courses at Joshua Tree National Park are designed to appeal to a cross-generational student body, so lessons may be as practical as learning to read a compass and map, as novel as mastering the art of stellar navigation, as thrilling as tracking an animal, or as demanding as scaling the park's tallest peak. Some guided hikes explain how this harsh land shaped the existence of the ancient people who lived here, while another brings students up to date on the very active gold-mining operations launched here in the 1890s and relaunched by necessity during the Depression. In a unique twist on standard botany lessons, foraging for edible desert plants is followed by a final exam that includes ways to prepare and eat at a firepit cookout. Another course looks at how snakes have adapted to an ecosystem of extreme temperatures and a scarcity of water. Many programs focus on the creative arts, including appreciation for and attempted re-creation of Native American pottery and baskets. The barren landscape magnifies the profile of anything that survives here, so painters and photographers have the inspiration—and instruction—to capture the park's unique beauty. *joshuatree.org/desert-institute*

More Field Institute Opportunities

NatureBridge: Four park units with overnight, hands-on environmental science programs for children and teens. Yosemite National Park, Golden Gate National Recreation Area, Olympic National Park, and Prince William Forest Park; *nature bridge.org*

Olympic Park Institute: K–12, customized programs, family programs, and teacher training; four hours to 10 days. Olympic National Park; *olympic parkinstitute.org*

Petrified Forest Field Institute: From archaeology and geology to paleontology and ecology; one-of-a-kind landscape best explored with a world-class guide.

Petrified Forest National Park; *petrifiedforestfieldinstitute.org*

Sequoia Parks Conservancy: Naturalist-guided night sky hikes, nature walks, backpacking trips, and winter activities; one hour to three days. Sequoia and Kings Canyon National Parks; *sequoiaparksconservancy.org*

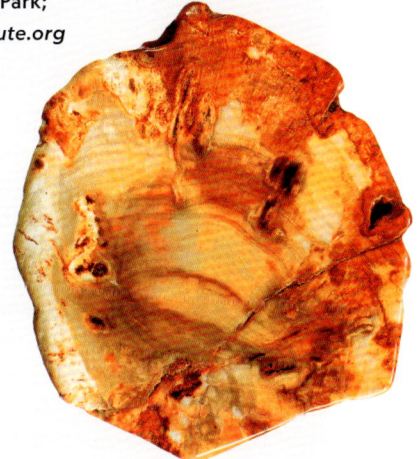

RANGER PROGRAMS

No matter which park unit you visit, you're sure to leave knowing more than when you arrived. And many times that is thanks to rangers—folks who are passionate about their parks and want everybody to understand what humans and nature have conspired to create. Check out these one-of-a-kind programs that greatly enhance the learning experience.

1 Carlsbad Caverns
NATIONAL PARK
New Mexico

Every evening a few hundred thousand bats peel themselves off the ceiling at Carlsbad Cavern to take off for dinner. From Memorial Day weekend through October, just before sunset, a ranger standing near the natural amphitheater created by the cave entrance shares information about bats and the importance of protecting them. With a little luck, the 30-minute talk will end just as the first bats appear from deep within the cave, a small trickle slowly turning into an aerial stampede as thousands swarm into the sky. No cell phone or devices of any kind are allowed, and attendees are cautioned to remain quiet, as voices and the high-frequency pitch of flash cameras can distract the bats. As summer passes the size of the maternal colony grows, the baby bats having grown enough to take flight and hunt for themselves. The **Bat Flight Program** is free, and space in the amphitheater is first come, first served.

2 White Sands
NATIONAL PARK
New Mexico

It's normal to have a lot of questions about the world's largest gypsum dune field. In fact, the rangers at White Sands National Park expect and enjoy them. Ask away on one of the park's daily **Sunset Strolls,** a leisurely, family-friendly exploration of the dunes with a focus on the park's mind-boggling geology and plant and animal life. Timed to coincide with the fall of the sun against the mountains, the one-hour walk is approximately one mile in length. For a deeper dive into gypsum geology and the dry lake bed considered the dunes' birthplace, the three-hour **Lake Lucero Tour** is presented monthly from November through March. The rest of the year, escape the desert sun with the **Full Moon Hike,** a chance to explore the dunes under the glow of the moon. Because of the nature of the programs and extreme conditions, the reservation-only Lake Lucero and Full Moon hikes are not recommended for children.

3 Gettysburg
NATIONAL MILITARY PARK
Pennsylvania

Credit the stories woven into the Battle of Gettysburg that keep this national park atop the popularity lists. Supplementing the ranks of uniformed rangers each summer are **living historians** who appear in period dress while adapting the persona of someone related to the battle. In addition to an impressive daily total of more than a dozen ranger programs, guests may join

A park ranger explains the geological history of White Sands National Park.

A park ranger checks the harness of one of Denali's sled dogs before a meal.

a rotating cast of characters who share firsthand tales of their experiences here. There may be the expectant mother who saddled up to show Union soldiers local roads, the *New York Times* reporter who covered the battle only to find his dying son in the field, a sketch artist on assignment, a reporter from the *Times* of London, or troops coming to Gettysburg to remember their fellow soldiers. You can always learn something from a ranger, but a completely new level of learning is reached when you suspend your disbelief and listen to the stories of people who were actually there.

4 | Denali
NATIONAL PARK & PRESERVE
Alaska

A lot of things distinguish Denali from other national parks, among them that it's the only national park with a kennel of sled dogs. In fact, Denali has had working dog teams for more than 100 years. Multiple

daily sled dog demonstrations happen in the summer season, and although you can visit the kennels on your own, watching rangers interact with these skilled canines is worth planning ahead. Hop on a free shuttle from the Denali Visitor Center—no reservations are required. The 30-minute programs feature an engaging—and, yes, downright cute—demonstration of mushing, the traditional mode of travel in Alaska, plus time for questions and a visit with the fuzzy huskies.

5 Lava Beds
NATIONAL MONUMENT
California

When the California summer becomes too hot, it's a welcome relief to find somewhere to get out of the sun—like inside a 55°F lava tube that snakes beneath a national park. Daily during peak season, rangers at Lava Beds National Monument are ready to get down to business, which means getting their groups belowground. Unlike limestone caves, which evolve over hundreds of thousands of years, lava tubes are created relatively quickly and from a powerful geological force. The two dozen tunnels created here offer three levels of difficulty, so there will always be a balance of opportunities. On a **ranger-led lava tube exploration,** guests don gloves and helmets, carry flashlights, and descend into the catacombs for about an hour. In easier caves you can follow a lighted and paved trail; in more difficult tubes you get a quick introduction to the serious challenges of caving. This is also an activity that can be done without a ranger leading the way, if desired.

Being a Park Ranger Is Cool

Not everyone has a job they really and truly enjoy, but one could wager that most park rangers do. Beyond working in some of the most beautiful "offices" in the country, most park rangers are fulfilling a passion to work outdoors and promote their own love of nature. Such is the case with Dan Ng.

Ng was formerly the chief interpreter (that is, naturalist) at Bryce Canyon National Park in Utah. Inspired by a visit to Yellowstone National Park in Wyoming as a child, he took a detour from his first career path of dentistry and pursued a degree in biology, following it up with one in park recreation management. He explains why he's invested more than a quarter century with the National Park Service.

"I love the outdoors and scenery and animals and I've been able to work at Yellowstone, Big Bend, Grand Teton, Golden Gate, and the Grand Canyon. I love interacting with the public and protecting and promoting the values of the Park Service.

"When you get to work at a place where people love to vacation, that tells you a lot about the opportunities we have. People love to visit us, and they enjoy where they are. They want to be here. Compare that to what I'd experience if I were a dentist.

"Plus, as a park ranger, you don't have to worry about what you wear. There's a lot of tradition in the uniform, and you take pride in wearing it. Visitors see the uniform and they know that we can help them understand the park and we're there to provide help if they need it."

What's not to love?

Dressed for the experience, anyone can purchase cave maps at the visitor center and explore the chilly tubes solo.

6 | Mammoth Cave
NATIONAL PARK
Kentucky

Travel beyond the regular guest experience and dive deep into a challenging and chilly (mid-50s) subterranean world at Mammoth Cave. Among a dozen different themed excursions, the wildest option is undoubtedly the **Wild Cave Tour.** How wild? Decked out in caving gear (helmet, gloves, kneepads, headlight), guests follow the ranger deep into sections of the cave that lack lights and amenities (this may not be the first adventure choice for those who are claustrophobic). But thoughts of finding a restroom (reached at an underground lunchroom about two hours into the experience) may be put on hold as novice spelunkers deal with more immediate concerns, such as moving from a relatively large section of cave that reduces to the size of a kneehole in a desk before reducing again to narrow passages as small as just nine inches high. There are restrictive size and age requirements for this daylong excursion, but before you considering signing up for the five-mile jaunt, be certain that you can handle the pressure of living like a mole—once inside, it's hard to get out.

Mount Rushmore's auditorium hosts outdoor events.

7 | Mount Rushmore
NATIONAL MEMORIAL
South Dakota

At nearly any time of day, the spectacle of Mount Rushmore is already awe-inspiring. Then comes nightfall and the spectacular 45-minute **Evening Program.** Held from around late May through September, the presentation is one of the most popular ranger-led programs at any national park. At the spacious amphitheater, patriotic music plays as dusk starts to shade the four faces on the mountain. The program begins when a ranger strides to the center of the stage to explain the story behind the sculpture, take questions from the audience, and then introduce the short film "Freedom: America's Lasting Legacy." Next, to the musical accompaniment of "America the Beautiful" and sing-along favorite "The Star-Spangled Banner," the faces are slowly illuminated. Additionally, summertime visitors won't want to miss the Lakota, Nakota, and Dakota Heritage Village that spotlights the customs and traditions of the area's local Native American communities.

8 | Bryce Canyon
NATIONAL PARK
Utah

Bryce Canyon is a geological wonder. One of the park's most popular ranger programs is the **Full Moon Hike,** typically offered monthly.

This hike shines a new light on the hoodoos—the park's iconic towering spires. Depending on the night, two hike options may be available: one that descends into the canyon (think steep and rocky trails) and another that meanders along the plateau's rim. The approximately two-hour experience offers the best of all worlds: Bryce is seen in an unusual natural state, with the full moon illuminating the canyon, while rangers infuse the adventure with lessons in geology, wildlife, and astronomy. Due to popularity, the visitor center holds a lottery prior to the hike. All participants must wear hiking shoes, and children under the age of eight are not permitted. Can't get a ticket or have a younger crowd in tow? Bryce has a number of other ranger programs, including geology talks, kids' programs, and a night sky telescope program.

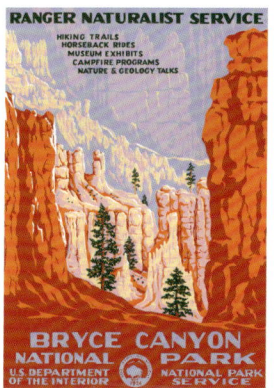

 **Canyon de Chelly**
NATIONAL MONUMENT
Arizona

Located in northeastern Arizona, Canyon de Chelly is entirely contained within the Navajo Nation, and who better to share the story of the canyon than local guides, many of whom are Native Americans. These are more than just nature walks. Depending on how much time is available, hikes range from simple **three-hour treks** to demanding **weeklong expeditions** that explore more remote reaches of the canyon. For guests who'd prefer not to go on foot, guides also lead **horseback and jeep tours** that can expand the roaming range as well as awareness of the region's history, from natural and geological history to facts about the Navajo and ancestral Puebloan people.

10 | **Fossil Butte**
NATIONAL MONUMENT
Wyoming

While Fossil Butte is not a large national park unit, it quite likely has some of the oldest exhibitions. Preserved across and below this sagebrush-filled landscape are an inestimable number of fossils from the Eocene epoch—in cosmic terms a relatively recent 50 million years ago. On summertime weekends a ranger leads guests in the popular **Fossil Quarry Program.** After a brief uphill hike, guests reach an area where researchers are combing the rock to find the fossilized remains of dragonflies, fish, snakes, turtles, and other prehistoric creatures and foliage. Learn what paleontologists do and witness their ongoing research—all fossils found remain in the park's collection.

Field Day: Citizen Scientist Programs

As the saying goes, the best way to learn is to do. National park visitors can get into the field—and contribute helpful data—by participating in one of the park system's many citizen scientist programs.

Citizen scientist (also called community scientist) projects are those that engage public volunteers in conducting scientific research, whether through designing experiments, collecting data, or analyzing results. The National Park Service uses information gathered from visitors—often via sources as simple as a mobile app—to help manage its resources.

A FEATHERED HOLIDAY

There are a few sure things when you join Bryce Canyon National Park's Christmas Bird Count: You'll be joined by passionate and skilled birders, you're certain to learn, and your findings will go toward conservation efforts.

The rest is up to chance, weather, and the whims of nature—and that's what makes it so fun. As the longest running citizen scientist program in the country, the Christmas Bird Count was started in 1901 by Audubon Society members who preferred to census birds rather than hunt them. Now an international event, more than 2,600 counts took place in 2022, lending important data points to further our understanding of bird biology and the impact of climate change.

FOR THE BIRDS

Ranger Peter Densmore, Bryce Canyon's visual information specialist, remembers his first Christmas Bird Count, in 2016, fondly. "I went with a coworker who was similarly new to the experience, and we were paired with some skilled birders. We learned so much—where to look for birds, what kind of trees have food in the winter, how to identify the species we spotted. It opened up an awareness for what birding can do to connect you to everything around you." Densmore was hooked. He's been directing the annual count ever since.

Densmore recognizes that while the Christmas Bird Count is an important data collection exercise, the impact doesn't stop there. "While the event's data is scientifically useful—it is regularly cited in Environmental Protection Agency reports—even more so than that, this is an outreach event."

Bryce Canyon specifically encourages beginners to bird, reaching out to local schools as well. "Birds are a natural starting point for connecting with nature. They are an easy friend to make on that journey," Densmore says. "And like so many things, depending on where you enter, typically you end up broadening out from there. If you're into the birds, you may end up getting into plants; from the plants, the insects, and so on."

THE MOST WANTED LIST

Across the country in Tennessee, Todd Witcher and his team at Discover Life in America (DLIA) are on a mission to catalog every species in Great Smoky Mountains National Park—no small feat for a place world-renowned for its biodiversity. Currently the director of the nonprofit, Witcher has spent his career in conservation and education, and the last 16 years at DLIA. "I've always thought that in order to have people care about conserving biodiversity, they have to be more attached to it. And the best way to do that is to engage people in the science of it all."

Founded 26 years ago, DLIA runs BioBlitz programming in the park, hosts local courses and workshops, and offers the regional middle school iScience project.

Various citizen scientist programs take place at Yosemite, including for birding and pollinators.

Its flagship effort is the All Taxa Biodiversity Inventory, a joint effort with Great Smoky Mountains National Park to identify and catalog all its living species. The public can contribute via the iNaturalist app, with emphasis placed on DLIA's Smokies Most Wanted list, which prioritizes 100-plus species of birds, plants, and insects that are fairly simple to identify and for which scientists need more data points.

Witcher encourages people to take their time: "We go slow, we turn over rocks, and we look at those things that people don't typically stop to look at." If you spend your time looking for a bear and you don't see a bear, you might be disappointed, he says. "But it's just as thrilling—maybe even more so—to spot lichen or a special bark or something on our Most Wanted list. The more people

care about the little things, the more they're going to care about the big things—about conservation, biodiversity, and the big world around them."

JOIN IN

There are dozens of citizen scientist projects in national parks, such as the Snowshoe Hare Pellet Plot Project at Denali National Park and Preserve; the Western Monarch Milkweed Mapper at Grand Canyon National Park; the Otter Spotter at Great Smoky Mountains National Park; the Pika Patrol at Rocky Mountain National Park; and the Saguaro Census at Saguaro National Park.

If you'd rather get going on your own, apps like eBird and iNaturalist are self-directed and can be started whenever you're ready.

10 BEST

OUTFITTED ADVENTURES

For those who'd rather leave the planning to the experts, outfitters provide ready-made park tours. With an emphasis on outdoor adventure, most trips feature hiking, biking, boating, or all three. They range from budget camping trips where everyone pitches in to luxury journeys with lodges and fine dining.

1 Climate Ride
CYCLING, HIKING, PHILANTHROPY

Montana-based nonprofit Climate Ride challenges people to get active for a really good cause: our planet. Multiday fundraising events merge **physical adventures with philanthropic commitments,** resulting in unforgettable trips. The six-day Glacier Ride cycling event spans Glacier National Park and Canada's Waterton National Park, while the six-day Death Valley National Park Ride is a spring or fall tour of unparalleled desert. If hiking is more your speed, the Climate Hike Glacier traverses the Crown of the Continent with a daily distance of eight to 20 miles. Climate Ride handles all the trip details so participants can focus on fundraising,

training, and fun. More than 100 nonprofit organizations receive grants based on the distribution choices that riders and hikers select, making protecting Mother Earth a fun and actionable merging of passions. *climateride.org*

2 National Geographic Expeditions
WILDLIFE, PHOTOGRAPHY, HISTORY, CRUISING

From Kenai Fjords and Yellowstone to the Channel Islands and Bryce Canyon, traversing national parks with National Geographic Expeditions is every bit as extraordinary as you'd imagine. You can browse itineraries by type (such as **expedition cruises** or **land-based** or

family-focused journeys) and by interest (**photography, wildlife encounters, history and archaeology,** or **hiking,** to name just a few). Given National Geographic's reputation for incredible images, the photography tours are among the most popular. The **chromatic adventure** can include early morning photo shoots of wildlife in Yellowstone's Lamar Valley, access to lesser known areas of national parks, as well as workshops led by professional photo instructors able to take even amateurs to the next level. Also popular are **expedition cruises,** which nimbly glide guests into remote destinations as the specialty ships navigate narrow waterways. Trips are led by a top-notch expedition team with marine biologists, historians, and seasoned photographers and include sites such as the Columbia and Snake

Tourists explore Alaska's wilderness with National Geographic Expeditions.

Rivers along Lewis and Clark's legendary route. *nationalgeographic expeditions.com*

3 | St. Elias Alpine Guides
ALASKA-FOCUSED OUTDOOR ADVENTURES

Few do Alaska like McCarthy-based St. Elias Alpine Guides. Deep-wilderness backpacking, rafting wild and scenic rivers, climbing the continent's highest peaks, backcountry ski touring, and even kite skiing on remote ice fields—these folks do it all. While most of their adventures take place in and around Wrangell–St. Elias National Park and Preserve, they also venture into Denali and the Chugach Mountains. Many of the company's most extreme adventures (like scaling 18,000-foot Mount St. Elias) are designed for experienced hikers, climbers, or backpackers craving a **world-class wilderness challenge.** But the company also offers easy one-day glacier hikes and river rafting out of McCarthy, as well as relatively **easy four-day trips** to places like Skolai Pass and Iceberg Lake that sample a small portion of the Wrangell wilderness. Participants fly in by bush plane (landing in a grassy meadow or rocky glacial field), pitch their tents at some jaw-dropping location, and spend the days doing day hikes on glaciers, lakeshores, and wildlife trails and the nights

The National Geographic Society awards nearly 600 grants to researchers and conservationists around the globe every year.

curled up inside the tent, sipping hot chocolate and swapping tales of the day. *steliasguides.com*

4 | Austin Adventures
LUXURY OUTDOOR ADVENTURES

In 1985, Austin Adventures launched as one of the world's first **boutique adventure travel** outfits. Its co-founders pioneered a new type of travel that combined adrenaline and luxury, with small groups led by experienced guides with a mandate to experience places like a local would. Nearly 40 years later, that's still the trademark. Austin Adventures' multisport jaunt through Glacier National Park is typical of what you can expect on one of their trips: a mosaic of hiking backcountry trails over granite and ice, rafting the unfettered Flathead River, biking the Going-to-the-Sun Road, and retiring at night in the comfort of one of the park's legendary lodges. The company offers journeys in more than a dozen parks, with trips

designed for families, couples, solo travelers, and more. *austin adventures.com*

5 | Country Walkers
GROUP AND SELF-GUIDED WALKING TOURS

Walking is the best way to absorb the geography and culture of a place. And that's what Country Walkers is all about: **immersion via steady footsteps.** This is ambling even for those who don't do it as part of everyday life, either as part of a group or self-guided using Country Walkers' maps, notes, and inside tips. The guided trips cover seven different parks from coast to coast. The Olympic Peninsula adventure is par for the course, a leisurely weeklong wander through most of the park's highlights, including the tide pools and seamounts of Kalaloch and the temperate rainforest of the Hoh River region. Nights are passed in the comfort of the park lodges. The hiking is rated easy to moderate, with everyone putting in a reasonable number of magical miles per day. *countrywalkers.com*

6 | Road Scholar
EDUCATION, OFF-THE-BEATEN-PATH SITES

A nonprofit organization dedicated to **lifelong learning opportunities for older adults,** Road Scholar

provides interesting low-cost travel opportunities with an educational angle. Formerly known as Elderhostel, the group rebranded its programs "Road Scholar" in 2009 to emphasize the combination of journey and learning that characterizes every trip. National parks are among their more popular programs. In fact, Road Scholar visits more parks than any other travel outfitter in the country, including off-the-beaten-path destinations like Assateague Island National Seashore in Maryland and Virginia, Canyon de Chelly National Monument in northern Arizona, and Capitol Reef National Park in central Utah. Trips often combine more than one park, like a southwestern Colorado adventure that incorporates Great Sand Dunes, Mesa Verde, Black Canyon of the Gunnison, and Colorado National Monument. *roadscholar.org*

7 | O.A.R.S. Nationwide
RAFTING & FLOAT TRIPS, WILDLIFE, HISTORY

America's largest white-water outfitter organizes **aquatic adventures in several national parks** from its base in the San Francisco Bay Area. A lauded itinerary is the 24-day John Wesley Powell Retrace, where participants take reinforced dories and rafts to follow the footsteps of Powell's 1869 expedition through the Green and Colorado Rivers. In addition to exploring remote river canyons, there's plenty of time for casual swims, side hikes, and wildlife-watching. Rafters camp along sandbanks at night and take their hard-earned meals over open fires among soaring scenery. O.A.R.S. also organizes a combined Yosemite hiking and Tuolumne River rafting trip, as well as multiday float trips in such parks as Dinosaur National Monument. *oars.com*

8 | REI Co-op Adventure Travel
GUIDED OUTDOOR EXPERIENCES, CAMPING, CRUISING

The adventure travel arm of the well-known outdoor equipment store organizes guided trips—**hiking, biking, paddling, and more**—in and around many national parks. Its portfolio includes a multisport and bear camp adventure in Alaska's Kenai Fjords, a Glacier Bay cruise and kayak combo, a family adventure trip to Washington's San Juan Islands, four days of **camping and kayaking** on Yellowstone Lake, and a Great Smoky Mountains voyage with amenities like cushy tents and fantastic food. Led by highly experienced guides, most of these trips are rated moderate and in many cases no previous experience is required. Camping is normally at remote campsites, in tents or alfresco (camper's choice), with meals prepared by both guides and guests. *rei.com/adventures*

O.A.R.S., a family-owned and -operated company since it opened in 1969, specializes in rafting and outdoor adventures.

9 Natural Habitat Adventures
WILDLIFE, PHOTOGRAPHY, CONSERVATION

As its name suggests, Natural Habitat specializes in **close encounters of the animal kind** all around the globe. In the United States, the Colorado-based adventure travel company throws out some pretty wicked wildlife trips. One unique trip revolves around photographing the winter wonders of Yellowstone and Grand Teton. A new tour called Alaska Bear Camp promises rustic luxury—and bear thrills—in renowned Lake Clark National Park.

Conservation and exploration go hand in hand with Natural Habitat, which works with the World Wildlife Fund to access locations largely inaccessible to tourists. *nathab.com*

10 Backroads
CULTURAL IMMERSION, MULTISPORT ADVENTURES

Californian Bill Hale started Backroads in 1979 as a way to create "epic experiences" for people. While all of Backroads' excursions are active experiences, Hale says an even bigger part of the Backroads experience is **cultural immersion.** The company offers hundreds of trips each year, including **multisport outings** in a number of national parks. One top trip is a Crater Lake multiadventure in the Cascades Range of Oregon. The itinerary includes a pedal around Crater Rim Drive with jaw-dropping views of the deep caldera; whitewater rafting on the Deschutes River; and a hike along the high-desert trails of Smith Rock. Among the overnight stops on the six-day tour are Crater Lake Lodge and a top-notch resort. Other Backroads itineraries feature Bryce Canyon, Zion, Joshua Tree, the Grand Canyon, Yellowstone, and many more. *backroads.com*

Rail Journeys to National Parks

Once upon a time luxury trains provided transport to the gateways of many national parks. While only a few of the vintage lines remain, new ones have been added to the rails. Here are some highlights:

Alaska Railroad: This rail line runs between Anchorage and Fairbanks, traveling up the eastern edge of Denali National Park and Preserve. It stops at a station only 100 yards from the park's visitor center. Another branch runs from Anchorage to Seward, gateway to Kenai Fjords National Park. *alaskarailroad.com*

Amtrak Vacations: Amtrak has trips to 18 national park units, including the Glacier National Park Getaway round-trip from Chicago and the Peaks to Pacific Journey between Denver and San Francisco that includes a side trip to Yellowstone. *amtrakvacations.com*

Grand Canyon Railway & Hotel: This historic railway runs from Williams, Arizona, to the South Rim of the Grand Canyon. Since the last spike was driven in 1901, the line has carried numerous well-known figures to the canyon, including Teddy Roosevelt, John Muir, and Clark Gable. *thetrain.com*

10 BEST PARKS FOR

LITERARY PILGRIMAGES

In the decades immediately after the Revolutionary War, the United States existed in the cultural shadow of Europe. As the 19th century progressed, homegrown writers made their pen marks in the world of literary arts. The following national parks showcase the lives of some of our nation's finest authors.

1 New Bedford Whaling
NATIONAL HISTORICAL PARK
Massachusetts

Thousands of American students have written essays on *Moby-Dick* and the symbolism of Ahab's pursuit of the white whale. The novel ranks among the great achievements of modern literature and contains beautiful writing and indelible scenes and turns of phrase. **Herman Melville** based much of his 1851 story on his own experiences on a whaling ship. He embarked from New Bedford, Massachusetts— which he described as "perhaps the dearest place to live in, in all New England"—in 1841 and returned to Boston in 1844. At New Bedford Whaling National Historical Park, learn how fortunes were made from whale oil and relive the voyages of whalers. The park's Seamen's Bethel has been a house of worship for mariners since 1832. Melville attended services here and later wrote about the chapel's marble memorials to sailors lost at sea. The pulpit in the shape of a ship's bow was added in 1961, based on Melville's description of a pulpit in *Moby-Dick*.

2 Edgar Allan Poe
NATIONAL HISTORIC SITE
Pennsylvania

Debate about **Edgar Allan Poe's** literary qualities—not to mention his personal character—began during his lifetime, intensified at his early death (at age 40), and continues today. There can be no doubt, though, that few American writers have been as influential as the Boston native known as the master of the macabre. His chilling horror tales are still avidly read, and stories such as "The Murders in the Rue Morgue" have led many to credit him with the invention of the detective story. Poe's chronically underfinanced life forced him to move often. The house that is now Edgar Allan Poe National Historic Site was his Philadelphia home for about a year beginning in 1843. You are greeted by a sculpture of a raven, a reminder of Poe's renowned poem "The Raven" (in which the bird utters the ominous word "Nevermore"). Exhibits include recordings of Poe's works narrated by actors such as Vincent Price, Basil Rathbone, and Christopher Walken.

Theodore Roosevelt's Sagamore Hill home on Long Island holds a wealth of the president's personal possessions.

3 | Sagamore Hill
NATIONAL HISTORIC SITE
New York

Famously described by historian Henry Adams as embodying "pure act" for the decisive force of his personality, **Theodore Roosevelt** arguably did more in his lifetime than a dozen ordinary people. The 26th U.S. president, Roosevelt was also a soldier, rancher, conservationist, adventurer, voracious reader, and respected author. He published a historical study, *The Naval War of 1812,* while in his early 20s; it was just one of some 18 books he would write on subjects including history, biography, nature, and government, as well as his autobiography. Sagamore Hill, on Long Island near the town of Oyster Bay, was Roosevelt's home from 1884 until his death in 1919. Once called his Summer White House, the rambling, 23-room Victorian home is full of original Roosevelt family furnishings and personal items, including his Rough Rider hat from the Spanish-American War, his gold-plated shaving kit, hunting trophies, and a ring containing a strand of Abraham Lincoln's hair. The Sagamore Hill Nature Trail winds through a forest of oaks and tulip trees along the same path Roosevelt and his family took on many swimming and camping excursions.

4 Minute Man
NATIONAL HISTORICAL PARK
Massachusetts

True, Minute Man National Historical Park is best known as the location of the "shot heard round the world." But this site in eastern Massachusetts preserves more than just Revolutionary War attractions. Concord visitors can tour what has been locally labeled the Home of Authors, which celebrated authors **Louisa May Alcott, Nathaniel Hawthorne,** and **Margaret Sidney** all called home in the 19th century. The Alcotts, residents from 1845 to 1852, called the house "Hillside." It was here that Louisa and her sisters enjoyed many of the adventures that the author mirrored in her book *Little Women*. The family was said to have had a close relationship with neighbors Ralph Waldo Emerson and Henry David Thoreau. From 1852 through 1869 the home passed to Nathaniel Hawthorne, author of such works as *The Scarlet Letter, House of the Seven Gables,* and *Twice Told Tales*. Hawthorne renovated the home, which he named "The Wayside," and added a three-story tower for his study. The last private family to own the storied site was the Lothrops. Harriett

 *Louisa May Alcott and her family were prominent abolitionists and their Massachusetts home, now known as The Wayside, is also part of the National Underground Railroad Network to Freedom program administered by the National Park Service.*

Lothrop (pen name Margaret Sidney) was a children's book author and preservationist whose dedication to Concord's historic places made their future protection in the national park possible. Guided tours of The Wayside happen seasonally.

5 North Cascades
NATIONAL PARK
Washington

The striking landscape of North Cascades National Park—glacier-sculpted mountains dotted with alpine lakes, waterfalls, rocky streams, and the largest collection of glaciers in the lower 48—has the power to inspire anyone, but it holds a special place in the American literary movement known as the Beat Generation. Three Beat writers—novelist and poet **Jack Kerouac,** poet and environmental

activist **Gary Snyder,** and poet **Philip Whalen**—all worked for a time as fire lookouts on mountain peaks here in the 1950s. Kerouac, perhaps the most famous of the three for books such as *On the Road,* was stationed atop 6,102-foot Desolation Peak, now in Ross Lake National Recreation Area, part of the North Cascades park complex. His experience was reflected in his novel *Desolation Angels*. Snyder worked on 8,127-foot Crater Mountain, in Pasayten Wilderness just east of the national park, and on 6,106-foot Sourdough Mountain in the park's North Unit, where Whalen also worked. All three peaks can be climbed by physically fit hikers; none require technical mountaineering skills.

6 Carl Sandburg Home
NATIONAL HISTORIC SITE
North Carolina

Known as the "poet of the people," **Carl Sandburg** was the son of an immigrant railroad worker and dropped out of school after eighth grade to work and travel the country—experiences that shaped his political beliefs and writings. He won Pulitzer Prizes for both poetry and a biography of Abraham Lincoln. In 1945, Sandburg and his wife, Lilian, bought a house and property in Flat Rock, North Carolina, primarily so she could have a place to raise her prize-winning dairy goats. The house was always full of music and friends—not to mention more than 15,000 books. Sandburg died here in 1967 at the age of 89. You can see thousands of family items, from teapots to clothing to guitars on which Sandburg performed folk songs. The Park Service maintains a small herd of goats, and rangers demonstrate milking and cheesemaking seasonally. The park includes more than five miles of trails through the woods and fields, where Sandburg felt it was important "to go away by himself and experience loneliness" at times.

7 | Eugene O'Neill
NATIONAL HISTORIC SITE
California

A life punctuated with sorrows provided background material for the plays of **Eugene O'Neill,** winner of four Pulitzer Prizes and the Nobel Prize in Literature. Born in a hotel on New York City's Broadway, O'Neill suffered from alcoholism and depression, married three times, and had difficult relationships with his children. When O'Neill and his third wife, Carlotta, bought a 158-acre ranch near Danville, California, in 1937, he hoped he could live and work there permanently. He and Carlotta lived in the house, called Tao, for seven years, during which he wrote the plays *The Iceman Cometh, Long Day's Journey Into Night,* and *A Moon for the Misbegotten.* Illness

Longfellow House boasts manicured gardens that are open year-round.

and the circumstances of World War II forced the O'Neills to leave in 1944. O'Neill never completed another play and died in Boston in 1953. Private vehicles are not allowed at the historic site, but there is a shuttle bus from Danville. After a guided tour of the home, there's time for a walking tour of the grounds as well.

8 | Dayton Aviation Heritage
NATIONAL HISTORICAL PARK
Ohio

The first African American to be widely acclaimed within literary fields, **Paul Laurence Dunbar** produced more than 400 works in his short career, including 12 books of poetry, four novels, four short story collections, and lyrics to many

songs. He was known for his use of fanciful metaphor and conversational dialect, a skill that earned him international renown. In downtown Dayton, Ohio, the Wright-Dunbar Interpretive Center has exhibits on Dunbar (as well as aviation pioneers Orville and Wilbur Wright), including his newspaper *The Dayton Tattler,* which was printed there. The Paul Laurence Dunbar House, where the poet worked before his 1906 death at the age of 33, was the first national historic landmark established in honor of an African American.

9 | Longfellow House– Washington's Headquarters
NATIONAL HISTORIC SITE
Massachusetts

For a time in the mid-19th century **Henry Wadsworth Longfellow** was arguably the most famous living writer, not just in the United States but worldwide. Poems such as "The Song of Hiawatha," "Paul Revere's Ride," "The Courtship of Miles Standish," and "Evangeline" earned him enough money that he resigned his teaching position at Harvard University and devoted himself to writing and translation. Longfellow spent the rest of his life in the house on Brattle Street in Cambridge, Massachusetts, a gift from his father-in-law in 1843. Guests who stopped by included Nathaniel Hawthorne, Ralph

Waldo Emerson, Charles Dickens, and Oscar Wilde. The house had gained fame even before the poet's residence, though: In 1775, during the Revolutionary War, George Washington used it as the head-quarters of his Continental Army for nine months. Today, the national historic site offers small-group tours of the house seasonally, allowing you to see Longfellow's library, study, bedrooms, family portraits, and paintings by artists such as Gilbert Stuart and Albert Bierstadt. The annual Longfellow Summer Arts Festival celebrates poetry, music, and community, and the site's for-mal gardens are open year-round.

10 Fort McHenry
NATIONAL MONUMENT & HISTORIC SHRINE
Maryland

On September 13, 1814, American lawyer and amateur poet **Francis Scott Key** was under British deten-tion on a ship in the Baltimore harbor, conducting negotiations related to the ongoing War of 1812. That night, American military post Fort McHenry came under heavy bombardment from British ships, and as night fell Key was unsure of the battle's outcome. "By the dawn's early light," though, he saw the "star-spangled banner" still flying over the fort—a sign that the British attack had failed. Key was inspired to write a poem he called "Defence of Fort M'Henry." Set to the melody of a British drinking song, it became a popular patriotic tune and, in 1931, was officially designated the national anthem of the United States with the name "The Star-Spangled Banner." Today, Fort McHenry is a national monument and historic shrine, the only National Park Service site with that designation. You can take a self-guided tour to see military memorabilia, an electric battle map, barracks, commander's quarters, guardhouse, and powder magazine. Flag changes take place twice daily, weather permitting.

Literary Role of National Parks

A number of other authors have written both fiction and nonfiction works associated with national parks.

Colin Fletcher: Fletcher's *Man Who Walked Through Time* chronicles the author's 1963 adventure hiking the length of the Grand Canyon.

Peter Matthiessen: *Shadow Country,* which won the National Book Award in 2008, combines three novels by Matthiessen set in Florida's Ten Thousand Islands area, now part of Everglades National Park.

John Muir: Though best known as a naturalist and conservationist, Muir wrote many influential books and essays, including *The Yosemite* and *Travels in Alaska.* He is remembered at Muir Woods National Monument and John Muir National Historic Site, both in California.

Nevada Barr: Barr has written a series of mysteries featuring fictional national park ranger Anna Pigeon, including *Track of the Cat* (set in Guadalupe Moun-tains National Park) and *High Country* (Yosemite National Park).

Tennessee Williams: His Pulitzer Prize–winning 1947 play, *A Streetcar Named Desire*, was shaped by the cultural melting pot of New Orleans, whose heritage is the focus of one unit of Jean Lafitte National Historical Park and Preserve.

DISCOVERING HISTORY

The Stone House, at what is now Manassas National Battlefield Park, weathered two Civil War battles.

INDIGENOUS SITES

These sites celebrate the rich stories and cultures of Indigenous communities throughout the United States. While much of the culture from these civilizations has been lost to the impacts of time and colonization, locations across the country memorialize everything from important archaeological sites to artifacts to sacred grounds, with increasing efforts made to preserve these important legacies. All will leave you awed and inspired.

1 | Mesa Verde
NATIONAL PARK
Colorado

Designated a national park in 1906, Mesa Verde was the first primarily archaeological unit in the National Park System. Unquestionably one of the world's most important archaeological sites (as well as among the most visually striking), it was declared a cultural **World Heritage site** in 1976. More than 4,000 individual archaeological sites exist within this southwestern Colorado park, the most famous of them a series of large, beautiful **cliff dwellings** set into canyon walls. Though this plateau was populated by the ancestral Puebloan people from about A.D. 550 to 1300, it was only during the last century of occupation that they built and resided in the elaborate cliff dwellings. The structures here were abandoned in the late 13th century for unknown reasons. Plan your trip to Mesa Verde during the period from late spring through fall, when all facilities are open, including the most impressive cliff dwellings: Spruce Tree House, Cliff Palace (the largest cliff dwelling in North America), Balcony House, and Long House. Some can be seen only on ranger-led tours, while smaller ruins such as Cedar Tree Tower can be explored on your own. A visit to the Chapin Mesa Archaeological Museum is a first-stop must. Watch an orientation film, see dioramas representing daily life in an ancestral Puebloan village, and learn details about Mesa Verde from exhibits and artifacts.

2 | Chaco Culture
NATIONAL HISTORICAL PARK
New Mexico

Archaeologists recognize this Native American site in northwestern New Mexico as having been the major regional center for the ancestral Puebloan culture. Chaco's political and economic influence from A.D. 850 to 1250 extended to communities many days' journey away. Visitors know the park as one of the largest and most striking collections of Native American structures in the Southwest. One highlight is **Pueblo Bonito,** a Chacoan great house, which was a very large multistory public building with formal ceremonial rooms known as kivas. Pueblo Bonito once rose four stories high and contained more than 600 rooms and 40 kivas; it remains a sacred place to several Native American tribes. Five other major archaeological sites are located around the nine-mile **Canyon Loop Drive,** including **Casa Rinconada,** which contains the largest kiva in the park. Many of the structures at Chaco Canyon seem to have been aligned to observe astronomical

Cliff Palace at Mesa Verde National Park is one of the most stunning examples of ancestral Puebloan architecture.

events, a tradition continued in today's park with the Chaco Night Sky Program presenting astronomy programs from April to October. The park was designated an International Dark Sky Park in 2013.

3 | Gila Cliff Dwellings
NATIONAL MONUMENT
New Mexico

One of this park's attractions is its setting in the rugged Gila National Forest, adjacent to a 558,000-acre wilderness area. Traveling to the site through the mountains of southwestern New Mexico creates the proper mood for a trip back in time. There's mystery, too: No one knows why the people of the Mogollon

culture built **stone-and-wood structures in cliffs** here around A.D. 1270, only to abandon the site after just a few decades. Seeing the dwellings, set in shallow caves in the cliffs, requires taking a trail that climbs 175 feet from the valley floor. Guided tours of the cliff dwellings are offered regularly. Don't miss the short **Trail to the Past,** which leads to pictographs and to a single isolated Mogollon dwelling in a small canyon.

4 | Hopewell Culture
NATIONAL HISTORICAL PARK
Ohio

This intriguing park in south-central Ohio showcases the Native American culture that dominated the

forested regions of eastern North America from 200 B.C. to A.D. 500. The Hopewell culture encompassed political and spiritual beliefs of various tribes and was characterized by the construction of tall burial mounds and large earthworks in geometric patterns. For example, a parallelogram-shaped earthwork encloses 111 acres, and a 13-acre rectangular enclosure contains at least 23 mounds. The **Mound City Group, Seip Earthworks,** and **Hopewell Mound Group** complexes feature self-guided walking trails. The park's museum displays some of the thousands of artifacts found in the earthworks, including knives, pipes, animal effigies, masks, arrowheads, copper ornaments, pottery, and tools.

5 | Ocmulgee Mounds
NATIONAL HISTORICAL PARK
Georgia

A prehistoric Native American site representing more than 12,000 years of continued human habitation, Ocmulgee Mounds has been home to at least four different Indigenous cultures since the last ice age. Start at the visitor center for a brief orientation film and a chance to peruse thousands of artifacts from the peoples of the Macon Plateau. The historical park protects seven mounds of varying sizes that have been used for sacred ceremonies and commerce.

Learn about thousands of years of history and see the seven historical mounds along eight miles of trails at Ocmulgee Mounds.

Eight miles of trails—most a mile or less one way—take guests through wetlands and forests and to historic mounds like the **Great and Lesser Temple Mounds. Earth Lodge,** just a short stroll from the visitor center, is a reconstructed council chamber with an original floor that carbon-dates to 1015. In the future, Ocmulgee may be upgraded to a national park, which would make it the first for Georgia, and dramatically expand its protective acreage. The Ocmulgee Indigenous Celebration takes place here each September and features artists, dancers, musicians, storytellers, and crafters.

6 | Canyon de Chelly
NATIONAL MONUMENT
Arizona

The attractions of this park in northeastern Arizona are threefold: ruins of **ancient dwellings perched on cliff ledges, spectacular walls of red sandstone** rising hundreds of feet above valley floors, and a community of Navajo families whose heritage dates back 2,000 years. Canyon de Chelly is administered cooperatively by the Park Service and the Navajo Nation, with access more limited than in most parks. Two roads wind along canyon rims, offering superb vistas, and one publicly accessible trail leads down to the canyon floor. Otherwise, you must be accompanied by a ranger or Navajo guide. The 18-mile South

> *The country's largest archaeological dig happened at Ocmulgee Mounds between 1933 and 1936, with more than 800 workers excavating 2.5 million artifacts.*

Rim Drive and the 17-mile North Rim Drive pass sites such as **Mummy Cave Ruin,** one of the park's largest structures, and the amazing spectacle of **Spider Rock,** an eroded pinnacle rising 800 feet above the canyon floor. At several places you can see the pastures, fields, and traditional hogan houses of modern Navajo farm families on the canyon floor.

7 | Fort Union Trading Post
NATIONAL HISTORIC SITE
Montana & North Dakota

For nearly four decades in the mid-1800s, Fort Union had an annual average of 25,000 buffalo robes and $100,000 in merchandise trades. Constructed by the American Fur Company at the request of the Assiniboine Nation, the commercial zone was considered a prime example of peaceful coexistence, and it was here that the Northern Plains tribe would exchange furs for goods like knives, beads, blankets, alcohol,

and cookware. Today's park hosts an archaeological reproduction of the **trading post** as well as a visitor center at **Bourgeois House,** the former residence of the head trader. In the summer and early fall, stop by the Trade House to hear from living-history interpreters, or plan a visit around the park's Fort Union Rendezvous, Indian Arts Showcase, or Living History Weekend.

8 | Trail of Tears
NATIONAL HISTORIC TRAIL
Georgia to Oklahoma

Treaties and tensions were already ignited against Indigenous peoples when, in 1828, gold was discovered in Georgia. European-American settlers, eager for the tribal homelands on "rich" soil, passed the Indian Removal Act in 1830, forcing the relocation of eastern Native Americans past the Mississippi River into what was then called Indian Territory (present-day Oklahoma). Of the more than 16,000 Cherokee, Choctaw, Seminole, Creek, and Chickasaw who endured the westward route via land and river, an estimated quarter perished. Winding 5,043 miles across nine states, the **Trail of Tears National Historic Trail** links federal, state, tribal, and locally owned sites that interpret these stories of deep tragedy and improbable survival. Marked by blue-and-white signage, dozens of sites include water routes, museums, commemorative parks, historic ferries, and more. The trail's website has interactive maps for each state along the route.

9 | Knife River Indian Villages
NATIONAL HISTORIC SITE
North Dakota

The nomadic lifestyle of many Plains Indians means that few physical traces of their communities remain. This park interprets the history of the Hidatsa and Mandan people, Northern Plains Indians who built earthen lodges. The Lewis and Clark expedition passed through this area in October 1804, encountering a thriving community of around 4,500 people who welcomed them and let them build a fort where they could spend the winter. While here, Lewis and Clark hired Toussaint Charbonneau, a French-Canadian fur trapper living with the Hidatsa, as an interpreter; his wife, Sacagawea, accompanied him. Later, the villages suffered greatly from smallpox, a European disease against which they had no immunity. The Mandan population was reduced by 90 percent and the Hidatsa by half. In 1845, they moved upriver along the Missouri. The park preserves a reconstructed **earth lodge,** and walking trails wind through **village sites** where depressions of other lodges can be seen.

10 | Walnut Canyon
NATIONAL MONUMENT
Arizona

Set in a beautiful canyon near Flagstaff, Arizona, this park protects elaborate structures of rock, mud, and wood built on alcoves set back into near-vertical sandstone walls. The people who lived here from A.D. 1100 to 1250 are known by archaeologists as the Sinagua culture. The name comes from the early Spanish explorers' description of the nearby San Francisco Mountains. (*Sin agua* means "without water," which isn't exactly true.) Ruins of Sinagua pueblos can also be seen at other Arizona sites such as Montezuma Castle and Tuzigoot National Monuments. Building their homes into cliff walls may have had several advantages for the Sinagua, including safety from attack, protection from weather, and moderation of temperature extremes. Descend into the canyon via the steep 0.9-mile **Island Trail** to see 25 rooms of the **cliff dwellings,** with more visible across the canyon.

EARLY COLONIAL SETTLEMENTS

The explorations of Christopher Columbus sparked a land rush in the Americas, with Spanish, French, English, and Russian powers, among others, scrambling to establish colonies and monopolize the riches of the New World. Proof of their impact can be found in parks commemorating the nation's colonial era.

1 | Roger Williams
NATIONAL MEMORIAL
Rhode Island

Fleeing religious intolerance in England, Puritan minister Roger Williams arrived in Massachusetts in 1631. Speaking out against the narrow-minded ways of his fellow colonists, Williams was convicted of sedition and heresy and, in 1635, fled the colony in order to avoid arrest. He established a settlement among the Narragansett west of the Seekonk River, christening the new colony **Providence** and dedicating it to both religious freedom and separation of church and state—two principles on which the United States was later founded. Located in downtown Providence, Rhode Island, the 4.5-acre memorial was founded in 1965 and celebrates the theologian and guiding light of American idealism. The visitor center includes displays and a short film on Williams's life; the grounds include outdoor exhibits set around what was once the town's public commons.

2 | Cabrillo
NATIONAL MONUMENT
California

Perched at the entrance to San Diego Bay, this small but popular national monument commemorates the first European to navigate the western coast of America, as well as California's maritime heritage and unique coastal ecosystems. Juan Rodríguez Cabrillo had already earned fame as the captain of Hernán Cortés's crossbowmen during the Spanish conquest of Mexico, and then as a gold miner and shipbuilder in Guatemala, before agreeing to lead the first Spanish voyage along the California coast in 1542. Cabrillo and his three ships sailed north along the coast, going as far as today's Russian River, north of San Francisco; he would die on the return trip from complications of an injury suffered in a skirmish with Indigenous warriors. His expedition laid the groundwork for the Spanish settlement of California and a string of Franciscan missions that would eventually stretch between San Diego and Sonoma. The park also includes the iconic **Old Point Loma Lighthouse** (1855)—replete with 19th-century furnishings—and a number of historic **coastal defense batteries.** Terraces provide astonishing bird's-eye views of San Diego Bay and the vast Pacific. Trails lead through pristine coastal chaparral vegetation and along sandstone cliff tops above tide pools.

3 | Sitka
NATIONAL HISTORICAL PARK
Alaska

Although the Russians were late starters in the European bid for North America, their presence in Alaska endured from 1741 to 1867, when the territory was sold to the United States. Sitka National Historical Park commemorates the site of the most significant skirmish between Russian forces and Alaska's Indigenous people—the 1804 **Battle of Sitka**—as well as the settlement the Russian fur traders built there afterward. The region's Russian and Native American heritage is explored at the park visitor center and the adjacent **Southeast Alaska Indian Cultural Center.** Trails lead to totem poles and the historic battleground on a peninsula between Sitka Sound and the Indian River. Separate from the main park, the Russian **Bishop's House** sits in the middle of modern Sitka town. Built in 1843, the imposing log structure is one of the

Historic Places of Worship

Faith was one of the leading reasons why Europeans colonized America, and our parks now safeguard some of the nation's oldest places of worship.

Concepción, San José, San Juan, and Espada (San Antonio, Texas): Franciscan friars built these four Spanish colonial churches in the early 18th century with the goal of converting the Indigenous populations to Christianity. Strung together in San Antonio Missions National Historical Park, the missions blend Moorish, Gothic, and baroque features. All are active Roman Catholic congregations.

Gloria Dei–Old Swedes' Church (Philadelphia, Pennsylvania): The single oldest church in the park system, this historic site opened in 1700 on the site of an even older Lutheran chapel by descendants of the New Sweden colony (1638–55). Services are held in the small brick structure Sunday mornings; however, the congregation is now Episcopal.

Old North Church (Boston, Massachusetts): This 1723 church is now part of Boston National Historical Park. Sexton Robert Newman flashed two lanterns from its steeple on the night of April 18, 1775, to warn local patriots that the British were coming. An active Episcopal congregation, services are held on Sunday. The Georgian-style church is also home to the oldest church bells (1744) in North America.

Touro Synagogue (Newport, Rhode Island): Now a national historic site, this 1762 synagogue is the nation's oldest place of Jewish worship and the only surviving colonial-era synagogue. The congregation was created in 1658 by Sephardim fleeing the Spanish Inquisition. The Orthodox Congregation Jeshuat Israel renders Shabbat services on Friday evening and Saturday morning; there is separate seating for men and women, and proper attire is required.

The foundations of El Morro, a triangular Spanish colonial bastion in San Juan, Puerto Rico, date back to 1539.

oldest remaining Russian colonial structures in North America. In addition to living quarters restored with period furnishings, the house also boasts the richly decorated **Russian Orthodox Chapel of the Annunciation.** The nearby St. Michael's Cathedral is a faithful reproduction of the 1848 original that burned down in the 1960s.

 San Juan
NATIONAL HISTORIC SITE
Puerto Rico

San Juan, Puerto Rico, the second oldest European settlement in the New World after Santo Domingo, was settled by Spanish colonists in 1508—only 16 years after Columbus "discovered" America. By 1521, they had relocated from the Spanish Main to a small offshore island that offered better living conditions and defensive possibilities. Soon after, the Spanish began construction of the first in a series of sandstone fortifications that compose the heart of today's national historic site, as well as the oldest European structures in the entire National Park System. **Castillo San Cristóbal** is the largest of these, a sprawling citadel that took

See inside a Paspahegh home at the historic Jamestown settlement, part of Colonial National Historical Park.

more than 150 years (1634–1790) to complete. But the hulking **Castillo San Felipe del Morro,** known as El Morro, is the famous triangular bastion overlooking the harbor mouth that fell only once in battle during its 400-year history. El Morro also bears the distinction of firing the first American shot of World War I—a 1915 salvo against a German warship. Daily ranger talks and walks (in English and Spanish) relate the history of the forts and the Spanish in Puerto Rico. Or you can explore the meandering ramparts on their own, trekking the top of the walls or the waterfront Paseo del Morro trail.

5 | Colonial
NATIONAL HISTORICAL PARK
Virginia

This park bookends the British colonial era in the New World, from the establishment of Jamestown (1607) to the landmark battle of Yorktown (1781) that ended British rule in the 13 Colonies. Perched on either side of the Virginia Peninsula, Colonial National Historical Park comprises two anchor segments connected by the 23-mile **Colonial Parkway,** which meanders through heavily wooded countryside and connects to **Colonial Williamsburg, historic Jamestown,** and the **Yorktown battlefield** (collectively referred to as the Historic Triangle). The settlement at Jamestown eventually grew

▶ *First performed on July 4, 1937, The Lost Colony is the longest-running outdoor symphonic drama in the United States. Only two seasons have been canceled: 1944, due to World War II, and 2020, due to the COVID-19 pandemic.*

into the first capital of the Virginia Colony, which would play a pivotal role in the American Revolution 169 years later. One of the more popular units in the National Park System, this park offers a wide range of activities from living history tours of Jamestown to interpretive talks on the Yorktown battlefield.

6 | De Soto
NATIONAL MEMORIAL
Florida

While Juan Rodríguez Cabrillo was sailing the California coast and Francisco Vásquez de Coronado was trekking the Southwest, another intrepid Spaniard, Hernando de Soto, was on a meandering four-year journey through the American Southeast. Far more than the other two expeditions, De Soto's venture would forever change the lives of the Indigenous people with whom he came into contact.

De Soto landed somewhere near present-day Tampa Bay, Florida, with a legion of 600 men and more than 200 horses in 1539. They set off northward on an odyssey that would take him across the Great Smoky Mountains and all the way to the banks of the Mississippi River, where De Soto died from a fever. The marathon expedition and its aftermath are explored in a short film and displays at the Bradenton, Florida, memorial visitor center. A living history exhibit called **Camp Uzita** operates during the winter season, with rangers and volunteers dressed as 16th-century Spaniards and Native Americans. Ranger-led kayak tours of the park's mangrove swamp and shallow coastal waters are offered in the warmer months.

7 | Coronado
NATIONAL MEMORIAL
Arizona

In 1540, Francisco Vásquez de Coronado marched 339 Spanish soldiers and more than a thousand native allies from Mexico City into what is now Arizona. Coronado and his entourage trekked to Kansas and back on a futile two-year search for the Seven Cities of Cibola and their legendary riches. They found instead a flourishing ancestral Puebloan culture and natural treasures like the Grand Canyon. Located on the Arizona-Mexico border, this park reflects on the significance of the 16th-century expedition and its lasting impact on the Southwest. The visitor center displays Spanish colonial armor and other artifacts, as well as a short film about the expedition. Hike the monument's eight miles of **mountain and valley trails,** through high desert, grassland, and woodlands, to see a landscape little changed since the 16th century. Save time to explore **Coronado Cave,** a 600-foot undeveloped cavern that was likely first used as human shelter more than 8,000 years ago.

8 | Fort Raleigh
NATIONAL HISTORIC SITE
North Carolina

Sponsored by Sir Walter Raleigh in 1585, the first English settlement in the New World was the ill-fated **Roanoke Colony,** established on the coast of today's North Carolina, an area that had been home to Carolina Algonquian for centuries. The venture failed, and Raleigh dispatched a second voyage in 1587, a group of roughly 116 men, women, and children who populated what came to be called the "Lost Colony" because of their ensuing disappearance. No definite proof has ever been found as to why or how the colony vanished. The modern park includes a **visitor center** and **museum,** an **archaeological dig site** that yielded a number of significant 16th-century artifacts, **Elizabethan gardens,** and a reconstruction of the **earthen fort.** The park's most

Salinas Pueblo Missions was once occupied by the Spanish and Puebloan peoples.

renowned feature is the **Waterside Theater,** where the Roanoke Island Historical Association presents an adaptation of Paul Green's *The Lost Colony* on summer nights.

9 | St. Croix Island
INTERNATIONAL HISTORIC SITE
Maine

With a charter from France's Henry IV in hand, French explorer Pierre Dugua de Mons led a 1604 expedition to the Bay of Fundy and established a small colony on St. Croix Island in the middle of the river of the same name. Among the 74 settlers were cartographer Samuel de Champlain and Mathieu de Costa, a skilled translator and the first known Black man to set foot in Canada. Nearly half the group perished in the winter, and come spring, the French moved their first North American colony to Port Royal, Nova Scotia. An **interpretive trail** (on the Maine coast opposite the island) spins the tale of these intrepid souls, and during the summer rangers give history talks. Note that St. Croix Island itself is closed to the public due to its fragile ecosystem.

10 | Salinas Pueblo Missions
NATIONAL MONUMENT
New Mexico

Attempting to spread Christianity in the New World, Spanish Franciscan friars arrived in the upper Rio Grande region in the late 15th century. By the 1620s they had made their way to the remote Salinas region on the eastern side of the Manzano Mountains, named after the area's salt lakes *(salinas)* and salt trade. **Abó** (1622) was the first of a small string of missions the Franciscans created among the local Tiwa and Tompiro peoples. But owing to persistent famine, drought, and Apache raids, as well as extreme distance from other Spanish colonial hubs, the friars were gone within 75 years. The ruins of their wondrous red-rock churches can be seen in the park's three historical segments—**Quarai, Abó,** and **Gran Quivira**—linked by the **Salt Missions Trail** (N. Mex. 55). The site's visitor center is located in Mountain Air, New Mexico.

BATTLEFIELDS

Battlefields bring you face-to-face with conflicts: both the ones that founded our country and the ones that later threatened to divide the young nation. The following national parks honor the impact of the battles of the Revolution and Civil War, Little Bighorn, and World War II. The history is unmistakable; the heroism, inspirational.

1 Boston
NATIONAL HISTORICAL PARK
Massachusetts

Many of the ideas and actions that sparked the American Revolution came about in Boston and nearby Charlestown, and it's that spirit that ties together the variety of historic sites that compose this park. Landmarks and two visitor centers are spread along the 2.5-mile **Freedom Trail** between Boston Common and Bunker Hill. The **Boston Massacre Site** marks one of the earliest episodes, the 1770 slaughter of five unarmed civilians by British troops. Thousands of colonists gathered at **Old South Meeting House** in 1773 to protest "taxation without representation" and launch the Boston Tea Party. Paul Revere kept his eye on the **Old North Church** to see if British forces were advancing by land or sea in April 1775. Two months later, the war's first large-scale battle was waged at **Bunker**

Hill. The park's **Charlestown Navy Yard** pays homage to Boston's rich maritime heritage, from the Revolution through World War II. Preservation work is under way across the park. Check the website for updated closure information.

2 Guilford Courthouse
NATIONAL MILITARY PARK
North Carolina

Few other battles in U.S. military history proved so fortuitous to the losing side as the 1781 clash at **Guilford Courthouse.** The battle in North Carolina Piedmont country was the culmination of a 16-month British campaign to retake Georgia and the Carolinas. In an effort to stymie that plan, Nathanael Greene amassed 4,400 colonial troops at Guilford on March 15. On the other side were Gen. Charles Cornwallis and 1,900 highly trained Redcoats

and Hessian allies. Ninety minutes later—after an intense British bayonet charge—the fighting had ended and the Americans were in tactical retreat. But the British suffered 28 percent casualties. Severely weakened, Cornwallis adjourned to Yorktown to lick his wounds, a strategic blunder that would lead to the end of the war just seven months later. Films, exhibits, and original artifacts at the visitor center bring the bloody clash to life; the battlefield itself is best explored as a cell phone–guided driving tour. A battle anniversary wreath-laying ceremony happens here each March.

3 Colonial
NATIONAL HISTORICAL PARK
Virginia

Yorktown was the last major land battle of the Revolutionary War. In the summer of 1781, British commander Charles Cornwallis had encamped along the York River in order to resupply and plan an attack against rebel forces in Virginia. After a French fleet routed the Royal Navy in the Chesapeake Bay, Cornwallis found himself suddenly trapped between the river and a large French-American army under the command of George Washington. A three-week siege convinced Cornwallis and his 8,300 troops that the situation was hopeless, and the British surrendered on October 19. It was another year before the Treaty of Paris officially ended the war, but Yorktown was the final British hurrah. The battleground serves as the eastern anchor of the eclectic park. Many of the original redoubts and batteries remain, explored along the seven-mile **Battlefield Tour Road** and nine-mile **Encampment Tour Road.** The visitor center offers a short orientation film on the siege and audio tours of both routes, as well as information on historic house tours and interpretive programs.

4 | **Minute Man**
NATIONAL HISTORICAL PARK
Massachusetts

Nobody knows for sure who fired the "shot heard round the world" on Lexington Common, but that first bullet on April 19, 1775, sparked a chain of bloody events that made war between Britain and

▶ *The Gettysburg Cyclorama that hangs in Gettysburg National Military Park is 377 feet long, 42 feet high, and weighs 12.5 tons.*

the 13 Colonies inevitable. Minute Man preserves several key battlegrounds, as well as the heritage of the Patriot militia that defeated elite British troops. The Redcoats had been dispatched to the area to seize rebel arms and supplies. After a skirmish in Lexington, they marched to Concord to complete their mission, only to be met by colonial minutemen—militiamen trained to fight on a moment's notice—at North Bridge. Routed, the British began a long and costly retreat to Boston. The five-mile **Battle Road Trail** follows their route between Concord and Lexington, winding past historic houses, skirmish points, and the spot where Paul Revere was captured.

5 | **Perry's Victory & International Peace Memorial**
NATIONAL MEMORIAL
Ohio

"We have met the enemy and they are ours," wrote Commodore Oliver Hazard Perry after his celebrated naval victory during the War of

1812. The enemy in this case was a British fleet trying to slip into U.S. waters from Canada. And the place was the southern shore of Lake Erie, an archipelago off the Ohio coast that now serves as the setting for a park that commemorates both war and peace. The **memorial column** rises 352 feet above South Bass Island and overlooks the spot where the furious maritime battle raged on September 10, 1813. The Americans prevailed, capturing all of the enemy warships and ending the threat of British invasion via the Great Lakes. You can hop an elevator to an open-air observation deck near the top of the monument. In summer, rangers give interpretive talks on the battle, the War of 1812, and other topics. Living-history demonstrations unfold on summer weekends, with rangers and volunteers clad in period military uniforms and civilian garb.

6 | **Gettysburg**
NATIONAL MILITARY PARK
Pennsylvania

Most historians consider the series of clashes that took place in southern Pennsylvania on July 1–3, 1863, to have been the pivotal events of the Civil War. Coming from heartening victories in Virginia, Confederate Gen. Robert E. Lee hoped that fresh triumphs in Union territory might demoralize Northern politicians and pressure them to grant independence to the South to avoid further bloodshed of a continuing war. Instead, Lee was forced to retreat after three days of fighting that resulted in 51,000 total casualties for both sides, making **Gettysburg** the bloodiest battle. Never again would the Confederacy reach so far into the North or come so close to winning the war. At its expansive visitor center, Gettysburg National Military Park provides an excellent and evocative look at this defining moment in American history. Purchase tickets for the museum, the film entitled "A New Birth of Freedom," and the historic **Gettysburg Cyclorama** (a 360-degree painting depicting Pickett's Charge, the climactic event of the battle)—all are must-sees. Near the visitor center is the **Soldiers' National Cemetery,** where President Abraham Lincoln delivered the famed Gettysburg Address on November 19, 1863.

You can take a self-guided tour of the battlefield (using park maps), hire a Licensed Battlefield Guide, or choose among the many daily ranger programs.

7 | Manassas
NATIONAL BATTLEFIELD PARK
Virginia

"Manassas: End of Innocence" is the name of a film shown at the visitor center in this Northern Virginia park,

Many of the major military parks, like Gettysburg, stage an annual reenactment of their key battle, bringing history to life.

Appomattox Court House National Historical Park

Confederate Gen. Robert E. Lee surrendered his troops at Appomattox Court House on April 9, 1865.

In early April 1865, Confederate Gen. Robert E. Lee found his forces exhausted, surrounded, and outnumbered, and reluctantly concluded that surrender was his only option. He sent a message to Union Lt. Gen. Ulysses S. Grant, and the two men met in the parlor of a private home in the town of Appomattox in south-central Virginia. Their meeting essentially ended the long, bloody Civil War. Today the park's centerpiece is a reconstruction of the home of Wilmer and Virginia McLean, where the generals met and set in motion the reunion of a long-divided nation.

The parlor is furnished with both original and reproduction items. Original structures within the park include houses, cabins, offices, stores, and the 1819 Clover Hill Tavern. The visitor center is in a reconstruction of the county courthouse, and exhibits include the pencil used by Lee to make corrections to the terms of surrender. A variety of hiking trails route to several additional sites associated with the meeting.

and with good reason: Both Union and Confederate forces entered the first major land battle of the Civil War on July 21, 1861, imagining a quick and easy victory. Northern troops thought they would quickly vanquish the ragtag Southerners, while the "Rebs" believed that one skirmish would cause the North to grant them independence. The Union was soundly defeated and forced to flee to Washington, D.C.; the Confederates learned that the North had no intention of ending the fight quickly. Both sides realized that the war would be long, difficult, and bloody. A larger battle at the same location in late August 1862 also ended in a victory for the South, and marked the high-water mark of Southern military success. A **hiking trail** near the visitor center focuses on the first battle, while sites of the second battle can be seen on a **driving tour** of the area. A popular feature is an equestrian statue of **Gen. Thomas "Stonewall" Jackson,** who got his nickname when another Confederate general supposedly pointed out "Jackson standing like a stone wall" in the face of Union attack.

8 | Antietam
NATIONAL BATTLEFIELD
Maryland

Civil War buffs revere **Antietam** for the preservation of its battlefield, allowing you to accurately visualize events of September 17, 1862. On that date, the Army of the Potomac, led by Maj. Gen. George B. McClellan, held off a push into western Maryland by Confederate forces led by Gen. Robert E. Lee, who had to retreat after a battle that had momentous consequences: President Abraham Lincoln had been waiting for a Union victory before declaring freedom for enslaved people in the South. Five days after Antietam, he issued the preliminary Emancipation Proclamation. From that point forward, the Civil War's aim was both to preserve the Union and end slavery. Among park attractions is a self-guided, 8.5-mile **auto tour** that takes visitors to 11 significant stops within the battlefield. The **Pry House Field Hospital Museum** (limited hours), located in the house that served as McClellan's headquarters, features a re-creation of a wartime operating theater. And plan to spend time at the visitor center, which has been extensively restored to feature interpretive exhibits.

9 | Little Bighorn Battlefield
NATIONAL MONUMENT
Montana

Waged June 25–26, 1876, the **Battle of the Little Bighorn** was the culmination of several hundred years of westward expansion and Native American retreat. The battles were waged by well-known personalities: Sitting Bull, Crazy Horse, and Lt. Col. George Armstrong Custer. Nobody knows for certain why Custer ordered the ill-fated attack rather than wait for reinforcements. Perhaps he underestimated the number of Lakota, Cheyenne, and Arapaho encamped in the valley of Montana's Little Bighorn River or perhaps it was merely hubris on the part of a military leader. Regardless, Custer and more than 260 of his men perished. The battlefield can be seen via the five-mile self-guided **auto tour,** or on board a **mini-coach guided ride** narrated by members of the native Crow tribe. Places where soldiers in the Seventh Cavalry fell are marked by white headstones erected in 1890. They are most abundant atop **Last Stand Hill,** where Custer's command was surrounded and obliterated. Red granite pillars mark the known spots where Cheyenne and Lakota warriors fell in battle.

10 | War in the Pacific
NATIONAL HISTORICAL PARK
Guam

One of the last stepping stones across the Pacific to the Japanese mainland, the island of Guam was the scene of fierce fighting between Allied troops and the Japanese Imperial Army in the summer of 1944. The park is unique in that it honors not just the Americans who fought and died on Guam, but also Allied troops from eight other nations, the Japanese troops who fell here, and the Chamorro (Guam natives) who were killed or wounded during the conflict. The T. Stell Newman Visitor Center has movies, exhibits, and artifacts relating to the **War in the Pacific,** as well as maps and information on visiting the hundred historic buildings, bunkers, caves, and memorials that make up the park. The **Agat Beach** and **Asan Beach** units, where Allied forces came ashore in 1944, are easy to reach via Maritime Drive (Route 1). Some of the more secluded units—like **Mount Alifan,** site of the Japanese command post—are much more difficult to reach owing to steep topography and thick vegetation. The park's free audio tour allows visitors to access photos and narration via an app.

The Marine Landing Monument at War in the Pacific National Historical Park commemorates the U.S. marines who stormed the beach during the Battle of Guam.

10 BEST PARKS FOR

CAPITAL ATTRACTIONS

Washington, D.C., has more than lived up to the vision of George Washington, architect Pierre-Charles L'Enfant, and others who took an active part in its planning and creation. The Park Service manages a variety of the capital's parks, monuments, and memorials, many of which have become national symbols and global icons.

1 National Mall
NATIONAL MALL & MEMORIAL PARKS
District of Columbia

A stroke of urban planning genius, the National Mall runs nearly two miles between the Capitol building and the Lincoln Memorial. The broad, grassy passage is often described as America's front yard. Inspired by the gardens of Versailles, French-born architect and engineer Pierre-Charles L'Enfant included the grand avenue in his 1791 blueprint for the new American capital. From political demonstrations and rock concerts to presidential inaugurations and outdoor art exhibits, the Mall has hosted tens of thousands of events, including some of the most memorable in U.S. history. Dozens of landmarks are set within its bounds, from presidential shrines like the **Washington Monument** to poignant military tributes like the **World War II Memorial,**
Korean War Veterans Memorial, and **Vietnam Veterans Memorial.** The Mall also borders great storehouses of culture and knowledge like the Smithsonian Institution museums and the National Gallery of Art. With the **Tidal Basin** and **East Potomac Park** included within its official boundaries, the park is also recognized for its natural beauty.

2 Theodore Roosevelt Island
NATIONAL MEMORIAL
District of Columbia

One of the guiding lights of conservation, Theodore Roosevelt was responsible for the creation of myriad national parks and monuments, the U.S. Forest Service, and 50 wildlife refuges. The 88-acre Theodore Roosevelt Island in the middle of the Potomac River commemorates the 26th president's legacy. **Hiking trails** and **boardwalks** wind through lush forest and marshland, where deer, beaver, and fox roam. The **memorial plaza** features a bronze statue of the ex–Rough Rider and granite slabs inscribed with some of his memorable quotes about nature and conservation. Native Americans used the island as a seasonal fishing village and later patriot-politician George Mason transformed the isle into a private estate. Other than a brief period during the Civil War when Union troops were stationed there, the island has been uninhabited since the 1830s, when the Mason family vacated.

3 Thomas Jefferson Memorial
NATIONAL MEMORIAL
District of Columbia

There likely isn't a more sublime building in all of Washington than this memorial to the third U.S. president, its Ionic columns and dome reflected

Exquisitely sited and designed, the Thomas Jefferson Memorial commemorates America's third president.

in the Tidal Basin or silhouetted by sunset over the Potomac. Thomas Jefferson was the architect of so many American ideals, in particular those enshrined by the Declaration of Independence. Designed by John Russell Pope and inspired by the Pantheon in Rome, the white marble memorial was dedicated in 1943.

The interior is graced by a 19-foot **bronze statue of Jefferson,** the wall engraved with passages from his writings on liberty, equality, and freedom of religion. Rangers lead walking and bike tours, but in many respects this is a memorial best explored alone, reflecting on the inspiration and influence of the man.

4 | The White House & President's Park
PARK
District of Columbia

Perched on the northern edge of the Mall, this unit of the park system includes the **Chief Executive's mansion** and **surrounding green space and monuments.** Tours of this famous residence need to be booked well in advance through your local member of Congress. The visitor center presents a 14-minute film on the history of the White House, as well as exhibits on the first families and other presidential topics. The best place to snap photos is the pedestrian-only stretch of Pennsylvania Avenue and adjoining Lafayette Park on the north side, a public space where there is almost always someone or some group practicing their First Amendment right to free speech. The Ellipse Park on the south side is where the National Christmas Tree glimmers each year.

5 | Chesapeake & Ohio Canal
NATIONAL HISTORICAL PARK
District of Columbia, Maryland & West Virginia

This **184.5-mile waterway** facilitated transportation along the Potomac River between the District of Columbia and Cumberland, Maryland. Gouged out between 1828 and 1850, the canal opened the upper Potomac basin to waterborne commerce with the rest of the East Coast. The first 10 miles of the Chesapeake & Ohio hug the eastern bank of the Potomac. Seven visitor centers stretch the length of the canal's many miles, offering displays, maps, and information. The Georgetown Visitor Center is a

▶ *Zero Milestone, a four-foot-high granite marker in President's Park south of the White House, designates the north and south meridians of the District of Columbia. It was originally intended to be the point from which all road distances were measured.*

starting point for rides on reproduction canalboats that journey through the bygone warehouse district (seasonal). Rangers also lead interpretive walks on a variety of local topics. A nice getaway from the busier nearby towns, the meandering C&O towpath between Georgetown and the Maryland border is especially good for hiking and biking.

6 | Lincoln Memorial
NATIONAL MEMORIAL
District of Columbia

Prepare for a surge of emotion the first time you ascend the stairs at this memorial and see Daniel Chester French's **immortal statue of Abraham Lincoln** gazing down. The Lincoln Memorial is a masterful blend of architecture and ideas.

Henry Bacon's timeless design, completed in 1922, features 36 Doric columns representing the 36 states of the union at the time of Lincoln's assassination. The interior walls bear passages from the 16th president's Gettysburg Address and 1865 inaugural speech, as well as allegorical murals by Jules Guerin that reflect Lincoln's lofty ideals and momentous accomplishments. The structure rests on an emotive site at the western end of the Mall, its facade glimmering in the Reflecting Pool. The memorial has long been a place of pilgrimage for those seeking freedom, justice, and truth, most notably Dr. Martin Luther King, Jr., who delivered his "I Have a Dream" speech here in 1963.

7 | George Washington
MEMORIAL PARKWAY
District of Columbia, Maryland & Virginia

This winding roadway connects more than 20 scenic, nature, recreation, and historic areas on both sides of the Potomac in the Washington, D.C., metropolitan area. The parkway was conceived in the 1930s as a tribute to the nation's Revolutionary hero and first president, a leafy route through areas that Washington often traversed by horse or carriage. It has since evolved into so much more, an eclectic urban green space where you can participate in a dozen activities on any given day, from bird-watching, biking, and boating to outdoor concerts and ranger-guided history walks. Among the park's varied segments are the **U.S. Marine Corps Iwo Jima Memorial, Clara Barton National Historic Site, Theodore Roosevelt Island, LBJ**

Memorial Grove on the Potomac, Dyke Marsh Wildlife Preserve, and 800-acre **Great Falls Park.** The parkway also provides access to adjacent landmarks like Arlington National Cemetery and the C&O Canal. Regular interpretive programs are available, such as popular bird walks at Dyke Marsh.

8 Rock Creek
PARK
District of Columbia

From urban playground to commuter corridor, Rock Creek Park is many things to many people. Despite its modern vibe, this park is actually **one of the oldest federally operated parks,** set aside in 1890 as a

Presidential Park Honorees

John Adams and John Quincy Adams: A Boston-area park preserves the childhood home of the second and sixth presidents.

Martin Van Buren: A national historic site protects the eighth president's Hudson Valley estate.

Abraham Lincoln: Three park units document the migration of the Lincoln family across the Midwest.

Ulysses S. Grant: The general turned president is remembered at Grant's Tomb in New York City and the family farm near St. Louis.

James A. Garfield: The Ohio home of the 20th president, not far from Cleveland, is a park unit.

Theodore Roosevelt: Several park units recall the nature-loving 26th president.

William Howard Taft: The 27th president spent his childhood at a house in Cincinnati, Ohio.

Herbert Hoover: An Iowa site explores the rural childhood and Quaker upbringing of the 31st president.

Franklin D. Roosevelt: The 32nd president and his wife, Eleanor, are buried on the grounds of the Hudson Valley home where Roosevelt grew up.

Harry S. Truman: The Missouri home and farm where the "People's President" lived for more than 50 years is now a park.

Dwight D. Eisenhower: The 34th president spent the last decades of his life at a farm near Gettysburg, Pennsylvania.

John F. Kennedy: The Brookline, Massachusetts, house where Kennedy was born is a historic site.

Lyndon B. Johnson: A park in the Texas Hill Country preserves the home and ranch of the 36th president.

Jimmy Carter: The 39th president's hometown Plains High School, local train depot, and boyhood farm make up this site in southwest Georgia.

"pleasure ground" for the American people. More than twice the size of New York's Central Park, Rock Creek Park encompasses more than 1,700 acres and stretches roughly 10 miles from north to south, fat at the top and increasingly skinny at the bottom as the creek meanders toward the Potomac. The park has many diversions: more than 25 miles of hiking trails and 13 miles of equestrian paths; a boating center where kayaks, canoes, and rowing shells are available for rent; a public golf course; and 25 tennis courts. A seven-mile bike path that starts at **Beach Drive** loops along winding, closed roads and multiuse trails. The **nature center** offers all sorts of exhibits on park flora and fauna, as well as a planetarium for ranger and night-sky programs.

9 | National Capital Parks–East
NATIONAL PARKS, PARKWAYS & STATUARY
District of Columbia

A tribute to historical preservation and environmental rebirth, this collection of **14 parks** gives visitors a compelling reason to explore a part of the nation's capital that until recently was far off the tourist trail. The park's 8,000 acres encompass museums and nature areas, historic golf courses and farmland,

The iconic Washington Monument stands at the center of the National Mall in Washington, D.C.

hiking trails and even campgrounds all within a dozen miles of Capitol Hill. National Capital Parks–East is rich in African American history: the **Mary McLeod Bethune Council House,** the **home of author Carter G. Woodson,** and **Cedar Hill,** home of famed abolitionist Frederick Douglass. Among its outstanding nature areas are **Kenilworth Park and Aquatic Gardens,** famous for both its water lilies and wildlife that includes beaver, otter, fox, and more than a hundred bird species. Animals of a different ilk are the focus at **Oxon Hill Farm,** where activities include milking

cows, feeding chickens, and scenic wagon rides. Recreational offerings include hiking the eight miles of wooded trails at **Greenbelt Park** in suburban Maryland.

10 | Washington Monument
NATIONAL MEMORIAL
District of Columbia

The Washington Monument soars above the District of Columbia as a tribute to the first president. In 1836, architect Robert Mills won a competition to design a lasting memorial to the former commander in chief. But lack of funding and disagreement over the architectural merit of Mills' design—an **Egyptian-style obelisk** surrounded by a Greek-style colonnade with a rooftop statue of Washington driving a chariot—forced the government to scale back his grand vision. Political squabbling and the Civil War also took their toll, and ultimately it was nearly 50 years before the monument was finally completed in 1884. The bottom third of the 555-foot tower is finished with a slightly different-colored white marble than the upper two-thirds, one consequence of the delay-plagued construction. An elevator whisks you to the observation deck and its consummate views of the nation's capital, plus Virginia and Maryland.

FORTS

Simple wooden stockades or defensive earthworks were the first permanent structures built by colonists and settlers for the purpose of protecting against wild animals, attacks, and raids. Forts were also vital to the American westward push, erected as fortified trading posts and U.S. Army outposts, and stand now as fascinating glimpses into the past.

1 | Castillo de San Marcos
NATIONAL MONUMENT
Florida

In contrast to other U.S. bastions, San Marcos is a typical Iberian design, a hulking **masonry star fort** similar to those found throughout the Spanish colonies, but rare on mainland North America. Construction began in 1672 and continued for more than 20 years as the Spanish sought to protect St. Augustine—their Florida capital—from pirates and rival European powers. Lacking quality stone, in a stroke of genius the engineers used coquina—conglomerate shell similar to limestone—as the primary building material. When the British laid siege in 1702, the coquina walls easily absorbed the shock of the cannonballs and Spain won the day. Over the next 200 years, the citadel had several masters, including the British (1763–84) and

the Confederacy (1861–62). Its most contentious era was the 1870s, when it housed Native American prisoners of war, most notably Cheyenne warrior, artist, and theologian David Pendleton Oakerhater.

2 | Gulf Islands
NATIONAL SEASHORE
Florida

There's no doubt that beautiful beaches draw visitors to Gulf Islands National Seashore, but several mainland harbor forts in this park showcase a more bold and combative time. Worth exploring, **Fort Barrancas** and **Fort Pickens** each have visitor centers, ranger-guided tours, and plenty of wildlife. Fort Pickens, begun in 1829 and used through the end of World War II, is considered both a monument to evolving military technology and a memorial to past injustices, as both enslaved

workers and prisoners of war are credited for building and repairing the fort. Fort Barrancas, located within Pensacola Naval Air Station, was completed in 1844 on the bluffs of Pensacola Bay. Six million bricks constructed this fort atop the ruins of previous British, French, and Spanish forts, making a mighty structure that received extra support from the Advanced Redoubt fort and faced combat during the Civil War. Both Fort Barrancas and Pickens were added in recent years to the National Underground Railroad Network to Freedom.

3 | Fort Laramie
NATIONAL HISTORIC SITE
Wyoming

Fort Laramie played a pivotal role in Manifest Destiny as a **trading post, military stronghold,** and **way station** for thousands of people heading west on the California,

Fort Jefferson, one of the largest 19th-century coastal forts built for defensive purposes, never actually saw battle.

Oregon, and Mormon Trails. Fur trader William Sublette founded the fort in 1834 on a strategic site near the confluence of the Laramie and North Platte Rivers, in today's southeast Wyoming. By the late 1840s the federal government had purchased the fort and set about transforming Laramie into a forward base for confronting and pacifying the Plains tribes. In 1890, with the threat over, Fort Laramie, now looking more like a small town than a military post, was abandoned.

Eleven of the original structures have been restored and decorated with period furnishings and artifacts. The visitor center is housed in an 1884 commissary storehouse. Overlooking the grassy parade ground are the comfy Captain's Quarters, the New Guardhouse (with a collection of artillery pieces), and two-story "Old Bedlam"—named after the famous English asylum because of the rambunctious behavior of the bachelor officers who bunked there.

4 Dry Tortugas
NATIONAL PARK
Florida

Fort Jefferson is the focal point of secluded Dry Tortugas, situated in the Gulf of Mexico about 70 miles from Key West. Set on Garden Key, one of seven keys that make up the park, the fort was planned as a post for defending Florida and the Gulf Coast. Construction started in 1846 but was never fully completed; more than 16 million bricks went into the hexagonal ramparts that surround the 11-acre citadel. Fort Jefferson remained in Union hands during the Civil War and never fired a shot in anger. During most of its active service, the fort served as a military prison or quarantine station. Activities at the fort, which can be reached via boat or seaplane from Key West, are limited to the visitor center and a self-guided tour. Beyond the fort, Dry Tortugas offers a wide array of outdoor adventures, including scuba diving and snorkeling, camping, fishing, and birding.

5 | Fort McHenry
NATIONAL MONUMENT & HISTORIC SHRINE
Maryland

Constructed in 1798 to defend Baltimore from seaward attack, **star-shaped Fort McHenry** was no different than dozens of other forts along the Atlantic seaboard—until September 13, 1814, when British Royal Navy warships appeared at the harbor entrance. They pounded McHenry for 25 hours before deciding the fort would not fall. Watching from a nearby truce ship was Washington lawyer Francis Scott Key, who shortly after penned a poem called "Defence of Fort M'Henry" as an ode to the brave American defenders. Set to the tune of a popular British drinking song, the ditty rose in popularity through the 19th century but didn't become the official U.S. national anthem until

▶ *During the Civil War, several civilian prisoners were brought to Fort Jefferson, including Samuel Mudd, the physician who set the leg of John Wilkes Booth and was later convicted as part of President Abraham Lincoln's assassination.*

1931. Park activities range from flag ceremonies and interpretive programs to morning bird walks in the nearby wetlands.

6 | Fort Stanwix
NATIONAL MONUMENT
New York

Strategically located about halfway between New York City and Canada, Fort Stanwix played pivotal roles in both the French and Indian War and the American Revolution. Named after British Gen. John Stanwix, who oversaw its construction in 1758, the **wooden fort** was also guardian of the Oneida Carry, a vital portage link on the trade route between the Hudson Valley and the eastern Great Lakes region. After the French and Indian War, the fort stood abandoned until 1776, when colonial troops garrisoned it in order to block a major

British invasion route from Canada and aid the Patriots' Oneida allies. Redcoats besieged Stanwix in the summer of 1777, but the nearby Battle of Oriskany gave the defenders the diversion they needed to counterattack and lift the siege. The fort later featured in relations between settlers and the upstate Indian nations but then fell into disrepair. It was painstakingly rebuilt in the 1970s on its original site in the middle of what is now modern Rome, New York. In addition to three short trails—one of them along a portion of the Oneida Carry—the national monument offers guided tours, historic weapons demonstrations, and living-history programs, especially during the summer months.

7 | Fort Sumter & Fort Moultrie
NATIONAL HISTORICAL PARK
South Carolina

Charleston's historical prominence is as a port city, and this park unit is named for two of its harbor forts. Construction on the squat **Fort Sumter** started after the War of 1812 and was still unfinished when, in the first shots (and bloodshed) of the Civil War, Confederate guns opened fire on the fort on April 12, 1861. A 34-hour Rebel barrage ended in surrender of the Union troops inside. The only way to reach the island is by private boat or concessionaire ferry. The Visitor Education Center renders an

Sunrise burns off the fog surrounding the Golden Gate Bridge at Fort Point.

excellent introduction to the political, social, and economic factors that sparked the Civil War. Once on the island, you can join a ranger-led tour of the fort that includes the parade ground, the fort museum, and the rooftop of Battery Isaac Huger with its display of historic flags. The park's other namesake fortress, **Fort Moultrie,** was the first on Sullivan's Island. Initially built of palmetto logs and sand during the colonial period, it was attacked by British ships in 1776, but the city was saved from occupation. The structure was reconstructed multiple times over the next hundred years, especially following heavy damage from the Civil War. Today, Moultrie is restored to showcase a brick iteration circa 1809.

8 Golden Gate
NATIONAL RECREATION AREA
California

It has been said that California's history is a study in overlapping cultures, including Indigenous, Spanish, Mexican, U.S. military, and more, with the culture of defense impacting all of them. Much of Golden Gate National Recreation Area's 82,000 acres were once used by the U.S. Army, and today, **37 natural, recreational,** and **historic spots** are preserved in the expansive San Francisco Bay–area park unit. Among them are almost every type of coastal protection constructed in America from the mid-1800s through the mid-1900s, so a visit to the sites is like a tour through time,

technology, and military methodology. Significant spots in the park include **Forts Baker, Mason, Point, Funston, Barry,** and **Cronkhite,** as well as **Battery Chamberlin, Crissy Airfield, Battery Wallace,** and the **Presidio.** Today the forts are studies in both preservation and reuse. At the Presidio, an active military installation and barracks turned military intelligence school now features a historic hotel, outdoor activities, and exhibits. Fort Mason, once the San Francisco Port of Embarkation, now hosts art facilities, galleries, and shops, as well as the park visitor center.

9 Fort Vancouver
NATIONAL HISTORIC SITE
Oregon & Washington

Fort Vancouver was born as a commercial operation, the bustling headquarters and supply depot of Britain's Hudson Bay Company in western North America. Founded in 1824 as a fur-trading post in the lower Columbia River Valley, the fort grew into the hub of a vast trade network and the most populous "town" on the Pacific coast between Alaska and Mexico, home to around 600 people from more than 35 ethnic and tribal groups. Hudson Bay Company continued business at the fort long after Oregon Country became part of the United States in 1846. The U.S. Army moved in when the British left in 1860 and stayed until 1948, when the fort became part of the park system. The park has two units: the **Fort Vancouver site** in Vancouver, Washington, and the **McLoughlin House site,** which preserves two pioneer-era homes in Oregon City, Oregon. Much of the fort has been rebuilt. Ranger, audio, and self-guided tours help visitors experience the park. The fort stages several special events each year as well, including an annual Memorial Day celebration and the Brigade Encampment in June.

10 | Fort Smith
NATIONAL HISTORIC SITE
Arkansas & Oklahoma

This southwest Arkansas park preserves the remains of two forts. The first, **Fort Smith** (1817–24), was built by the U.S. Army at the highest navigable point on the Arkansas River but was soon abandoned; the stone foundations are visible on a bluff at Belle Point. A second, larger fort was started in 1838 as a U.S. Army post on the edge of the recently created Indian Territory. It later served as a training ground for the Mexican War and was occupied by both sides during the Civil War. By the early 1870s, Fort Smith had been transformed into a **U.S. federal court and jail,** one that prisoners dubbed "Hell on the Border." During his 21 years on the bench, infamous Judge Isaac C. Parker sentenced 160 men to death. The visitor center occupies the former barracks-courthouse-jail building; an eerie reconstruction of the gallows stands behind. A riverside path leads to the Trail of Tears Overlook, which commemorates the forced removal of Cherokee, Choctaw, Chickasaw, Creek, and Seminole tribes to Indian Territory in the 1830s.

Forts on National Trails

Given their function as military bastions and trading posts, many of the fortifications on our list are found along the Park Service's National Trails System.

Fort Smith is near the western end of the Trail of Tears National Historic Trail, which follows the two routes that thousands of Native Americans from southeast tribes walked from their homelands to Indian Territory, today's Oklahoma.

Fort Vancouver lies on the Lewis and Clark National Historic Trail—the path that the explorers took from the Mississippi River to the Pacific coast—and is near the western end of the Oregon National Historic Trail that so many 19th-century settlers trod.

Fort Stanwix in upstate New York lies astride the North Country National Scenic Trail, a meandering route that runs all the way between the Hudson River and North Dakota.

Fort McHenry offers a historic port-of-call on the watery Captain John Smith Chesapeake National Historic Trail through the tidelands of Virginia, Maryland, and Delaware.

Fort Laramie overlooks four national historic trails that follow the same route through much of Nebraska and Wyoming: Oregon, California, Mormon Pioneer, and Pony Express.

All Aboard

The romance of the rails is one of our country's lasting legacies. And while technology and time have moved us toward other modes of transportation, national parks protect the history and heritage of trains.

DONE. This one word, telegraphed across the country on May 10, 1869, marked a turning point for industry and travel when it announced the completion of the first transcontinental railroad line at Promontory Summit in northern Utah. It was here that members of both the Union Pacific and Central Pacific Railroads hammered a golden spike, symbolically connecting the Atlantic to the Pacific.

Today that buzz of excitement remains at Golden Spike National Historical Park, where reenactments mark the momentous event and *Jupiter* and No. 119, Victorian-era reproductions of the steam locomotives, make demonstration runs. Open year-round, the 1.5-mile Big Fill Loop Trail walking path traces the train grade. You can spot drill marks where workers blasted rock and take in the Big Fill and Big Trestle. Two auto tours—the East is two miles; the West is seven miles—also traverse the cuts, fills, and culverts of the legendary track.

Thirty-five years before the golden spike was hammered, another historic event occurred. It took 10 inclined planes and endless ropes to pull passengers and freight up and down the steeply graded Allegheny Mountains of Pennsylvania in the mid-19th century. Called the Allegheny Portage Railroad, it linked Philadelphia and Pittsburgh, ran from 1834 to 1854, and finally solved a major obstruction to the nation's western growth. Allegheny Portage Railroad National Historic Site honors the breakthroughs of the era with exhibits, remains of the system, a former traveler's tavern, and a replica Engine House No. 6. A worthwhile 30-minute drive from the site's Summit Level Visitor Center is the Staple Bend Tunnel, the first railroad tunnel in America, now preserved for hiking and biking.

At Pennsylvania's Steamtown National Historic Site, the era of the steam engine, plus the people who built, repaired, and rode its rails, is honored with living-history programs, exhibits, and attractions like an active roundhouse. The park, located in Scranton, encompasses 40 acres of the Delaware, Lackawanna, and Western Railroads. Three-mile rides travel the rail yards, go over the Lackawanna River, and stop at the University of Scranton before returning to the historic site. For a different perspective, guests can reserve a ticket for a brief caboose ride within the park. Longer excursions travel beyond Steamtown, filling days with adventures to see craft shows, fall foliage, and the Poconos.

A RIDE FORWARD

Ohio's Cuyahoga Valley National Park is often viewed aboard the Cuyahoga Valley Scenic Railroad (CVSR). Popular as a leisure ride, the two-hour sightseeing excursion takes guests along the Cuyahoga River with multiple seating options and a café car. Active bikers, hikers, and paddlers can use the CVSR as a one-day shuttle following a trip down the river or the Ohio & Erie Canalway Towpath Trail. The Valley Railway dates back to 1880, when it was used for coal and passenger travel. In 1972 the line became a scenic excursion route, and since then other itineraries have been added, including the festive North Pole Adventure, the wine-themed Grape Escape, and patriotic Veterans Day rides.

Under one of the darkest skies in the lower 48 states,

A train ride aboard the Alaska Railroad's flagship locomotive is a unique way to take in the splendor of Denali.

the Nevada Northern Railway sets out on summer evenings for a guided exploration of the heavens. Its Great Basin Star Train doesn't travel into or out of its eponymous national park, but it is hosted by the park's cleverly named Dark Rangers. Pulled by a vintage diesel locomotive, the Star Train departs around sunset so passengers can see the sun sink over Steptoe Valley. Away from the lights of town, the conversant rangers lead engaging star-gazing sessions.

Park Service guides also hop on Amtrak's trains through its Trails & Rails program, which presents passengers a "traveling show-and-tell." The *Coast Starlight* travels from Seattle to Portland along the stunning Cascade Mountains while volunteer guides from Klondike Gold Rush National Historical Park share stories of Indigenous tribes and the impact of groups who began arriving in the late 1700s. From San Jose to Paso Robles, California, the tales from Juan Bautista de Anza National Historic Trail come alive as park staff recount how the train travels the route once taken by Anza's expedition.

While not built specifically to service parks, lines like the Grand Canyon Railway remind visitors that railroads played a pivotal role in history. Begun in 1901, this train predates Grand Canyon National Park's establishment and was actually used to bring water and supplies during the construction of its facilities. Today, the commercial railway embraces Western-style fun with musicians and costumed cowboys aboard its daily route from Williams, Arizona. It's not hard to imagine its impact along the way: engaging the nostalgia of adults and, just perhaps, inspiring fresh wonder in youths.

MONUMENTS & MEMORIALS

Ranging from global icons to faraway battlefields and paying homage to both natural disasters and human tragedies, national parks in the form of monuments and memorials stir the soul in their remembrance of American heroes or the ideals on which this nation was founded.

1 Johnstown Flood
NATIONAL MEMORIAL
Pennsylvania

One of the only units of the park system that revolves around a natural disaster, this memorial in southwestern Pennsylvania recalls an 1889 deluge that took the lives of more than 2,200 people and revolutionized the way that Americans respond to major calamities. On the afternoon of May 31, rain-filled Lake Conemaugh burst South Fork Dam, sending an estimated 20 million gallons of water downstream in a wave that reached 40 feet in height. About an hour later, the deluge smashed into **Johnstown,** a burgeoning steel town and railroad hub, washing much of the town away. Clara Barton and her newly formed American Red Cross spearheaded the relief effort. A visitor center shows a riveting film about the disaster. Paths lead through the old lake bed and remains of the dam. In summer, rangers lead tours of flood-related sights. Every May 31, the victims are remembered in a ceremony during which 2,209 luminaries are displayed on the grounds.

2 Pearl Harbor
NATIONAL MEMORIAL
Hawaii

As President Franklin D. Roosevelt declared in 1941, December 7 lives in infamy not only as the date of Japan's shocking attack on the Pacific Fleet base in Pearl Harbor but also as the day that set the United States on a path to enter World War II. In just a few brief hours more than 20 battleships and vessels were sunk or damaged, and more than 2,400 people killed. Visitors can see artifacts and photographs of the attack, hear recorded oral histories, engage with interpretive exhibits, and watch a documentary film. And while the memorial recognizes the impact of this event, it also emphasizes the magnitude of peace and reconciliation. Reservations are required for the popular Park Service–facilitated **USS Arizona Memorial Program,** which includes a round-trip Navy-operated shuttle boat ride to the *Arizona.* Neighboring independent historic sites include the Pacific Fleet Submarine Museum, Battleship Missouri Memorial, and Pearl Harbor Aviation Museum.

An overhead view reveals the remnants of the USS Arizona below Hawaii's blue-green waters at Pearl Harbor National Memorial.

3 Mount Rushmore
NATIONAL MEMORIAL
South Dakota

"American history shall march across that skyline," quipped Gutzon Borglum when he first saw the Black Hills in the early 1920s. The Danish-American sculptor spent the next 17 years turning that vision into a larger-than-life reality, the faces of four presidents—Washington, Jefferson, Lincoln, and Roosevelt—carved in granite on a South Dakota mountainside. With Mount Rushmore, Borglum successfully crafted a lasting monument to the American spirit and the country's ideals of freedom and liberty. His artwork has become a symbol of the United States around the globe. The **Avenue of the Flags** (representing the U.S. states and territories) leads from the parking area to the excellent Lincoln Borglum Museum, named after the artist's son, who supervised completion of the memorial after his father died in 1941. The **Presidential Trail** takes you to wooden viewing platforms directly beneath the famous faces, as well as the studio where Borglum rendered plaster scale models of the presidential faces.

4 Flight 93
NATIONAL MEMORIAL
Pennsylvania

"A common field one day. A field of honor forever." So marks this emotive memorial recalling the sacrifice

▶ *Dedicated in 1993, the Vietnam Women's Memorial honors the 265,000 military and civilian women who served around the world during the Vietnam War.*

of those aboard United Flight 93 on September 11, 2001. One of four airplanes hijacked that day by terrorists, Flight 93 was on its way from Newark to San Francisco when four men commandeered the cockpit and diverted the Boeing 757 toward Washington, D.C. Authorities believe they planned to crash the plane into the White House or U.S. Capitol. Learning of the other hijackings by cell phone, passengers fought to regain control of the flight. During this struggle, the aircraft crashed into a field in Somerset County, Pennsylvania, killing everyone. The monument's design centers around a **memorial** at the impact site that features a wall inscribed with the names of the 40 passengers aboard the flight. The 93-foot **Tower of Voices** with 40 wind chimes was built as a visual and audible reminder of their heroism. Hiking trails and ranger programs add to the site's year-round appeal.

5 Gateway Arch
NATIONAL PARK
Missouri

Erected in the 1960s along the west bank of the Mississippi River in St. Louis, the 630-foot-tall **Gateway Arch** is a soaring monument to the energy, curiosity, and ambition that stoked America's westward expansion in the 19th century. Designed by famed Finnish-American architect Eero Saarinen, the stainless steel arch is a giant upside-down version of a catenary curve—the ideal shape of a hanging chain. You can ride a tram to the top for a view across St. Louis and the Mississippi Valley. Underground, the free **Museum at the Gateway Arch** pays tribute to the city's founding and its role in America's frontier mosaic. The **Old Courthouse** forms another section of the park, dedicated in 1828 and the scene of numerous landmark trials, including the first two sessions of the 1847 Dred Scott case that had such a profound impact on slavery.

6 Vietnam Veterans & Vietnam Women's Memorial
NATIONAL MEMORIAL
District of Columbia

While the war itself was no doubt one of the most divisive eras in our nation's history, there's no question that the monument dedicated to the lives lost in Vietnam is incredibly moving. Just off Constitution Avenue near the Lincoln Memorial, the **Vietnam Veterans Memorial** consists of a V-shaped wall of polished black granite inscribed with the names of more than 58,200 U.S. military personnel who were killed or missing after the conflict. Directories guide visitors to the location of names, and rubbings, often done with charcoal or pencil, are popular mementos. Other significant sites at the memorial include the **Three Servicemen Statue,** the **Vietnam Women's Memorial**—the only memorial to military women on the National Mall—as well as a flagpole flying both the United States and MIA-POW flags. Site rangers and volunteers provide regular programing at the memorial, as well as special observances for Memorial Day and Veterans Day.

7 | Statue of Liberty
NATIONAL MONUMENT
New York

World-renowned icon of freedom and democracy, beacon to immigrants all around the globe, **Lady Liberty** is one of our planet's most recognizable landmarks. A gift from the people of France to the United States, the Statue of Liberty Enlightening the World was created in Paris by sculptor Frédéric-Auguste Bartholdi and engineer Gustave Eiffel, stuffed into 214 crates, and shipped across the

Atlantic in a French warship. It's now hard to imagine given her famous green patina, but Lady Liberty was a dark copper brown when reassembled on Bedloe's Island in 1886. She is literally draped in metaphors: The seven points of her crown represent the seven seas and seven continents; the tablet in her left hand is inscribed with the date of American independence; her torch lights the path to freedom. Visitors can ogle the New York skyline and harbor through the lofty "jewel" windows of her crown. The national monument also includes the federal immigration station on Ellis Island, where more than 12 million people entered the United States between 1892 and 1954.

World War II Incarceration Site Memorials

The National Park Service marks several significant World War II-era forcible incarceration sites.

Amache National Historic Site (Colorado): Authorized by President Joe Biden in 2022 and currently managed by the nonprofit Amache Preservation Society, this new park unit features a museum dedicated to the 10,000 people, mostly U.S. citizens, who were incarcerated here from 1942 to 1945.

Tule Lake National Monument (California): The largest of the Japanese American confinement sites, Tule Lake housed nearly 30,000 people in prisonlike environments. Later this site was known as Camp Tulelake, where Japanese Americans and German prisoners of war faced extreme conditions.

Aleutian Islands World War II National Historic Area (Alaska): The Unangan (Aleut) people hunted for seals, whales, sea lions, and more from their traditional bidarkas (shown here). More than 800 Unanagans were imprisoned for two years in southeastern Alaska when their homeland—called a "back door to America"— became a battleground.

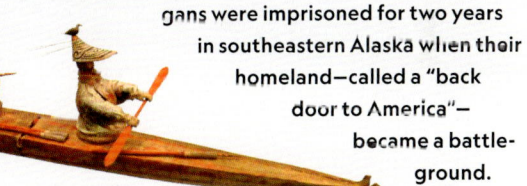

8 | Independence
NATIONAL HISTORICAL PARK
Pennsylvania

Dubbed the most historic square mile in the United States, Independence National Historical Park memorializes symbols of freedom and democracy at sites like **Independence Hall** and the **Liberty Bell.** Start at the Independence Visitor Center at Sixth and Market Streets, where rangers will help you customize a visit to your availability and interests. Philadelphia was the hub of liberation activity in the late 18th century, with such visionaries as John Adams, George Washington, Patrick Henry, Benjamin Franklin, Thomas Jefferson, and John Hancock meeting here to plot the direction of the country's freedom. More than two dozen sites are open to the public, with guided tours available at many. Independence Hall's "Great Essentials" exhibit is a can't-miss, as it features copies of the Declaration of Independence, the U.S. Constitution, and the Articles of Confederation. Liberty Bell Center and **Congress Hall** are steps away. The Park Service's free app has five self-guided audio tours on topics like civil rights and Benjamin Franklin, while an interactive map allows exploration of many public sites at your own pace. Young visitors will enjoy signing the Declaration of Independence, ringing the Liberty Bell, and more in a separate Junior Ranger app.

On view at Independence National Historical Park, the Liberty Bell reads, "Proclaim Liberty Throughout All the Land Unto All the Inhabitants thereof."

9 | Manzanar
NATIONAL HISTORIC SITE
California

In 1942, more than 110,000 Japanese Americans and Japanese immigrants were forcibly moved from their homes to remote areas of the country where incarceration camps were rapidly and fearfully created at the start of World War II. **Manzanar War Relocation Center** was one of the 10 camps, and today this well-preserved park has a self-guided driving tour as well as an interactive visitor center that sets the historic scene before visitors explore chilling

Block 14. Here, a mess hall, women's latrine, and barracks reveal the harsh realities faced by the men, women, and children forced to live at Manzanar. Save time to peruse the Children's Village and the Japanese gardens—of the more than 100 original plots made during the incarceration, a few have been uncovered and stabilized.

10 | Federal Hall
NATIONAL MEMORIAL
New York

The mention of Wall Street these days channels thoughts of the stock exchange, investment banks, and financial institutions. But in the late 1700s, this Manhattan district was the birthplace of American government. **Federal Hall National Memorial** marks the location of George Washington's inauguration, the first meeting of Congress, and the first home of the Supreme Court. Once the location of New York's City Hall and later the original Federal Hall, the current building, the U.S. Custom House, was completed in 1842. Visitors can view exhibits about Washington's presidency, including the Bible he used for his inaugural oath, and see models of City and Federal Halls. Self- and ranger-guided tours are available year-round, and dates for reenactments can be found online.

AFRICAN AMERICAN CIVIL RIGHTS

At these sites, visitors learn of the decades spent fighting for African American equal rights—from the time of enslavement to the present—and meet the heroes of that worthy quest. From locations that launched historic court cases to personal homes of legendary figures, these sites illuminate crucial memories in our nation's fraught history.

1 | Brown v. Board of Education
NATIONAL HISTORICAL PARK
Kansas

In a 1954 case generally known as *Brown* v. *Board of Education,* the U.S. Supreme Court ruled that schools segregated by race violated the constitutional principle of equal protection. It was the beginning of the end of segregated schools in the United States, and one of the most important landmarks in the struggle for African American rights. Brown v. Board of Education National Historical Park is located in Topeka, Kansas, where in 1951 an African American girl named Linda Brown was refused enrollment in an all-white school near her home. The setting is the former **Monroe Elementary School,** once a school for African American students (and attended by Linda Brown). Exhibits include a film entitled "Race and the American Creed" and galleries on the themes of education, justice, and the legacy of the historic legal case. Adjacent to the park, the Landon Nature Trail connects to the Santa Fe Trail and Oregon Trail, both historic national trails.

2 | Charles Young Buffalo Soldiers
NATIONAL MONUMENT
Ohio

One of the most respected military leaders of his time had many odds stacked against him, the most significant of which was his skin color. Charles Young was born to enslaved parents in 1864 and yet graduated from the United States Military Academy at West Point. He was a lauded leader of the African American troops referred to as "buffalo soldiers" for their dark hair and fierce fighting. These troops defended our country while they still faced much persecution and inequality in civilian life. Ultimately, Young achieved the highest ranking of any African American officer in the Army and was posthumously promoted to brigadier general. Interestingly, he also was the first Black man to serve as a superintendent of a national park, which he did for Sequoia and General Grant (now Kings Canyon) in California in 1903. **Young's Ohio home,** once used as a stop on the Underground Railroad, was declared a national monument in 2013 and is undergoing restoration work to return it to its 19th-century form.

The 1965 Selma-to-Montgomery March played a pivotal role in the granting of voting rights to African Americans.

3 Selma to Montgomery
NATIONAL HISTORIC TRAIL
Alabama

On March 7, 1965, a predominantly African American group in Selma, Alabama, began what they intended to be a peaceful walk to protest racial discrimination. But marchers were soon brutally attacked by police. Scenes from "Bloody Sunday" were shown on television and in newspapers around the world, causing outrage. Two weeks later, marchers successfully completed a five-day walk to the Alabama State Capitol in Montgomery, 54 miles to the east. By the time they reached their goal, the 4,000 people who had begun had grown to 25,000, a gathering addressed by famed civil rights leader Dr. Martin Luther King, Jr. The Selma-to-Montgomery Voting Rights March helped ensure the passage of the federal Voting Rights Act of 1965, which assisted Black citizens in gaining political representation. Travelers on this historic trail can visit sites including the **Martin Luther King, Jr., Walking Tour** and the **Edmund Pettus Bridge** (where the police attack occurred) in Selma; the **Lowndes Interpretive Center,** located midway between Selma and Montgomery; and the **Dexter Avenue King Memorial**

Baptist Church, the **Rosa Parks Museum,** and the **Civil Rights Memorial** in Montgomery.

4 | Birmingham Civil Rights
NATIONAL MONUMENT
Alabama

A four-block area of downtown Birmingham, Alabama, marks both a devastatingly brutal past and the promise of a hopefully enlightened future. The Southern city, arguably one of America's most segregated, became the topic of international headlines, known for its violent civil rights struggles of the 1960s. It was here that police turned dogs and water hoses against peaceful protesters and where Black children were killed in a Baptist church bombing. Hope came in the form of civil rights leaders such as Martin Luther King, Jr., Rev. Fred Shuttlesworth, and Rev. Ralph David Abernathy, who met at the **A.G. Gaston Motel** to plan their nonviolent campaign. Seven points of interest now mark the national monument, including such sites as the previously mentioned motel, which is being restored to its 1963 appearance; **Sixteenth Street Baptist Church,** the site of the Ku Klux Klan bombing; the **Birmingham Civil Rights Institute;** and **Kelly Ingram Park,** where the Children's March led to the arrest of 1,000 child picketers. Sculptures, cell-phone tours, interpretive

▶ *Dr. Carter G. Woodson's home, a national historic site, was the headquarters for the Association for the Study of African Life and History. His work established what we now celebrate as Black History Month.*

signs, and world-class exhibits are available for visitors to this new national park unit, with more plans and restoration under way for the coming years.

5 | Booker T. Washington
NATIONAL MONUMENT
Virginia

Born into slavery in 1856, Booker T. Washington once wrote, "I had the feeling that to get into a schoolhouse ... would be about the same as getting into paradise." A dedicated believer in the power of education, Washington worked tirelessly to gain more opportunities for African Americans. At the age of 25, he was named the first president of Tuskegee Institute (now Tuskegee University) in Alabama. Booker T. Washington National Monument is located at the site of the tobacco

plantation where Washington was born, 25 miles southeast of Roanoke, Virginia. The visitor center contains **exhibits** on his life and legacy. The short **Plantation Trail** passes reconstructions of buildings like those that would have been on the farm, including a replica of the kitchen cabin where Washington lived as a boy. The park also includes a picnic area, garden, and a farm where sheep, pigs, horses, and chickens are raised.

6 | Nicodemus
NATIONAL HISTORIC SITE
Kansas

"Go to Kansas" was the recommendation given to formerly enslaved people at the end of the post–Civil War Reconstruction period. Advertised as a promised land of opportunity, the free state seemed quite enticing, and in 1877 the town of Nicodemus was founded by a group of African American settlers, with trades varying from farming to banking and pharmacy. The community thrived and the population soared to 600 before a railroad line bypassed the town, forcing development out of Nicodemus. Today, the national park unit is the **oldest and only remaining Black settlement** west of the Mississippi River. Five historic buildings, including Township Hall and the African Methodist Episcopal Church, make up the historic site and represent

A plaque marks the sidewalk before Martin Luther King, Jr.'s birthplace in Atlanta, Georgia.

the settlement's core values of faith, self-government, education, family, and business. Although the town's current population is less than 20, descendants of the original settlers flock in each summer for a festive homecoming.

7 | Martin Luther King, Jr.
NATIONAL HISTORICAL PARK
Georgia

Born in Atlanta, Georgia, in 1929, Martin Luther King, Jr., became one of the most important leaders of the 20th century. His passionate advocacy of equal rights for African Americans, while emphasizing the importance of nonviolent protest, helped shape the civil rights movement of the turbulent 1960s. Felled by an assassin's bullet in 1968, he nonetheless set in motion changes that profoundly reformed American society. Martin Luther King, Jr. National Historical Park in Atlanta's **Sweet Auburn neighborhood** interprets several locations connected with King. The visitor center includes exhibits and audiovisual presentations on King and the civil rights movement. The 1922 **Ebenezer Baptist Church** is where King's grandfather and father preached, where King was baptized and ordained as a minister, where he joined his father as co-pastor in 1960, and where his funeral was held. The **King Center** includes exhibits on King's life and work as well as his grave site, an eternal flame, and a reflecting pool that serves as a place for contemplation. The house where King was born and where he lived until age 12 can be visited only on a ranger-led small-group tour. The tours, when available, are popular; arrive early to secure a spot.

8 | Tuskegee Airmen
NATIONAL HISTORIC SITE
Alabama

Racial discrimination was still a fact of life in the U.S. military as World War II began. African Americans were generally relegated to non-combat roles, and none of them were allowed to be a pilot in the Army Air Corps (the precursor to today's Air Force). This historic site in eastern Alabama honors the hundreds of men who participated in an experimental training program to become pilots, bombardiers, navigators, instructors, and maintenance staff. During combat the Tuskegee Airmen distinguished themselves for skill and bravery, helping break down barriers to integration in the military. Visitors to the site can see **interpretive films** and, from Wednesdays to Saturdays, tour a museum in a **restored hangar** that contains historic aircraft, model airplanes, audiovisual presentations of personal recollections from airmen, as well as exhibits on the unit's accomplishments in and out of combat.

9 | Frederick Douglass
NATIONAL HISTORIC SITE
District of Columbia

As national and international momentum in favor of abolition grew in the mid-19th century, Frederick Douglass wrote and spoke eloquently on the topic, using his own life experience as inspiration. Born into slavery in 1818, Douglass learned to read despite his lack of formal schooling and escaped to

Tuskegee Institute

A small university in east-central Alabama boasts an honored legacy in African American history. In 1881, Tuskegee Institute (now Tuskegee University) hired as its first president a young teacher named Booker T. Washington, who became one of the most respected educators in the country's history. Among the staff he employed was George Washington Carver, a renowned botanist who worked to improve the lives of poor farmers. During World War II, Tuskegee trained African American military pilots who gained fame as the "Tuskegee Airmen." Some campus buildings were designed by Robert R. Taylor, the first African American graduate of the Massachusetts Institute of Technology. In recognition of these and other accomplishments, the school has been designated a national historic site and is part of the newly formed Alabama Black Belt National Heritage Area.

At Tuskegee University, a statue of Booker T. Washington illustrates him lifting the "veil of ignorance."

Tours of Harriet Tubman's home (at right) are offered through the nonprofit Harriet Tubman Home, Inc.

the free North. His 1845 autobiography brought him fame; he became a newspaper publisher and served in various government positions, including U.S. minister to Haiti. **Cedar Hill,** his estate in Washington, D.C., is now Frederick Douglass National Historic Site, where you can see exhibits on Douglass's life, and tour the house (with original furnishings and personal items) and nine-acre grounds. Special events honoring Douglass's impact, such as the annual Oratorical Contest, take place at the home.

10 | Harriet Tubman Underground Railroad
NATIONAL HISTORICAL PARK
Maryland

By car, the 125-mile **Harriet Tubman Underground Railroad By way** winds past more than **three dozen sites** significant to the Underground Railroad and its best known "conductor." Along Maryland's Eastern Shore and through Delaware, these humbling and bucolic landscapes are the ones Tubman saw as she led herself and at least 70 enslaved people north to freedom. Begin with the visitor center's exhibits, audiovisual program, research library, and legacy garden, then download the self-guided audio byway tour to explore the gardens, churches, cabins, and courthouses that tell the tale of this incredibly courageous passageway. Other nearby attractions, such as the **Harriet Tubman Museum** and the **Visitor Center at Sailwinds Park,** both in Cambridge, Maryland, have additional exhibits on Tubman's extraordinary life and commitment to justice.

10 BEST PARKS THAT

HONOR TRAILBLAZING WOMEN

Whether recognizing their ingenuity, sacrifice, leadership, or impact on history, the following national park units celebrate the enduring legacies of women. While the march to equality has been at times fraught, these women persisted and their stories still inspire.

1 Clara Barton
NATIONAL HISTORIC SITE
Maryland

While the symbol of the American Red Cross is now known nationwide, some may not realize that the organization's founding is due to the compassion, leadership, and foresight of one woman. Clara Barton began teaching at the age of 17 and was inspired to open the first free public school in her community. She moved to Washington, D.C., where she became proficient at fundraising and recruitment. During the Civil War, Barton earned the moniker "Angel of the Battlefield" for the supplies she gathered and the care she provided injured troops. Later, on a European tour, she learned about the caregiving efforts of the Red Cross, and in 1881 founded a branch in the United States. Barton's Glen Echo, Maryland, home

(from 1897 until her death in 1912) also served as **headquarters and warehouse for the organization.** The 14,000-square-foot house had 33 rooms; was well-stocked with supplies, food, and volunteers; and even boasted a vault with $3,000 in cash in case there was ever an immediate need for relief funds. In

1975, Barton's home became the first national park unit to be dedicated to a woman, and today it's open on weekends for guided tours.

2 Rosie the Riveter/ WWII Home Front
NATIONAL HISTORICAL PARK
California

Her image is unmistakable with its muscled arm, determined face, and telltale red-and-white polka-dot bandanna symbolizing wartime solidarity and, later, a push for feminism and equal rights. Rosie the Riveter was in many ways the image of America's critical home front effort during World War II. An estimated 18 million women entered the workplace—many in defense plants providing ships and weapons—while men were away on active duty. The impact of the war in Richmond,

California, is especially significant, and guests can get an overview at the visitor center housed in the historic **Oil House of the Ford Assembly Plant.** Most weeks feature Rosie Fridays (call ahead to confirm), with WWII Home Front workers available to share stories and answer questions. The park also partners with the City of Richmond and the Rosie the Riveter Trust to support other sites like the Liberty ship–shaped **Rosie the Riveter Memorial.**

3 | Lowell
NATIONAL HISTORICAL PARK
Massachusetts

Lowell, Massachusetts, is known as the city where American industrialization began. But for the young women that made up the textile manufacturer's first labor force, Lowell was also a launchpad for labor reform and women's activism. Called "mill girls," the 15- to 30-year-olds were initially enticed by the security of regular wages plus room and board. Ultimately, however, they faced long hours and grueling, often unhealthy, settings. Strikes were common in this era, with labor reformers promoting shorter workdays, increased pay, and work safety. It was said that in Lowell, the fight for better wages and working conditions helped the women gain skills they utilized for other initiatives, including the vote. Many mill girls became suffragists, and in

▶ *At Clara Barton National Historic Site visitors can witness the cleverly frugal ways Barton maintained such a large headquarters, like using muslin (traditional bandage material) rather than plaster for interior walls and ceilings.*

1910 the Annual Woman Suffrage Convention took place in Lowell. Today's **mills turned museums** feature exhibits and programs about the industrial revolution, including a tour of a **weave room** and **boardinghouse,** with a look at the human side of the work.

4 | Belmont-Paul Women's Equality
NATIONAL MONUMENT
District of Columbia

A walk down the Hall of Portraits in **Belmont-Paul Women's Equality National Monument** is a who's who tour of the fight for suffrage and the long-lasting work of the National Woman's Party (NWP). One of the oldest residential properties in Washington, D.C., the 200-year-old home on Constitution Avenue already had a storied history when

it was sold in 1929 to the NWP, who named it the "Alva Belmont House," honoring the work of the organization's former president and primary benefactor. Donated to the National Park Service in 2016, it was given its current moniker to honor Belmont as well as Alice Paul, the founder of the NWP. Initially focused solely on women's suffrage, the NWP's efforts for women's equality expanded after the ratification of the 19th Amendment into the social, political, and economic realms. Visitors to the monument can view historic suffrage banners used by picketers, Susan B. Anthony's desk, and the Florence Bayard Hilles Feminist Library.

5 | Women's Rights
NATIONAL HISTORICAL PARK
New York

A bronze sculpture titled "The First Wave" greets visitors to Women's Rights National Historical Park in downtown Seneca Falls. The piece depicts five visionary women—Elizabeth Stanton, Martha Wright, Mary Ann M'Clintock, Jane Hunt, and Lucretia Mott—who decided to hold a convention to discuss the social, civil, and religious rights and conditions of women. Held in the **Wesleyan Chapel** in Seneca Falls, the event gathered more than 300 attendees, 100 of whom signed their names to the Declaration of Sentiments. And while this was

At First Ladies National Historic Site, take a guided tour of the Saxton House, once home to Ida and William McKinley.

just the first of many conventions on women's rights that would take place in the coming century, the efforts of these leaders stirred future abolitionists, suffragists, and human rights advocates. In addition to the visitor center's museum exhibits and outdoor ranger talks, the Wesleyan Chapel is open daily and historic homes are open seasonally. Next to the chapel, the **Waterwall at Declaration Park,** a 100-foot water feature, is inscribed with the words of the Declaration of Sentiments. Consider planning your visit to coincide with Convention Days, a signature Seneca Falls event held annually to commemorate the 1848 Women's Rights Convention.

6 First Ladies
NATIONAL HISTORIC SITE
Ohio

No matter the political persuasion, the importance of the role of first lady of the United States has held true. Whether she's revered or criticized, impacted by the politics of the times or her perceived sociability, the mark of the president's spouse cannot be understated. Which is why, in 2000, a national park site was designated in Ohio to honor and preserve the contributions related to this iconic position. Visitors begin at the **Education and Research Center,** located in historic City National Bank. Home to the **National First Ladies' Library,** it's operated by a nonprofit group in partnership with the National Park

The Stone Cottage's exterior utilizes fieldstones from the estate's many stone walls at the Roosevelt homestead of Val-Kill.

Service and offers a curated experience of rotating exhibits, a film, and an archival collection of books, documentaries, and manuscripts. A block away is the **Saxton House,** the 1841 Victorian mansion where Ida Saxton, wife of President William McKinley, was raised and where the couple lived for more than a decade. Restored to its late 1800s splendor, the home's personal artifacts and images highlight past first ladies and their contributions to supporting both the president and the nation.

7 Eleanor Roosevelt
NATIONAL HISTORIC SITE
New York

Val-Kill, a country retreat property of Franklin and Eleanor Roosevelt, was not only a spot for the couple to relax and enjoy the outdoors, but also a place to entertain friends, family, press, activists, and other official visitors. Acres of gardens and trails grace the Hyde Park, New York, area where the Roosevelts enjoyed the outdoors, with tennis, horseback riding, swimming, and hiking among their favorite activities. At Val-Kill ("waterfall stream" in Dutch), Eleanor was able to pursue her interests of education, business, and advocating for the rights of minorities and the disadvantaged—her human rights efforts earned her the nickname "First Lady of the World." As the first Park Service site dedicated to a first lady, Eleanor Roosevelt National Historic Site offers tours of Val-Kill and Stone Cottage that give an overview of Eleanor's work. The one-mile **Eleanor's Walk trail** follows a path she walked daily on a road built by FDR in 1940.

"Rangers, Not Rangerettes"

Curious about the history of female park rangers? Here are four female park rangers who made their National Park Service marks early.

Esther Brazell (1916): The first woman park ranger documented by the National Park Service, Brazell began giving tours at Wind Cave National Park in the summer of 1916. Her father was the park's supervisor, and both Brazell and her brother worked for the Park Service.

Clare Marie Hodges (1918): Although her appointment letter noted her status as temporary, Hodges was no less impactful in her role at Yosemite National Park, which included mounted patrols, tourist registration, and the "power of arrest." After her tenure in the park she taught botany and horsemanship locally.

Mary Sullivan (1924): Hired at the age of 57, Sullivan worked summers for Glacier National Park, filling a recurring need by registering guests at the Polebridge entrance. Her husband, Thomas, was a ranger as well, which led to suspicion that her hire was one of convenience.

Marguerite Lindsley (1925): The first female park ranger at Yellowstone was, indeed, at home there. Lindsley's father was the park's interim superintendent during its transition from the Army to the Park Service. A naturalist with a degree in bacteriology, Lindsley led tours and ran a museum in Mammoth.

Marguerite "Peg" Lindsley (1923)

8 | Maggie L. Walker
NATIONAL HISTORIC SITE
Virginia

Community was always at the core of Maggie L. Walker's work. Born in Richmond, Virginia, Walker was a leader in a fraternal society when she established a newspaper, *The St. Luke Herald,* in 1902. Just a year later she opened the St. Luke Penny Savings Bank, becoming the country's first Black female to charter and serve as a bank president. Aiming to bring economic empowerment to the largely African American Jackson Ward neighborhood, Walker distributed coin banks to children, served on the board of directors for the NAACP, and launched a Girl Scout troop. The Walker family bought their **Jackson Ward home** in 1904 and immediately began making changes, adding central heating and electricity, then several bedrooms and porches. By the time the National Park Service purchased it in 1979, the home had 28 rooms and had housed four generations of the Walker family. The visitor center's 20-minute film, "Carry On," serves as an orientation to the park, and the ranger-guided experience ends with a tour of Walker's home, which has been restored to its 1930s appearance. Guests can also check out the Jackson Ward podcast tour on the Park Service app to learn more about this center of Black business and social life.

Learn about Maggie Walker's legacy at her home in Richmond, Virginia.

9 | Harriet Tubman
NATIONAL HISTORICAL PARK
New York

After emancipating herself at the age of 27, Harriet Tubman became enshrined in history for the valiant work she did guiding roughly 70 people to freedom by way of safe houses on the Underground Railroad. In 1859 she moved to Auburn, New York, a center of progressive thought as well as support for abolition and women's suffrage. There, she established a home for the elderly and continued her civil rights work. The National Park Service is currently rehabilitating the **Thompson AME Zion Church** in Auburn, which was Tubman's religious community for 22 years. Its eventual visitor center will share more of Tubman's story. Until then, visitors to the park unit can view **Tubman's residence and burial place,** each of which are run by independent partner organizations.

10 | Mary McLeod Bethune Council House
NATIONAL HISTORIC SITE
District of Columbia

Lifelong educator, public servant, presidential advisor, and unsung hero of the civil rights movement, Mary McLeod Bethune is best known for organizing the National Council of Negro Women (NCNW), an unprecedented achievement that brought the collective voice of Black women to the national stage. Created in 1935, the NCNW established its first headquarters in Bethune's **Second Empire row house** in the Logan Circle neighborhood of Washington, D.C. The site is now a memorial to her life and legacy, with guided tours that highlight the influence of her work. Visitors will learn the story of Bethune's passion for teaching, which led her to establish a school for African American girls and ultimately earned her a presidential appointment to federal director of the Office of Negro Affairs of the National Youth Administration.

10 BEST PARKS FOR
BARRIER-BREAKING HISTORIES

From urban parks to desert oases, these sites celebrate the groundbreaking history forged by our country's diverse and industrious populace. Whether marking a battlefield turning point that changed the course of our nation's history or the spot at which human rights made invaluable strides, these parks are natural—and essential—parts of the National Park System.

1 | Stonewall
NATIONAL MONUMENT
New York

Tensions were already high when police raided a New York City gay bar on a June night in 1969. Patrons, weary of the discrimination and harassment they'd experienced based solely on their sexual orientation, formed a spontaneous act of resistance at the Stonewall Inn in Greenwich Village, escalating to a riot that ultimately amounted to six days of demonstrations in the blocks surrounding the bar. National attention was immediate, and the momentum was measurable: **Stonewall inspired LGBTQ rights groups** in nearly every major city across the country and was a key turning point for the movement. Established in 2016 as the first national park monument dedicated to LGBTQ civil rights, Stonewall was also the first federal park to fly the symbolic rainbow flag. A new visitor center with interpretive exhibits and ranger-guided tours tells the story of the site, which includes the Stonewall Inn, Christopher Park, and the adjacent city block, while regular outdoor events ensure pride lives freely outside the site's walls as well.

2 | Golden Gate
NATIONAL RECREATION AREA
California

As diverse and dynamic as the people it honors, Golden Gate National Recreation Area was established to offer a national park to urban residents. Centuries of overlapping history bring together California's Indigenous populations, Spanish colonialism, the Mexican Republic, and the comparatively recent development of San Francisco. A member of the International Coalition of Sites of Conscience, the recreation area uses intentional programming and education to **connect history within the park with current human rights and social justice issues.** Take, for instance, sites like the Presidio, where African American buffalo soldiers and WWII Japanese-American Nisei soldiers balanced patriotism and prejudice. At Alcatraz Island, visitors weigh the civil liberty concerns of the federal penitentiary's treatment of Black and Indigenous prisoners. With 37 San Francisco Bay Area locations in the

Fire Island National Seashore is home to the tallest lighthouse on Long Island.

recreation area's jurisdiction, the opportunity for this kind of analysis and engagement is both prescient and bountiful.

3 | César E. Chávez
NATIONAL MONUMENT
California

Some things are frozen in time at Nuestra Señora Reina de la Paz (translated Our Lady, Queen of Peace), the headquarters of the farmworker movement. This includes the office and library of its leader, César E. Chávez, just as he left them when he died in 1993; the memorial garden where he and his wife, Helen, are buried; and the historic structures where workers gathered to organize their strikes and boycotts. And yet other aspects have continued to push toward modernity, specifically the operations of the National Chavez Center, which hosts corporate conferences, educational seminars, and special events that continue to share the legacy of Chávez's activism. Chávez founded the National Farm Workers Association (which changed its named to the United Farm Workers in 1966) with the mission to **reform working conditions and wages for agricultural workers.** Revered not only among fellow Latino Americans, Chávez was lauded for the efficacy of his nonviolent work for human rights, earning him a Presidential Medal of Freedom in

▶ *It's 182 steps to the top of the Fire Island Lighthouse, where 360-degree views reward the journey. Can you spot the New York City skyline? On a clear day it's just beyond the sight of the Robert Moses Bridge.*

1994. Monument visitors can explore these efforts, plus enjoy the memorial and desert gardens.

4 | Fire Island
NATIONAL SEASHORE
New York

Given its close-enough proximity to Manhattan, yet tempered by the relative wildness of mostly undeveloped beaches and wilderness, it's easy to see how the Fire Island seashore became **an enclave of tolerance** in the early 20th century, a time when living openly as an LGTBQ person was illegal. Fire Island National Seashore stretches 26 miles along the 32-mile-long barrier island, and it's here, among ancient maritime forests, historic landmarks, and a globally rare ecosystem, that artists, performers, and other New York socialites helped grow the communities of Cherry

Grove, which was called **"America's First Gay and Lesbian Town,"** and later Fire Island Pines. Brimming with wildlife, rolling dunes, sandflats, and wetlands, this seashore was then—and remains now—an idyllic spot not only for escaping the hustle and bustle but also for embracing one's full self.

5 | Chamizal
NATIONAL MEMORIAL
Texas

Wedged between the bustling urban centers of El Paso, Texas, and Ciudad Juarez, Mexico, Chamizal is a green oasis in the middle of what is otherwise unrelenting desert and urban sprawl. The memorial **commemorates a landmark 1963 treaty** between the United States and Mexico that solved a long-running border dispute caused by the natural meander of the Rio Grande over the previous century. American authorities relocated more than 5,000 people on the U.S. side and ceded 437 acres of South El Paso to Mexico. In turn, Mexico relinquished claims to Cordova Island in the middle of the river. Both governments shared the cost of cementing the river and building new ports of entry on either side. Chamizal also **celebrates the shared heritage of the border region** in murals, museum displays, and art galleries with revolving exhibits. The park features a wide variety of

A tour of Vicksburg National Military Park includes Battery De Golyer, one of the largest concentrations of Union cannons.

activities, including ranger-guided history and culture programs, children's arts-and-crafts and reading events, as well as live music and dance with a Southwest flavor.

6 | Vicksburg
NATIONAL MILITARY PARK
Mississippi

Though often overshadowed by the Battle of Gettysburg, which ended a day earlier, the long siege of Vicksburg, Mississippi, ranks among the Civil War's most important events. This park recognizes this **critical historical moment—the battle's outcome was an important step to spur the country toward peace.** President Abraham Lincoln famously called Vicksburg the "key" to victory and stated that "the war can never be brought to a close until that key is in our pocket." The 47-day siege ended with Confederate surrender on July 4, 1863. The park is one of the war's most accurately laid out battlefield memorial sites—you can stand in soldiers' footsteps and imagine the Union siege of the city, when opposing forces were so close they could call out to each other during lulls in the fighting. Park activities include living history programs, artillery and rifle firings, and interpretive talks, but the site should also be noted for its more than 1,300 monuments, markers, tables, and plaques, including stone and bronze work crafted by renowned American sculptors.

7 Women's Rights
NATIONAL HISTORICAL PARK
New York

While the 1848 Seneca Falls Women's Rights Convention is certainly the focal event celebrated at Women's Rights National Historical Park, the ripple effect of that New York gathering cannot be overstated. In calling for signatures to the Declaration of Sentiments, organizers were recruiting those who would "employ agents, circulate tracts, petition the State and national Legislatures, and endeavor to enlist the pulpit and the press in our behalf." In a matter of weeks other groups convened in Ohio, Pennsylvania, and New York, with a National Women's Rights Convention taking place in 1850. And that was just the start. Visit the park to explore Wesleyan Chapel, considered the **formal launch site of the women's rights movement,** and the homes of two founding members: Elizabeth Cady Stanton and Mary Ann M'Clintock, who hosted the planning session for the first Women's Rights Convention in her home in 1848. Ranger-led programs at the historical park draw connections among this work with that of the **abolitionist movement and the broader conversation on human equality.**

8 Glacier
NATIONAL PARK
Montana

Waterton-Glacier International Peace Park was the first of its kind—a collaboration between two countries recognizing that neither nature nor Indigenous heritage can be constrained by governments or politics.

International Sites of Conscience

Including Golden Gate National Recreation Area, 20 national park units have been recognized by the International Coalition of Sites of Conscience. According to the organization, "a Site of Conscience is a place of memory—such as a historic site, place-based museum, or memorial—that prevents ... erasure from happening in order to foster more just and humane societies today." Add these sites to your bucket list and check the index—many are mentioned in these pages.

Brown v. Board of Education National Historical Park
Delaware Water Gap National Recreation Area
Ford's Theatre
Grand Teton National Park
Keweenaw National Historical Park
Little Rock Central High School National Historic Site
Longfellow House–Washington's
  Headquarters
Lowell National Historical Park
Manzanar National Historic Site
Minidoka National Historic Site
Minuteman Missile National Historic Site
New Bedford Whaling National
  Historical Park
North Plains National Heritage Area
Statue of Liberty National Monument
Ste. Genevieve National Historical Park
Timucuan Ecological and Historic Preserve
Ulysses S. Grant National Historic Site
Whitman Mission National Historic Site
Women's Rights National
  Historical Park

While multiple U.S. national parks share a border with Canada, America's Glacier and Canada's Waterton Lakes share **unique ecological and cultural significance** (both are Biosphere Reserves and World Heritage sites). Discussions for a collaboration between the parks began in the 1930s. While both are administered independently of one another, their guiding principles and goals intersect and are reviewed by various parties, including public and Indigenous communities. The international park's combined area features the snowcapped Rocky Mountains, high-altitude lakes, bunchgrass prairie, cedar-hemlock forests, and threatened and endangered species. The park symbolizes the impact of a **longstanding, positive international alliance.**

9 Blackstone River Valley
NATIONAL HISTORICAL PARK
Rhode Island & Massachusetts

Once referred to as the "hardest working river in America," the Blackstone was home to the 18th-century water-powered factories that launched American manufacturing, inextricably linking the Rhode Island and Massachusetts valleys to the Industrial Revolution and rising tide of **immigrants who would go on to build America into the diverse country we now know.** Samuel Slater, himself an immigrant, became wildly successful with his textile mill, hiring full families—children included—to live and work in the mill villages. With increased demand for products came a wider call for laborers, and soon Irish and

French Canadian mill workers found jobs and community in the region. Another wave came at the turn of the 20th century, with immigrants arriving from Sweden, Portugal, Italy, Poland, and Ukraine in search of farmland and the Blackstone Valley's industrial opportunities. Partner groups manage and work with the Park Service on several sites in this two-state park, including multiple visitor centers and recreational areas worth seeing.

10 Touro Synagogue
NATIONAL HISTORIC SITE
Rhode Island

Touro Synagogue, the most historically significant Jewish building in the United States, is a **symbol of religious freedom for all Americans.** The congregation was founded in 1658, with the building dedicated in 1763, and Orthodox services are still observed today. In 1790 George Washington's letter "To the Hebrew Congregation in Newport" assured his commitment to freedom of religion, stressing that the federal government "gives to bigotry no sanction, to persecution no assistance." An annual public reading of the letter marks its significance. The visitor center explores **Rhode Island's earliest Jewish community and the authority of the First Amendment,** while walking tours of the Touro neighborhood bring history to life.

Blackstone River Valley is home to the first water-powered textile mill in America.

SCIENCE & INDUSTRY

Among the eclectic sites administered by the Park Service are many that recognize the labor, invention, and economic achievements that moved America into the industrial age and beyond. These parks allow a look behind the scenes at the nuts and bolts of how things work, from simple blacksmithing to the space age.

1 George Washington Carver
NATIONAL MONUMENT
Missouri

One of the best ways to get to know the life and legacy of George Washington Carver is quite simply to go on a walk. It's apropos, as guided daily tours of the Carver Trail loop through the woods, streams, and tallgrass prairie where a fascination with the natural world first inspired the internationally lauded botanist. Born into slavery and raised on a small Missouri farm, Carver overcame the challenges of poverty and prejudice to rise from neighborhood "plant doctor" to **renowned agricultural scientist, educator, and humanitarian.** While Carver is buried at Tuskegee Institute in Alabama, his Missouri monument, the **first national park established to honor an African American,** preserves the site of his childhood and welcomes visitors to explore his passions though a 19th-century classroom, a hands-on science lab, and art appreciation.

2 Thomas Edison
NATIONAL HISTORICAL PARK
New Jersey

How could we not include the man who literally changed the world with his inventions, yet modestly claimed that "genius is one percent inspiration and 99 percent perspiration"? Born in 1847, Thomas Edison invented or improved a long list of things, but he's most famous for three: the **phonograph, electric lighting, and the motion picture.** Called the "Wizard of Menlo Park" (in New Jersey, where he did his early work), Edison later moved to new, much larger laboratories at West Orange, New Jersey, now preserved as a national historical park. Edison proudly called his West Orange headquarters the "best equipped and largest laboratory extant ... facilities superior to any other for rapid and cheap development of an invention." His labs were said to be the model for future research facilities such as Bell Laboratories. You can tour the laboratories where Edison worked until his death in 1931, see interpretive videos, and admire examples of his inventions. Tours are also available of Glenmont, Edison's 29-room Queen Anne–style mansion and the estate grounds and outbuildings. The main lab tour is self-guided, but ranger-led programs are offered, too, delving deeper into various aspects of the Wizard's productive life.

3 Wright Brothers
NATIONAL MEMORIAL
North Carolina

Humans have dreamed of flying since our brains got big enough to feel jealousy toward birds. True

In North Carolina in 1903, the Wright brothers made the first successful flights with a heavier-than-air machine.

flight had to wait, however, until two bicycle-builders from Dayton, Ohio, turned their obsession into reality on the barrier island sand dunes of North Carolina. On December 17, 1903, at a place called Kill Devil Hills, Wilbur and Orville Wright succeeded in **the first sustained powered flights in a heavier-than-air machine:** A new age of transportation had begun.

The memorial to their achievement includes functional replicas of their 1903 Flyer (the original is in the Smithsonian National Air and Space Museum in Washington, D.C.) and a 1902 glider that the brothers used to test steering mechanisms before they tried motorized flight. The park features reconstructions, based on photographs, of their hangar and living quarters. You can

walk along the exact flight paths taken by the Wright brothers' plane on that day in 1903; the brothers took turns flying the plane, making four attempts to sustain flight, and Wilbur was at the controls for the final and longest attempt—852 feet, lasting 59 seconds. Private pilots can land their own small planes at the park on the 3,000-foot First Flight Airstrip.

4 | Saugus Iron Works
NATIONAL HISTORIC SITE
Massachusetts

In the 17th century, settlers of the Massachusetts Bay Colony needed a variety of manufactured goods to help them build and set up houses and clear and farm the land. When the goods they had brought from Europe wore out, they needed replacements, and beginning in 1646, they could obtain them from Saugus Iron Works, the **first complete ironmaking factory in colonial America,** churning out hoes, nails, saws, axes, hinges, skillets, pots, and much more. Today's national historic site features a blast furnace, forge, rolling mill, warehouse, and working waterwheels, all reconstructed based upon extensive archaeological excavations. A museum displays many of the artifacts found on the site. Although Saugus operated for only 22 years, it gave a vital boost to the U.S. iron and steel industries.

5 | Golden Spike
NATIONAL HISTORICAL PARK
Utah

Today, when it is possible to jet from Boston to Los Angeles in a few hours and make computer measurements smaller than an angstrom, it might be hard to imagine the excitement that swept across America on May 10, 1869, when a telegraph operator in northern

▶ *The Wright brothers didn't work alone. From sea to sky, members of the local Coast Guard were known to help Wilbur and Orville launch their gliders and flyer.*

Utah sent the simple word "DONE" along the wires. It meant that two teams working toward each other from east and west had met at Promontory Summit, completing the **first transcontinental railroad line** and making possible a trip by train from the Atlantic to the Pacific: no more wagon trains, no more weeks-long voyages around Cape Horn. Nothing before had so unified the United States, both practically and psychologically. At Golden Spike National Historical Park, you can see exhibits on the transcontinental railroad and the effort required to bring it to reality, and drive part of the original rail route via an auto tour that passes construction sites, hill cuts, and the Chinese Arch, a natural limestone arch named **in honor of the thousands of Chinese workers who labored on the rails.** Modern reproductions of the two steam locomotives that met at Promontory Summit hold demonstration runs from May to Labor Day.

6 | Keweenaw
NATIONAL HISTORICAL PARK
Michigan

Thanks to geology, the Keweenaw Peninsula of northern Michigan was the site of **the world's richest known deposit of 97 percent pure copper.** "Copper fever" struck the area in the 1840s, and by the end of the decade this region of Michigan's Upper Peninsula produced 85 percent of the nation's supply of the mineral. Mining continued until the 1960s, and today's park allows you to recall those boom years. Here you can tour underground mine shafts; see the Quincy Mine, where the world's largest steam-driven hoist descended 9,260 feet into the earth; and admire the results of some of the wealth created, such as an 1899 opera house and a 1908 mansion built by a mine owner. The Finnish American Heritage Center, on the campus of Finlandia University in Hancock, **recognizes the heritage of the great number of Finns** who came to work the mines in the 19th century.

7 | Lowell
NATIONAL HISTORICAL PARK
Massachusetts

An important period of American industrialization began in the northeastern Massachusetts town of Lowell in the early 19th century. A group of businessmen established a planned city using waterpower to run textile mills, visualizing it as a more humane place than the gloomy factory towns of England. Employing mostly immigrants and "mill girls" (young unmarried women from around New England), the mills fueled a population growth in Lowell from about 2,500 in 1826 to more than 33,000 in 1850, when the city was one of the world's leading textile producers. By the 1930s, various economic factors had caused almost all the mills to close,

but today some have been restored or renovated for purposes ranging from museums to art galleries. Witness the fascinating process of turning raw cotton into cloth on 19th-century looms, or take a tour of the canals that used water from the Merrimack River to run vast amounts of machinery. Not only is the Lowell park an educational and entertaining experience; it's a fine example of **a beleaguered community reinvigorating itself.**

8 | Minuteman Missile
NATIONAL HISTORIC SITE
South Dakota

The Cold War, the stalemate between the United States and the Soviet Union that inspired the phrase "mutually assured destruction," was one of the **defining historical themes of the 20th century.** Several states were home to Minuteman nuclear missiles, each stored underground in a concrete "silo," with a crew ready in case an

The Reds of Glacier

Captured in vintage photographs and on thousands of mid-century home movies are images of low-profile "touring sedans" that once ferried passengers throughout parks. At one time, hundreds of these vehicles rolled down scenic roads and to vista points at Yellowstone, Yosemite, and Bryce Canyon, but as a new car culture took over after World War II, these classics were slowly nudged aside.

But not at Glacier.

Credit the topographical demands of the Going-to-the-Sun Road, where sheer drops and tight turns demand that drivers keep both eyes on the road. Even in the late 1990s, when park officials examined ways to alleviate pollution, reduce development, and protect the park's environment, the cherished "Reds"

(so named for the vehicles' color scheme of a slick red body and sharp black accents) were considered such a fundamental part of Glacier's heritage that a drive was on to find ways to keep them in service.

The park finally found an answer—and a partner: the Ford Motor Company. With Ford engineers and designers at the wheel, among the changes the Reds were given in their makeover was a switch to liquid propane, which burns 93 percent cleaner than gasoline; a stronger and more modern chassis, which supported a new body made of lighter materials; and the addition of ergonomic seats, a modern instrument panel, and various energy-saving and environmentally friendly extras. What remained of the old bus? Perhaps the most important thing: Its appearance. Committed to preserving a design that is both classic and contemporary, both Glacier and Ford managed to retain the vehicle's low profile, its multiple doors, the convenient rollback canvas top, and, of course, the iconic color scheme. The Reds roll on.

Emergency War Order arrived with a command to launch. At this western South Dakota national historic site, you can tour the aboveground Launch Control Facility Delta-01 and the underground Launch Control Center, where crew members could remotely launch missiles from 10 silos. A few miles away is Launch Facility Delta-09 (a missile silo), with an inoperative Minuteman inside. Tours begin at a visitor contact station at exit 131 on I-90. Reservations are required—because of the cramped space in these sites, space on tours is very limited.

9 | Salem Maritime
NATIONAL HISTORIC SITE
Massachusetts

Relive the days when sailing ships set out on voyages around the globe at this park on the Massachusetts coast. Privateers (essentially government-sanctioned pirates) operated from Salem during the Revolutionary War, and in the decades after American independence, the port saw vast quantities of goods arriving from Europe and the Far East, making this **one of the new nation's most important trade centers**—and the home of one of America's first millionaires. The historic site features a modern replica of the tall ship *Friendship,* which was launched in 1797 and captured by the British in the War of 1812. Also part of the park are wharves

Docked at Derby Wharf, the replica Friendship of Salem *represents how an original 1797 vessel would have looked.*

dating back to 1762 and the 1819 Custom House, where author and Salem native Nathaniel Hawthorne worked from 1846 to 1848.

10 | Steamtown
NATIONAL HISTORIC SITE
Pennsylvania

It's easy to be fascinated by the sights, sounds, smells, and raw power of steam locomotives. **Vital to the growth and economic development of the United States,** but doomed by the greater efficiency of the diesel-electric locomotive, steam had largely disappeared from railroading by the 1950s. Nonetheless, a significant number

of steam-train enthusiasts keep that colorful era alive in many places around the country, especially at Steamtown National Historic Site in Scranton, Pennsylvania. Located within the railroad yard of the Delaware, Lackawanna, and Western Railroad, the park displays **a collection of passenger and freight cars and steam locomotives,** including impressive *Big Boy,* built in 1941 for the Union Pacific, and a 1903 freight engine (the park's oldest locomotive) built for the Chicago Union Transfer Railway Company. Steamtown offers train rides seasonally, but call or check online for dates; the schedule varies depending on staff and equipment availability.

PRESIDENTIAL FOOTPRINTS

Since George Washington took the first presidential oath of office in 1789, our Chief Executives have steered the nation through moments of achievement, conservation, tragedy, and more with compassion and incredible vision. Their imprints have been left on homes, monuments, battlefields, and even hardscrabble western landscapes.

1 | Gettysburg
NATIONAL MILITARY PARK
Pennsylvania

No one knew when Abraham Lincoln followed the lengthy two-hour speech of famed orator Edward Everett that he would make perhaps **the most well-known and impactful speech in American history.** Lincoln had been invited to Gettysburg by local attorney David Wills to appear at the dedication of a portion of a battlefield as a national cemetery. At the dedication, the "few appropriate remarks" he delivered became known as the **Gettysburg Address,** 10 sentences that summed up the nation's past, present, and what Lincoln hoped for its future. Within two minutes, the "United States *are* ..." became the "United States *is* ..." and there was no doubt that

with a Union victory, Lincoln would unite the nation. When the president sat down after his comments, few in the crowd realized that he had even made a speech at all—it was only after he provided copies to friends and the press that his words became a part of the American story. This is still hallowed ground, and the vivid stories of the battles and Lincoln's healing words are an essential stop for anyone interested in stories of war and peace.

2 | Lyndon B. Johnson
NATIONAL HISTORICAL PARK
Texas

In a place that prides itself on being big, it seems the Lone Star State was right on track when it produced Lyndon Baines Johnson, a president who seemed larger than Texas. As his political and personal fortunes increased, LBJ purchased a **ranch of nearly 1,600 acres in the heart of Texas Hill Country,** roughly 15 miles east of Fredericksburg. He donated it to the National Park Service in December 1972, a month before his death, but the ranch remained the home of his widow, Lady Bird, until she passed in 2007. Using the audio tour available on the Park Service app, visitors follow directions around the ranch, passing the one-room schoolhouse Johnson attended, well-stocked wildlife enclosures, the grave of the former president, and guided tours inside the main home, known as the **Texas White House.** (Renovations of the ranch are ongoing, so check with the park before your visit for

Site of a horrific battle, Gettysburg is where Lincoln delivered a now famous speech on the principles of human equality.

up-to-date tour information.) Don't miss the state park, either, where you can learn about the botany of Hill Country. Fitting, since Lady Bird was a dedicated conservationist who advanced nationwide beauti-fication projects, including planting roadside wildflowers—notably the ubiquitous bluebonnets that bloom in Texas fields each spring.

3 | Mount Rushmore
NATIONAL MEMORIAL
South Dakota

This fabled sculpture has been celebrated, decried, and parodied in media from postcards to dinner plates. But no matter the emotion, visitors who round the corner on U.S. 16A near Keystone forever remember their first sight of

Mount Rushmore's enormous four faces. Tribal nations have every right to take issue with the work's placement, on a spiritual site of prayer known as Six Grandfathers Mountain The sacredness of the site didn't prevent the United States from breaking its own treaty in 1877, setting off years of legal battles. In 1924, sculptor Gutzon

Borglum (himself controversial) was recruited to honor four legendary presidents, and 14 years of blasting, drilling, and chiseling followed. Dedicated in 1941, the mountain **reveals the faces of George Washington, Thomas Jefferson, Abraham Lincoln,** and **Theodore Roosevelt,** selected respectively for their contributions to the birth, growth, development, and preservation of our nation. About an hour before sunset each evening between May and September, patriotic music fills the valley. Following a short film on the history of the project, the four faces are gradually illuminated. The park now consults with 21 tribal nations in ongoing conversation related to the significance of this, and many other, important sites.

4 The White House & President's Park

PARK

District of Columbia

Perhaps **the country's most iconic government building** (the U.S. Capitol runs a close second), the White House had been called the President's Palace, Presidential Mansion, President's House, and, more often, the Executive Mansion,

Presidential Visits to National Parks

Many presidents have made a point of visiting the oldest and grandest national parks. Yellowstone has always been a favorite destination: Chester Arthur was the first to visit, in 1883, exploring the park on a horse. Jimmy Carter fished in Yellowstone Lake. Bill Clinton lunched at the Old Faithful Inn and visited geysers with First Lady Hillary Rodham Clinton. Gerald Ford even spent a summer (1936) there as a young ranger. In 1974, Joe Biden took his boys for a week in hopes of healing from the loss of Biden's first wife, Neilia Hunter Biden, and daughter, Naomi. Barack Obama and his family spent a 2009 summer vacation in Yellowstone, reprising part of a national park–hopping tour he did as a kid.

Not surprisingly, nature-loving Theodore Roosevelt paid several memorable visits to national parks. During his 1903 trip to Yosemite, TR was led into the wilderness by none other than John Muir. Sixty years later, John F.

Theodore Roosevelt and John Muir stand atop Glacier Point in 1903.

Kennedy traveled to the Yosemite Valley and stayed at the Ahwahnee Hotel.

Given its proximity to the nation's capital, Shenandoah has been a favorite of several presidents, including Herbert Hoover, who spent many a day wandering the woods around Camp Rapidan. Franklin D. Roosevelt dedicated the park in 1936 and returned later during his long presidency. With six parks to his credit, FDR also owns the record for most presidential visits.

George W. Bush jumped right in to trail-building in Rocky Mountain National Park. Harry Truman happily cavorted through the snows of Mount Rainier. Ronald Reagan went underground at Mammoth Cave. William Howard Taft argued with an aide over whether he could ride a horse into the Grand Canyon. And in a show of bipartisanship, Richard Nixon and Lyndon B. Johnson visited Redwood National Park together in 1969 to dedicate the Lady Bird Johnson Grove of big trees.

until Theodore Roosevelt stamped its name on stationery in 1901. The history that's happened here is astounding. Beyond wars and social programs and presidents steering America from fledgling nation to global superpower, this is a site where *life* has unfolded: Think of John and Abigail Adams (the *first* first family to live here) hanging their laundry to dry in the East Room, or the raucous celebration of Andrew Jackson's 1829 Inauguration Day when supporters came in and trashed the place. Then there's Theodore Roosevelt losing sight in his left eye when sparring with a boxer; Malia and Sasha Obama sliding down the banister of the solarium; and John F. Kennedy's humorous observation about living at the White House and serving as the nation's Chief Executive: "The pay is good and I can walk to work." Tour the famous building yourself by making a reservation via your congressional representative.

5 | First Ladies
NATIONAL HISTORIC SITE
Ohio

Since George Washington's inauguration, every United States president has had a partner. For most this was a spouse, but in instances where the president was unmarried or widowed, the role was filled by a sister, daughter, niece, or even wife of a cabinet member. Thus far,

Birders can spot more than 180 bird species in Theodore Roosevelt National Park, including western meadowlarks.

all have been called first lady, the name for a role that, while never declared an official government duty, has often required all manner of labor and service recognized worldwide. (The term itself came into use gradually before taking root in the early 20th century.) At First Ladies National Historic Site, administered in a partnership between the National Park Service and the National First Ladies' Library, visitors can **explore the nuance of this role**—from fashion to activism and every critique and commendation in between—with **rotating and permanent exhibits, a restored first family home,** and **extensive archives.**

6 | Theodore Roosevelt
NATIONAL PARK
North Dakota

On February 12, 1884, 25-year-old Theodore Roosevelt welcomed the birth of his first child, Alice. Two days later, as he was grieving over the death of his mother that morning, he also learned he had lost his 22-year-old wife (also named Alice) from an undiagnosed kidney disease concealed by the pregnancy. Overwhelmed, Roosevelt left his infant daughter with his sister and **escaped into the Black Hills of North Dakota to find peace and an outlet for his anguish.** Already an avid outdoorsman, his two years in the Wild West would later affirm his reputation as the "Cowboy President" and confirm, in his own words, that "it was here that the romance of my life began." In the 1880s, these were the Badlands and, by and large, they still are. The same wildlife—**bison, elk, deer, antelope, and wild horses**—Roosevelt knew (and often hunted) still roam the grasslands, and nearly **200 species of birds,** many cataloged, sketched, and studied by the professorial cowboy, still flitter across the 70,000-plus acres of "grim fairyland." It took a tragedy to compel Roosevelt to retreat to the Dakotas, but it took time here and inner strength to return to his daughter, remarry, and, at 42, become America's youngest president.

7 | Jimmy Carter
NATIONAL HISTORICAL PARK
Georgia

Jimmy and Eleanor Rosalynn Carter's long union carried them from small-town life in Plains, Georgia, to naval stations, Georgia's governor's mansion, and the White House, before ultimately returning the couple to Plains following Carter's presidency. Both were born and raised in Plains; Carter graduated from **Plains High School** in 1941, and three years later, Rosalynn was named class valedictorian and graduated as well. A year later, she began dating Carter, the older brother of a childhood friend, and in 1946 they married. Visitors to the small, rural community today will start their experience at this same high school, now the park visitor center and museum, to see video interviews and exhibits on the family. The 1888 **train depot, Carter's boyhood home and farm,** plus **campaign headquarters** are also part of the experience.

8 | Gateway Arch
NATIONAL PARK
Missouri

New Yorkers might jest about the 1626 trading of $24 worth of beads by Dutch colonist Peter Minuit to the Algonquin people for the island of Manhattan, but the rest of the nation could argue an even better swap happened more than a century and a half later with the trading

▶ *President Jimmy Carter grew up on his parents' peanut farm and took over management of the business when his father died. Peanut popularity grew significantly during Carter's 1976 presidential campaign, with a caricature of the crop seen around the world.*

of $15 million for more than 800,000 square miles of land, ultimately doubling the size of the nation. President Thomas Jefferson was the prime mover in the Louisiana Purchase of 1803, and the subsequent exploration of the new land and beyond revealed the extent of the extraordinary deal. To commemorate the purchase and mark a site near where explorers Lewis and Clark passed on their epic trek, the **Gateway Arch**—previously known as the Jefferson National Expansion Memorial—was completed in 1965. As the centerpiece for the Mississippi riverfront complex, the stainless-steel arch is a dynamic testament to Jefferson's vision. In a spacious subterranean area below the arch, **museums, theaters, and exhibits celebrate the efforts of pioneers and explorers**—and add a

special tribute to the construction crews who built the arch. Ride a string of cars inside the arch to an observation area below the 630-foot peak. Through the windows looking east are America's origins, and to the west the fulfillment of Jefferson's vision.

9 | George Washington Birthplace
NATIONAL MONUMENT
Virginia

The "Father of His Country" was born in 1732 to Augustine and Mary Washington in a house 80 miles downstream from Mount Vernon, where George would later make his adult home. His father's Popes Creek Plantation—on the southern shore of the Potomac River, near land settled by the Washington family in 1657—is long gone, but what has been preserved has been done to a level that could be recognizable to George Washington today. The house he was born in burned down in 1779; its appearance is unknown. A **"memorial house," based on designs of 18th-century well-to-do Virginia homes,** was built here in the 1930s to honor the 200th anniversary of Washington's birth and is filled with period pieces. A one-mile walking path leads to the grave sites of Washington's father, grandfather, and great-grandfather.

Tour historic Ford's Theatre to glimpse the box where Lincoln was fatally shot.

10 | Ford's Theatre
NATIONAL HISTORIC SITE
District of Columbia

On April 14, 1865, Good Friday, the Civil War was largely over. Robert E. Lee had surrendered to Grant at Appomattox, and President Abraham Lincoln was given something he had prayed for: a chance to govern states that were, once again, united. Looking forward to a well-deserved evening out, he and wife Mary headed a few blocks from the White House in Washington, D.C., to Ford's Theatre to see the comedy *Our American Cousin.* Shortly after 10 p.m., famed actor and Confederate sympathizer John Wilkes Booth snuck into the president's private booth and fatally wounded Lincoln with a single shot. When the president died at 7:22 the following morning, **Ford's Theatre was written into history.** Although more than a century and a half has passed, the tragedy of the evening still stings. Due to a multiyear restoration completed in 2009, the theater's interior evokes the age of Lincoln, and the **basement museum** displays fascinating exhibits, including a handbill from the play, Booth's diary, and the small pistol Booth wielded.

SLEEPING & EATING

Settle in at the beautifully rustic Lake McDonald Lodge after a day of boating on Glacier National Park's Lake McDonald (p. 273).

HISTORIC LODGES

Many parks feature accommodations that immerse visitors in history, transporting them back to the days when railroads opened new areas of the country to travel and architects used native materials to create designs now called "parkitecture." Some lodges lack more modern amenities, but as in all real estate, the attraction here is location, location, location.

1 Death Valley
NATIONAL PARK
California

Well before a national park was established at the hottest, driest, and lowest spot in the United States, a borax-mining company built a beautiful lodge on a hill at the mouth of Furnace Creek Wash. Local mines had closed, and the company was looking for a way to make money from the narrow-gauge railroad it had built to ship ore. Opened in 1927, the Inn at Furnace Creek was built in the Spanish mission style, with stucco walls, towers, arches, and a red tile roof. Thanks to the skills of its architects and landscape designers, the inn blended nicely with the desert surroundings, its colors matching the Funeral Mountains that serve as its backdrop. Following a $155 million renaissance, the inn has been transformed into a true luxury resort, now named **The Oasis at Death Valley,** with palm trees, gardens, and a spring-fed swimming pool. There could hardly be a more comfortable respite from one of the most hostile environments on Earth. As a bonus, guests can play a round at the world's lowest-elevation golf course, an 18-hole layout more than 200 feet below sea level. *oasisat deathvalley.com*

2 Mount Rainier
NATIONAL PARK
Washington

In the southwest section of Mount Rainier National Park, **Paradise Inn** is easy to get to now. But that wasn't true when early 1900s visitors to Mount Rainier desired a place to stay when they explored the Paradise region of the park. A few iterations of camp came first, but in 1916 Paradise Inn was constructed—at a cost of approximately $100,000—in the exposed timber frame style that was popular at the time. Interior timber was cut from Alaska cedars. When it opened in 1917, Paradise had 27 guest rooms, plus platform tents to house overflow guests. A 104-room annex was added just a few years later. Today Paradise remains quite popular among visitors, even with the charm of '20s-era lodging dictating the lack of televisions, phones, and internet (though there is cellphone coverage in the area). In the inn's dining room, enjoy favorites like bourbon buffalo meatloaf by the great fireplace, then settle in on the patio for one of the nightly ranger talks. *mtrainier guestservices.com*

3 Glacier
NATIONAL PARK
Montana

The commanding, five-story **Many Glacier Hotel** sits in a spectacular location in the northern part of

Spectacular Many Glacier Hotel has been welcoming guests since 1915.

Glacier National Park, surrounded by rugged glacier-shaped peaks such as the great promontory of Grinnell Point. Swiftcurrent Lake is just steps away. Opened in 1915 by the Great Northern Railway, the hotel was designed to attract tourists who would arrive by rail; today it's reached by a road that runs alongside Swiftcurrent Creek and Lake Sherburne. A national historic landmark, Many Glacier Hotel was designed with a Swiss-chalet theme, which is reflected in its Interlaken Lounge, Heidi's Snack Shop, and lakeside restaurant, Ptarmigan. Rooms here are both rustic and comfortable, keeping with the lodge's distinct era. Hikers will find lots of nearby opportunities on trails that range from flattish and easy to long-distance treks into the park's backcountry; guided boat tours leave from the hotel's dock. *glaciernationalparklodges.com*

4 | Grand Canyon
NATIONAL PARK
Arizona

Just after the turn of the 20th century, the Atchison, Topeka, and Santa Fe Railway built a spur line north from Williams, Arizona, to the Grand Canyon to haul copper from a proposed mine. As it turned out, there wasn't enough copper to be profitable. So the railroad hired architect Charles Whittlesey to build a hotel on the South Rim to draw tourists and hopeful train passengers. Opened in 1905, **El Tovar Hotel** has elements of a Swiss hunting lodge, which was a popular design theme of the time. Since then, the hotel, a national historic landmark, has hosted Theodore Roosevelt (twice), William Howard Taft, Albert Einstein, Oprah Winfrey, Bill Clinton, Sir Paul McCartney, and thousands of other guests who have marveled at the canyonside views. Though the lobby and other public areas look much as they did in 1905, the rest of the hotel was extensively refurbished in the early 2000s; all 78 rooms have a range of modern amenities (note, however, that the hotel has no elevators). For meals, El Tovar's cuisine is internationally lauded; arrive early to enjoy the lounge's veranda. No matter how appealing El Tovar is, however, nothing can compete with the grand spectacle just 20 feet from its front door. Free park shuttle buses take guests to shops, visitor centers, and scenic overlooks along the South Rim. *grandcanyonlodges.com*

5 Yosemite
NATIONAL PARK
California

The Ahwahnee's major attraction dates to before its opening in 1927: The hotel was specifically sited to provide views of Yosemite Valley, one of the most spectacular landscapes on Earth. With Half Dome, Yosemite Falls, Glacier Point, and other iconic natural landmarks just outside, the Ahwahnee gives guests a chance to explore the valley before day-trippers arrive. The hotel's impressive rock-and-wood facade holds a secret: The "redwood" is actually concrete, shaped and stained to look like

Phantom Ranch

Phantom Ranch dates from an era (1922 to be exact) when national parks, or more specifically, railroads near the parks, needed fine accommodations to attract passengers to their rail lines. The Santa Fe Railroad and Fred Harvey Company enlisted architect Mary Colter to design an oasis destination on the site of an old prospectors' camp for well-heeled, mule-mounted tourists—a place where they could relax, splash in Bright Angel Creek, and enjoy fine dining. The mule trip to the bottom of the canyon became *the* signature trip in Arizona's Grand Canyon National Park and remained so for decades.

Today, hikers, river runners, and mule riders make the trek to the complex of stone cabins and dormitories that lies in the shade of cottonwood trees. Hikers need to be in great shape to make the round-trip 10.5-mile journey to the canyon bottom; a family-style meal is a great reward. Reservations are made via lottery 15 months prior to travel. Note: Dormitories are currently closed for extensive repairs to the water facilities. *grandcanyonlodges.com/lodging/phantom-ranch*

wood. Designers wanted to lessen the possibility of fire, a constant danger in all-wood buildings before the days of sprinkler systems. More than 5,000 tons of stone and 1,000 tons of steel were brought to the park during construction. The Ahwahnee's design blends elements of art deco, Native American, Middle Eastern, and arts and crafts styles. The Great Lounge features a massive stone fireplace and floor-to-ceiling windows with stained-glass panels, and the unique Mural Room boasts wood paneling and a large mural of Yosemite flora and fauna. Other amenities include a heated outdoor pool, sweets shop, rooms with parlors, and cottages. *travelyosemite.com*

6 | Hawai'i Volcanoes
NATIONAL PARK
Hawaii

Volcano House has long ranked with the most distinctive lodges in the National Park System. Perched on the rim of Kīlauea Crater, the hotel has been in operation in one form or another since the mid-19th century and famously offers a splendid vista into the sometimes active crater. The present 1941 establishment replaced an 1877 building whose guests included Mark Twain and Franklin D. Roosevelt. Accommodation options range from cozy and historic to deluxe crater views. Proudly "free of modern

▶ *Fifty-seven years of guest registers from historic Volcano House, in Hawai'i Volcanoes National Park, have been digitized, showcasing illustrations, stories, poetry, and reports of volcanic activity from visitors to the site from 1865 to 1922.*

distractions," the inn provides bikes, guided tours, board games, and evening cookies more than 4,000 feet above sea level. *hawaiivolcanohouse.com*

7 | Crater Lake
NATIONAL PARK
Oregon

The fact that this Oregon lodge is only partly historic is a little disappointing to architecture buffs, but there's plenty of good news to offset that fact. Completed in 1915, **Crater Lake Lodge** has had a checkered history, and at one point was in danger of demolition. Structural problems led to it being extensively rebuilt before its reopening in 1995. The picturesque Great Hall is largely original, however; and the

lodge's overall rustic style is reflective of the 1920s, in keeping with other national park inns of the era. Guests can learn about the lodge's past in the Exhibit Room, just off the lobby, and savor the flavor of sustainably and locally sourced Oregon cuisine. Rooms are more up to date than those at many parks' historic inns, and there's one more priceless attribute: The lodge sits right on the rim of Crater Lake, overlooking pristine waters and sheer cliff walls—one of North America's most scenic natural wonders. *travelcraterlake.com*

8 | Shenandoah
NATIONAL PARK
Virginia

Set near the halfway point of 105-mile Skyline Drive in Shenandoah National Park, **Big Meadows Lodge** takes its name from the large, grassy ridgetop area where it was built in 1939. Local stone and wood (much of it from chestnut trees, now virtually extinct) were used in construction, and the result is a lodge that well matches its Appalachian Mountains environment. The original lodge had 29 rooms, but a variety of other rooms and cabins have been added since the facility opened; newer rooms tend to be larger and have more amenities. The park's Harry F. Byrd, Sr. Visitor Center is located in the Big Meadows area, and a wide array of hiking

Three-hundred-year-old live oaks drape the historic Greyfield Inn, built by the Carnegies in 1890, on Cumberland Island.

The inn's rustic design had a major influence on subsequent architecture in other parks. Enlarged in 1915 and 1927 (under Reamer's supervision), the inn now has 327 rooms. There is no air-conditioning, internet, or television, and the inn is open only seasonally, but it goes without saying that, for a legendary spot like Old Faithful, reservations book up in a snap. *yellowstonenationalpark lodges.com*

10 | Cumberland Island
NATIONAL SEASHORE
Georgia

In 1900, Thomas and Lucy Carnegie, of industrial fame, built what is now the **Greyfield Inn** for their daughter Margaret. The only commercial property on protected Cumberland Island, Greyfield's 200 acres are accessible via the *Lucy R. Ferguson* ferry. The property, converted to an inn in the 1960s, is still managed by the family today and now operates as an all-inclusive haven featuring three farm-to-table meals a day plucked from the Greyfield Garden. Accommodations include fireplaces, shaded verandas, a cozy library, period furnishings, and plenty of Southern hospitality. Add as many (or as few) activities as you desire, with marshland to ocean options ranging from naturalist tours, kayaking, biking, and birding to simply savoring 18 miles of quiet beachfront. *greyfieldinn.com*

and horseback trails are nearby, including easy access to the Appalachian Trail. It's not far to Lewis Falls, one of dozens of waterfalls in the park. *goshenandoah.com*

9 | Yellowstone
NATIONAL PARK
Wyoming

The entire area around Yellowstone's iconic Old Faithful geyser in northwestern Wyoming is a national historic district, with a major highlight being the 1904 **Old Faithful Inn,** itself a national historic landmark. One of the few remaining log hotels in the United States, and one of the largest log hotels in the world, the inn was designed by Robert C. Reamer with asymmetry meant to reflect the disorder of the natural world; its steeply pitched roof mimics mountain peaks. Seven hundred feet long and seven stories tall, Old Faithful Inn boasts a lobby with a 65-foot-high ceiling and a huge fireplace built of volcanic rhyolite stone. No matter how many photos visitors may have seen, they are always awestruck viewing this space in person for the first time.

BACKCOUNTRY STAYS

Many national parks feature housing that can only be reached by foot or boat, or by remote, little-traveled roads. Such is the case with each of the following accommodations, where guests get a roof over their heads and varying levels of comfort and services, but always a sense of connection with the beauty of the park surrounding them.

1 Great Smoky Mountains
NATIONAL PARK
Tennessee

LeConte Lodge isn't for everybody, but guests who appreciate its solitude, simplicity, and beautiful surroundings clamor to snag a reservation year after year. No roads lead to this Tennessee establishment, the only lodging in Great Smoky Mountains National Park; getting to the lodge requires a hike of at least five miles. (The lodge brings in supplies three times a week on llamas.) There's no electricity or telephone, and bath time means using buckets and wash basins (bring your own washcloth and towel). Kerosene lanterns provide light, and propane heaters take some of the chill out of the elevation of 6,360 feet. Hearty meals are served family style, and at night guests retire to one of the extremely basic cabins or lodge units. At LeConte Lodge, rustic really means rustic. But once here, guests enjoy birdsong, fabulous Great Smoky sunrises and sunsets, relaxing rocking chairs, dark star-speckled skies, and the glorious feeling that they've left the modern world far, far away. *leconte lodge.com*

2 Voyageurs
NATIONAL PARK
Minnesota

Accessible only by water and 15 miles from the nearest road, **Kettle Falls Hotel** is the only lodging option in Voyageurs National Park. But despite few choices, guests don't feel like they're settling. Except, that is, when they visit the hotel bar, famous for its sloped (but safely supported) floor. Arrive by boat or plane to this remote spot on the edge of the U.S.-Canada border, where accommodations range from simple hotel rooms to villas that sleep up to eight. Electronic amenities are largely nonexistent, and that's for good reason. Rent a boat, canoe, or kayak to explore Rainy Lake, then settle in for a drink on the porch and fresh-catch dining hall meals. Kettle Falls has a colorful past, from fur traders to gold miners to bootleggers—be sure to browse old photos and ask your hosts for the tales. *kettlefallshotel.com*

3 Ross Lake
NATIONAL RECREATION AREA
Washington

Ross Lake Resort has all the trappings of a rustic old fishing lodge, but plenty of nonfishing types have discovered the serenity of this seasonal resort's roadless setting on the south end of 20-mile-long Ross Lake. The 15 self-catering floating cabins are modest but well furnished. Bedding, a full kitchen,

A "Bino" Primer

Whether you're bird-watching from the back porch of a remote lodge or scanning the horizon from a high point on the trail, whatever is worth seeing in the parks is worth seeing well—which is why good binoculars might be the best travel investment you can make.

If possible, resist buying bargain basement. Cheap (under $100) binos will magnify, but their optics are poor. They cause eyestrain that you might not be aware of; rather, you just lose interest in looking through them. When you shop, spend some time outdoors comparing models, and hold a fixed gaze for at least 30 seconds for each. You'll note differences. If you're looking at 7×42s, the seven indicates magnification; an object will appear seven times closer than with

the naked eye. Figures of seven, eight, and 10 are common. Magnification of 10 may sound good, but it can be hard to prevent the image from looking shaky. The 42 is the diameter of the objective lens in millimeters. A larger number means more light-gathering ability—in other words, better low-light performance. That's important because the best wildlife viewing is early in the morning or at dusk. With compact binoculars, say, 8×20, you get shirt-pocket convenience, but just average low-light performance.

Lots of other factors figure in to binoculars' cost as well, such as fully multicoated lenses that sharpen the view and sturdy construction that holds barrel alignment. Almost always, higher priced binos merit their price tags, something you'll be grateful for upon seeing that elusive critter.

---

and a barbecue are provided, but guests must bring along all their own food. Access is by ferry from Diablo Lake—the resort truck meets the ferry and shuttles guests in on a two-mile private road. Guests can rent the resort's boats or kayaks, hike, or honor the heritage of the place by fishing for wild rainbow trout or bull trout. A popular day trip is to rent a boat, motor 14 miles up the lake to the Desolation Peak trailhead, and make the four-mile, 4,000-foot pilgrimage to the peak, where Beat Generation writer Jack Kerouac spent a summer as a fire lookout. Cabins can book up

quickly, so prepare to get on the waiting list a year in advance. *rosslakeresort.com*

4 | Chesapeake & Ohio Canal
NATIONAL HISTORICAL PARK
Maryland

Staying in one of the three houses in Maryland once occupied by Chesapeake & Ohio Canal lockkeepers is similar to sleeping inside a historic landmark, which is exactly the case. The **Canal Quarters** lockhouses, which stand right beside the canal towpath, are each decked out with

period furnishings and stocked with literature about the canal's past, so it's easy for guests to immerse themselves in a historic journey during their stay. Lockhouse 22, at Pennyfield, might provide the most authentic experience of its mid-1830s heritage—a quiet, secluded location with no electricity or plumbing. Guests park nearby and tote in everything they need. Lockhouse 49, at Four Locks, is also quiet (although it's near a popular Potomac River boat ramp), but it has electricity and is furnished in 1920s style—the era when the canal ceased operation. Lockhouse 6, in the Bethesda neighborhood of Brookmont, is more urban (indoor plumbing and electricity) and is furnished in a 1950s fashion. The C&O Canal Trust is planning to rehab more of the old canal quarters as part of its effort to preserve and interpret the historic canal. *canalquarters.org*

5 Glacier
NATIONAL PARK
Montana

Two backcountry chalets—**Sperry** and **Granite Park**—in Glacier National Park offer guests a dash of creature comfort minus grid-associated niceties. Guests will not get hot showers or electricity, but they will be served three squares a day at Sperry and have access to the facilities to cook their own meals at Granite Park. Most important, guests get the silence of the backcountry: Both chalets are set amid Glacier's signature jagged peaks and wildflower meadows, and both are reachable only by foot—or hoof, in the case of Sperry. The 17-room, 1913 Sperry Chalet rewards a 6.7-mile approach via Sperry Trail. Granite Park is a simple 12-room shelter right on the Continental Divide, 7.6 miles from Logan Pass on the Highline Trail. Both chalets make great base camps for day hikes. From Sperry, for example, it's a spectacular 3.5-mile hike to Sperry Glacier. *sperrychalet.com, graniteparkchalet.com*

6 Denali
NATIONAL PARK & PRESERVE
Alaska

Camp Denali rests on a ridge amid open expanses of tundra 90 miles deep into the park. When it was built in the early 1950s, the camp of hand-hewn spruce-log cabins stood on the edge of the park. When the park tripled in size in 1980, Camp Denali remained as a 67-acre inholding. The camp retains a sense of rusticity befitting its roots and setting even as it has developed luxe touches and an upscale eco-lodge vibe. Guests are treated to handmade quilts on the beds and fresh produce from the camp's own organic greenhouse, served in a lovely timber-frame dining hall. But they still need to walk outside to use a private outhouse and stroll a few minutes to use the modern communal showers. After a fly-in, the stay (three-, four-, or seven-night stints) at the camp is filled with hikes, drives, and canoeing with staff naturalists as well as occasional evening programs with guest naturalists and artists. Those programs are held in the camp's log lodge, which is also the central reading room, stocked with Alaska books, and gathering place. And

The aurora borealis dances across the night sky above Camp Denali in Denali National Park and Preserve.

then there's the view—the breathtaking presence of Denali and major peaks of the Alaska Range. *camp denali.com*

7 | Yosemite
NATIONAL PARK
California

Five nights of backpacking in the High Sierra would ordinarily dictate carrying a 50-pound pack, but hikers who book Yosemite's **High Sierra Camps** get a backcountry experience that entails toting nothing more than a daypack filled with some clothing layers and toiletries. Everything else is waiting at each of the camps, strategically located around a large loop. Each camp has tent cabins, cots with blankets, and a chuck wagon that doles out ample breakfasts, dinners, and to-go trail lunches. Most hikers start from the **Tuolumne Meadows Lodge** (8,500 feet) to acclimate to the altitude, and proceed counterclockwise. Plan ahead and cross your fingers: The trips are quite popular and a lottery determines availability. *travelyosemite.com*

8 | Wrangell–St. Elias
NATIONAL PARK & PRESERVE
Alaska

Kennicott Glacier Lodge is everything a seasonal lodge in the middle of the country's largest national park should be: remote (eight-hour

▶ LeConte Lodge predates the establishment of Great Smoky Mountains National Park. Jack Huff, its founder, began building the retreat in 1926, and the Huff family operated LeConte until 1960.

drive or two-hour flight from Anchorage to nearby McCarthy), quiet (no cars on lodge property), and gloriously situated, with views of the Chugach and Wrangell Mountains and the Kennicott-Root Glacier confluence so close guests can hear the trickling of glacial meltwater. To get there, guests drive to the end of Alaska 10, walk across the Kennicott River on a footbridge, and meet the lodge shuttle for the five-mile drive in. Built in 1987 to reflect the copper-mining heritage of surrounding Kennicott, now a ghost town preserved by the park, the 25-room lodge feels like a prospector's retreat replete with family-style meals. The lodge is a home base for hikes on Root Glacier, float trips on the Kennicott River, or strolls through the old mining town, which boomed between the early 1900s and 1938. The lodge offers three meals a day, shared bathrooms, an exquisite porch overlooking the glaciers, and a sense of congeniality

befitting such a remote yet comfortable hostelry. *kennicott lodge.com*

9 | Virgin Islands
NATIONAL PARK
Virgin Islands

Cinnamon Bay Beach & Campground's spot in the national park on the north end of St. John is a choose-your-own-adventure-style getaway, with bare campsites, eco-tents, and cottages all providing easy access to acclaimed Cinnamon Bay Beach. Amid lush tropical foliage, the feel is akin to sleeping in a tree house. Watersports rentals, art classes, and more make disconnection all the more fulfilling. At **Concordia Eco Resort,** on the southeast end of the island, the ambience is more arid, but the views of cliffs and ocean are just as dramatic. The 25 eco-cabanas and eight villas are well-stocked with amenities. Both camps serve meals and provide kitchen facilities for cooking, and the trails and beaches of the national park are right outside the tent flap. *cinnamonbayvi .com, concordiaecoresort.com*

10 | Haleakalā
NATIONAL PARK
Hawaii

Here's a very different version of a romantic Hawaiian getaway: snuggling into one of Haleakalā's three

Postcard-worthy views await on the pristine shores of Cinnamon Bay in the U.S. Virgin Islands.

remote hike-in cabins, built by the Civilian Conservation Corps in the 1930s. Each is high on the mountain where the night air is brisk and clear. Amenities amount to cooking facilities (propane stove) and utensils, plus firewood to fuel a woodstove. Guests bring their own sleeping bag and use a nearby pit toilet. The main attractions of the cabins are their unmatched locations. **Hōlua Cabin,** 3.7 miles down the Halemau'u Trail, is set in native shrubland, where petrels sing at night. **Kapalaoa Cabin,** 5.6 miles via Keonehe'ehe'e Trail, is in the middle of a cinder desert, a place to stargaze and tell ghost stories over a howling wind. It's also the most private of the three; the others have campgrounds nearby. **Palikū Cabin,** 9.2 miles via Keonehe'ehe'e Trail, is in a cloud forest and can be very wet. All three are very popular, especially on full-moon nights. *recreation.gov*

10 BEST PARKS FOR
CABINS

Cabins can be a compromise between the convenience of a hotel and the "roughing it" quality of camping. Some cabins offer very comfortable modern furnishings, and others only very primitive accommodations. All have in common a prime spot in stunning environs and the potential to be the exact kind of cozy hideaway you're in need of amid the wilds.

1 | Grand Teton
NATIONAL PARK
Wyoming

By the time these cabins came to overlook the Tetons at **Triangle X Ranch,** they had already had a history of their own. All 20 porch-clad log homes originated in Jackson Hole: Some were built with local logs hauled in by horses; others housed 1800s settlers before being moved to their Grand Teton National Park site. Comfortably cozy and suitably rustic, Triangle X cabins range from one to four bedrooms and feature private bathrooms. But accommodations aren't actually the main appeal of this all-inclusive vacation spot: Triangle X is a five-generation family dude ranch, with Western activities to match all interests and ages— horseback riding, wildlife viewing, photography, "sing for your supper" cookouts, square dancing, float trips, plus kids and teens programs, too. *trianglex.com*

2 | Catoctin
MOUNTAIN PARK
Maryland

Catoctin Mountain Park is set in beautiful mountain scenery, but it has the advantage of an interesting history, too, given that it was established to create jobs during the Great Depression. Within the park, **Camp Misty Mount**—a camp of wood-and-stone cabins that date from 1936—is a charming place to spend a week. The cabins' original architectural detail and crafts-manship are largely intact, and the accommodations are basic but clean. The cabins have cots, mattresses, and electric lights inside, and picnic tables and grill rings outside. Guests need to bring their own bedding, cookware, and food storage devices. There are dining facilities and bathrooms located centrally in the camp, and the small camp store sells charcoal, ice, and other basic provisions seasonally.

Guests desiring extra privacy should ask for one of the cabins located off the main loop. The changing leaves in Catoctin are beautiful and the crisp autumn evenings refreshing, so plan a visit for the fall. *recreation.gov*

3 | Kenai Fjords
NATIONAL PARK
Alaska

Kenai Fjords National Park, at the tip of the Kenai Peninsula, is still truly outback territory. The Ice Age lives on here—Harding Icefield, parent of dozens of glaciers, lies at the heart of the park, much of which is still untrammeled wilderness. For guests wishing to immerse themselves in this unspoiled setting, Kenai has three rustic public-use cabins, two designated for summer months and one for winter. The summer-use cabins are accessible only by floatplane, water taxi, private vessel, or charter boat; the cabins need to be reserved

A Harding Icefield shelter looks out over Exit Glacier in Kenai Fjords National Park.

well in advance of a planned visit. **Holgate Cabin** offers great views of Holgate Glacier from its front porch, and guests may see bears ambling over the cobble beach nearby in search of berries. **Aialik Bay Cabin** looks out over Aialik Glacier, and at low tide you can walk out over the bay for more than a mile and poke around in the tide pools. **Willow Cabin,** the winter-use cabin, is propane heated and accessible only by snowmobile, cross-country skis, snowshoes, or dogsled, but a stay there will intimately acquaint you with the stillness and pristine beauty of an Alaska winter. *recreation.gov*

4 | Cape Lookout
NATIONAL SEASHORE
North Carolina

Still largely undeveloped, secluded Cape Lookout National Seashore is perfect for travelers seeking a primitive beach experience. The park offers basic rental cabins on two of its three barrier islands. The **Great Island Cabin Camp** on the South Core Banks has individual units that each sleep between four and 12 and have hot water, basic furnishings, and a kitchen and private bath. At North Core Banks, the **Long Point Cabins** are duplex-style, sleeping six to a side. Each of these units has all the amenities of the Great Island cabins, and several also have air-conditioning. The duplex cabins share porches and decks, with nice views across

▶ *Camp David, the tranquil presidential retreat, is located within Catoctin Mountain Park but, as could be expected, it is not open to the public.*

the beach to the ocean. Note: These cabins were damaged in Hurricane Dorian but are expected to reopen. In keeping with the primitive character of Cape Lookout, the cabins on both islands are pretty spartan, so guests need to bring their own bedding, towels, cookware, and generators. Be sure to purchase an off-roading vehicle permit prior to arrival (it's required). *recreation.gov*

5 | Yukon–Charley Rivers
NATIONAL PRESERVE
Alaska

Traversed by the mighty Yukon River and encompassing the entire course of the smaller Charley River and other minor waterways, the park has a number of cabins available for public use, all of which are easily accessed from the Yukon River. These cabins are primitive, offering basic accommodations, but their real charm lies in their historical resonance. **Nation Bluff Cabin,** for instance, was built in 1934 by a trapper; **Slaven's Public Use Cabin** is

just 100 yards from Slaven's Roadhouse, a restored 1930s roadhouse and dog-drop location on the Yukon Quest International Sled Dog Race; and **Coal Creek Camp** was a 1930s gold-mining camp in which one cabin has been renovated for public use. In addition to their handy proximity to the Yukon River—the main means of accessing the different regions of the park, especially during summer—these cabins provide an interesting window into the land's history and past use. The no-fee cabins are occupied on a first-come, first-served basis. *nps.gov/yuch/planyourvisit/publicusecabins.htm*

6 | Olympic
NATIONAL PARK
Washington

All but five percent of Olympic National Park is officially designated wilderness, and not a single road actually passes through its heart. Guests wishing to explore the park's interior will need to do a lot of legwork—literally. For a restful haven after a day of hiking and climbing, rent a cabin at **Kalaloch Lodge** sitting high on a bluff in the 73-mile strip of the park's protected Pacific coastline. The Bluff Cabins overlook the Pacific, and while the Kalaloch Cabins sit a row back from the Bluff Cabins, they are only a few steps farther from the sandy coast. The units have varying levels of amenities, from fully outfitted kitchenettes and fireplaces to more basic furnishings. From the lodge, it's only about 30 miles to the famous 7,980-foot Mount Olympus and the subalpine forests and alpine meadows of the park's interior. *thekalalochlodge.com*

7 | Death Valley
NATIONAL PARK
California

While **The Oasis at Death Valley**—the zhuzhed-up iteration of the historic Furnace Creek Inn property in Death Valley National Park—provides an assortment of options for accommodations, among the most popular are the 80 cabins added in recent years. Called Ranch Cottages, these 400-square-foot desert havens are centrally located on the property, with restaurants, golf, a saloon, and general store all within a short scenic stroll. Aligned in snug rows, the cozy one-bedroom quarters have the choice of one king or two queen beds, and all feature shared front porches, a welcome respite from a day exploring the otherworldly grandeur of Death Valley. Worth perusing, the Borax Museum in the resort's inn gives a history of the property, including artifacts from its mining days. *oasis atdeathvalley.com*

8 | Isle Royale
NATIONAL PARK
Michigan

Isle Royale National Park, a remote island in Lake Superior about 15 miles from the coast of Ontario, Canada, can be reached only by boat or seaplane, and visitors who make the effort to get there usually stay a few days. The largest overnight spot on the island

A Lot of Work for Hand Soap

For all of its dazzling geological riches, including gold strikes in the Panamint Range, it was a mineral called borax that put Death Valley on the map and in the American consciousness. The 1880s were "borax rush" days in Death Valley and the Nevada desert. Consumer demand for the stuff was huge, as it was used to aid digestion, keep milk sweet, improve skin complexion, remove dandruff, and was even a purported cure for epilepsy and bunions. Later it was used in powdered soaps called 20 Mule Team Borax and Boraxo, whose manufacturer, the Pacific Coast Borax Company, sponsored a popular radio and television series called *Death Valley Days.*

Back in the original Death Valley days, the ore lay on the surface of the desert floor in a compound form known as "cottonball," and it only needed to be scraped up, refined, and hauled away. But hauling anything out of Death Valley was a daunting task. The nearest railhead was in Mojave, 165 miles away. The ore had to be hauled up and over the Panamint Range across sand and rocks on extremely primitive roads. The specially made wagons weighed 7,800 pounds empty, 73,000 pounds loaded. William T. Coleman, owner of the Harmony Borax Works on the site of the present-day Oasis at Death Valley, then came up with the idea of using 20-mule teams to haul the borax and the water needed for the grueling journey. Though the operation ran only from 1883 to 1889, the mule teams became a symbol of the pluck and persistence of Old West prospectors.

is **Rock Harbor Lodge** at Rock Harbor, which has cottages with kitchenettes and private baths. All units have studio-type living areas and electric heat. Canoes can be used to get to sights like Lookout Louise, one of the most beautiful overlooks on the island, and Raspberry Island, home to a living bog. There is also easy access to hiking trails that lead to Scoville Point, surrounded on three sides by Lake Superior, and 938-foot Mount Franklin. On the southwestern end of the isle, **Windigo Camper Cabins** on Washington Harbor has comfortable quarters with private decks and outdoor restrooms. *rockharborlodge.com*

9 Yosemite
NATIONAL PARK
California

Stays in California's Yosemite National Park are neither cheap nor easily booked. **The Redwoods,** located four miles inside the south entrance to the park, offers 120 vacation homes, from one-bedroom rustic log cabins to six-bedroom spacious modern homes. All are outfitted with kitchens and fireplaces, and are fully stocked with linens, bedding, and cookware. There are two markets within walking distance of The Redwoods, and the complex is beautifully situated between the Chilnualna Creek and the South Fork of the Merced River.

Shenandoah's Skyland Resort, at the highest point on Skyline Drive, offers stunning views of the park.

It's about a 30-minute scenic drive to Yosemite Valley. One catch—each of the homes at The Redwoods is privately owned, so the decor and level of comfort will vary, and generally speaking guests get what they pay for. But the location is superb, and these cabins and homes are a convenient, scenic base from which to ramble in the famous park.

10 Shenandoah
NATIONAL PARK
Virginia

Shenandoah National Park, which stretches along the ridgeline of the Blue Ridge Mountains, offers nice options for cabin lodging. All are situated along the park's famous Skyline Drive, which runs the length of the Virginia park. Both the seasonal **Big**

Meadows Lodge and **Skyland Resort,** at mileposts 51 and 41.7 respectively, offer comfortable cabins with electricity and private bathrooms, and meals are available at the nearby lodges. Skyland Resort also has three family cabins with living areas, separate bedrooms, and kitchenettes. **Lewis Mountain Cabins,** at milepost 57.5, has both single and double-room cabins. A stay here is a truly restful experience—early-rising guests may catch sight of white-tailed deer and other wildlife before they seek the refuge of the deep woods during the day. All the cabins have heat, electricity, private bathrooms, and bedding provided. There are outdoor grilling rings, but guests need to bring utensils and food storage devices. A nearby camp store sells plenty of provisions. *goshenandoah.com*

JUST-OUTSIDE-THE-PARK LODGING

America's national parks have long been distinguished for their grand (and often quirky) lodgings. Some of the most historic, lauded, and luxurious establishments are located right outside the park gates in the adjoining towns, wilderness, and rural areas that perpetuate the vibe of that particular park.

1 Acadia
NATIONAL PARK
Maine

The locals have been known to call this area the "quiet side" of Mount Desert Island, and with all the amenities already available at **The Claremont,** that's a good thing. Opened in 1884, this elegant waterside property at the cusp of Acadia National Park has views of Somes Sound, Cadillac Mountain, and the inn's gracious gardens. Restored rooms have luxurious linens and historic architecture, and cottages and grandiose homes offer an even more elevated experience. All guests get to enjoy yard yoga and croquet, beach cruisers, cabanas by the heated pool, plus the Botanica Spa and mixology classes in the bar. And when it's time to lace your boots and explore the national park, Acadia is just a five-minute hop, skip, and jump away. *theclaremonthotel.com*

2 Glacier
NATIONAL PARK
Alberta, Canada

Set against a backdrop of jagged snowcapped peaks, the **Prince of Wales** is one of several classic wilderness lodges developed by the Great Northern Railroad during the early part of the 20th century. Located in Waterton Lakes National Park in southern Alberta, the hotel is just over the U.S.-Canada border from Glacier National Park in Montana. Although named after the future King Edward VIII of England—who later abdicated and became the Duke of Windsor—the 1920s building is more Swiss Alpine than British Empire. The seven stories rise to a massive, green A-frame roof capped with a 30-foot bell tower. Don't miss afternoon tea in the dining room, featuring Earl Grey and scones with a fabulous view over Waterton Lake. The 86 rooms, decked out with dark wooden furniture and earthy fabrics, feature lake or mountain views. *glacierpark collection.com*

3 Golden Gate
NATIONAL RECREATION AREA
California

For unmarried U.S. Army officers stationed at the San Francisco–area Presidio in the early 1900s, home was the Georgian Revival–style

Pershing Hall. Next door, the Montgomery Street Barracks housed a cavalry troop plus two infantry and six artillery companies. Today, fortunate travelers can check into the **Inn at the Presidio** and **Lodge at the Presidio,** respectively, to enjoy elegant rooms in this historic setting. Complimentary breakfast and afternoon wine and cheese receptions are welcome amenities, while golf and bowling add to the appeal. The boutique hotels also enjoy a particularly special setting, with unobstructed views of the Golden Gate Bridge and Presidio forest. *presidiolodging.com*

4 | Mount Rushmore
NATIONAL MEMORIAL
South Dakota

The "Residence of Presidents" in downtown Rapid City, South Dakota, **Hotel Alex Johnson** has hosted six American heads of state over the last 90-plus years. The "historically hip" nine-story hotel still looks much as it did on opening day in 1929, with the redbrick facade topped by a faux Tudor crown and the hotel's trademark neon sign. Native American motifs—a tribute to the Plains Indians who have called the area home for thousands of years—dominate the lobby with its huge stone fireplace, massive chandeliers, and wood-beam ceiling. The historical ambience has been complemented by a thorough updating of the 143 guest

▶ *Guests of the Stanley Hotel can request a "spirited" room that has been known to have high paranormal activity, including 217, the Stephen King Suite, and Ghost Hunters favorites 401, 407, and 428.*

rooms, including modern bathrooms and comfy king beds. The Presidential Suite is a masterpiece of handcrafted woodwork and one-of-a-kind light fixtures. Other additions include a market on the ground floor and AJ's Wicked Salon & Spa with its array of massages, facials, and other treatments. Rapid City is closest to Mount Rushmore National Memorial, but four other park system units are less than an hour's drive from the hotel: Wind Cave, Jewel Cave, Badlands, and Minuteman Missile. *alexjohnson.com*

5 | Shenandoah
NATIONAL PARK
Virginia

A small luxury lodge on the eastern flank of Shenandoah National Park, the **Inn at Little Washington** is both a marvelous place to base a stay in the region and an incredible way to stimulate your taste buds. The town

of "Little" Washington (pop. 83) takes its name from the man who originally surveyed the area in 1749, none other than young George Washington. The inn didn't come along until 1978, when chef/owner Patrick O'Connell converted an old wooden gas station and garage into a quaint colonial boutique hotel with sweeping views of the Virginia countryside. The 23 guest accommodations, each unique, were created by London stage designer Joyce Evans and are furnished with pieces made on both sides of the Atlantic. Many boast private balconies or gardens. O'Connell has developed the property into a bastion of fine dining that draws gourmands from around the globe. His Michelin three-star restaurant—which utilizes ingredients from the inn's own small farm or other local purveyors—has garnered numerous accolades, including a remarkable Lifetime Achievement Award from the James Beard Foundation. *theinnatlittlewashington.com*

6 | Rocky Mountain
NATIONAL PARK
Colorado

Inventor and entrepreneur F. O. Stanley (famous for the Stanley Steamer automobile) arrived in Estes Park in 1903 hoping that Colorado's pure mountain air would cure his tuberculosis. Within months he was on the road to recovery, and he

Located in Estes Park, just outside Rocky Mountain National Park, the Stanley is Colorado's most celebrated hotel.

would summer in Estes Park the rest of his life. Stanley built a guesthouse for his well-heeled friends from the East, a huge wooden, whitewashed structure that would soon morph into Colorado's celebrated **Stanley Hotel.** Theodore Roosevelt, "Unsinkable" Molly Brown, Enrico Caruso, and John Philip Sousa are among those who stayed here during its heyday. A young writer by the name of Stephen King came in 1974 seeking inspiration; the book that came from that stay, *The Shining,* became a horror classic. Nowadays the renovated Stanley is the epitome of an elegant, old wilderness escape, the rooms modernized, but not at the expense of bygone character and charm. Nearly every window offers stunning Rocky Mountain views, and a park entrance gate is just 15 minutes up the road. The hotel's restored concert hall offers a slate of performances, from theatrical séances to a spooky concert series and a New Year's Eve ball. *stanleyhotel.com*

7 Yosemite
NATIONAL PARK
California

One mile from the Hetch Hetchy Entrance to Yosemite National Park, **Evergreen Lodge** straddles a ridge between two branches of the Tuolumne River. The Main Lodge (built in 1921), recreation buildings, and cabin clusters are scattered across 20 acres of old-growth forest in a part of the Yosemite region less explored by outsiders. Silence, solitude, and comfort are what Evergreen is all about, a place to get away from it all and back to nature without totally roughing it. Cabins feature private bathrooms and balconies, gourmet coffeemakers, Amazon Alexa–connected smart devices, and daily housekeeping service. And, just in case guests forget this *is* California, a massage cabana boasts deep tissue, hot stone, reflexology, and other treatments. Daytime activities run the full gamut, from excursions inside the national park to biking, hiking, fly-fishing, geocaching, and white-water rafting. Dinners at the lodge always feature a great selection of California wines. The action continues through the evening with s'mores around the outdoor fireplace, movies, or kids' crafts in Tuolumne Hall, or live music and pool at the Tavern. *evergreen lodge.com*

8 | Olympic
NATIONAL PARK
Washington

Opened in 1926, rustic **Lake Quinault Lodge** perches on the secluded southwest side of Olympic National Park overlooking the lake of the same name. The three-story main building, with its shingled

Classic Souvenirs

National parks have certainly sparked more than their fair share of key chains and snow globes, but they have also inspired mementos that now rank as endearing period pieces and even high art.

Many classic park souvenirs display a touch of class—like a sterling silver sugar shovel and nut scoop embossed with Yellowstone scenes from the 1920s, and a lady's compact featuring a photo of Glacier or a bronze serving tray embossed with an image of Crater Lake, both from the 1930s. Others reflect something of the park, for example, the Great Smoky Mountains miniature iron skillet ashtray.

Some mementos are clever, such as the 1920s-vintage Fred Harvey Grand Canyon miniature mail pouch with 12 black-and-white photographs inside. While others—like colorful wooden Isle Royale fishing lures from the 1950s—were intended to be useful.

And then there are those that just make you scratch your head: In the 1940s, Yosemite gift shops stocked wooden phone book covers with taffeta ribbons and a bear carved on top. Some early souvenirs literally gave away part of the park: Petrified Forest sold small chunks of petrified wood with tiny metal dinosaurs glued to the top.

Southwestern parks seemed to inspire several drink-related items. In the 1950s, a collectible frosted tumbler full of iced tea was exactly what you wanted to have in your hand in Death Valley or Joshua Tree. Among vintage Grand Canyon keepsakes are copper shot glasses and elaborate beer steins.

How much are these souvenirs worth now? It depends on age, condition, rarity, and sheer novelty. A Glacier mountain goat wood carving by artist John L. Clarke now fetches in excess of $1,500. First edition books by 19th-century national park explorers and surveyors can go for much more.

Rocking chairs welcome guests to enjoy Blackberry Farm's mountain view.

sides and steeply pitched roof, looks like something out of a Brothers Grimm fairy tale. Guests gather at night for dinner in the Roosevelt Dining Room and for chats around the great fireplace in the lounge. Accommodations include rooms with fireplaces, spacious lakeside rooms, and pet-friendly boathouse rooms in an annex even older than the main lodge. A broad lawn runs down to a lakeshore perfect for boating, fishing, or swimming. Hiking and biking are also popular with guests. An outdoor swimming pool and sauna and massage services round out the amenities. The Quinault Rain Forest section of the park is on the lake's north shore, easily reached by road or boat; Olympic's Queets Valley and Kalaloch Coast segments are nearby as well. *olympicnationalparks.com*

 ## Great Smoky Mountains
NATIONAL PARK
Tennessee

In the pastoral setting of Walland, Tennessee, two sister Relais & Châteaux resorts, **Blackberry Farm** and **Blackberry Mountain,** strike an effortless blend of mountain haven and refined luxury. The original property, 4,200-acre Blackberry Farm, is the place of epicurean dreams. Fly-fishing, sporting clays, hiking, and the spa help you work up a hunger, and meals, included

in the stay, are decadent occasions in themselves. In 2019, the Beall family launched Blackberry Mountain, where cabins, multibedroom homes, and even luxury tree houses serve as home base for wellness getaways featuring fitness, a naturopathic spa, and family activities. *blackberryfarm.com, blackberry mountain.com*

10 | Hawai'i Volcanoes
NATIONAL PARK
Hawaii

The boutique hotel **Kīlauea Lodge & Restaurant** started life in 1938 as a rustic YMCA outpost called Camp Hale O Aloha. For decades it hosted Hawaiian school kids on outings to Hawai'i Volcanoes National Park. Despite its conversion into a modern hotel, the lodge retains much of its old Civilian Conservation Corps ambience, especially in the main building with its Fireplace of Friendship constructed of stones and coins from 32 countries. Just a mile outside the park, the lodge is perfectly situated for close encounters of the volcanic kind. For guests who crave an urban escape, Hilo is only 30 minutes up the road. The rooms are all different; some of them have fireplaces and private balconies. Sunday brunch in the restaurant is a local tradition, as are the dinners prepared by the resort's internationally influenced chefs. *kilauealodge.com*

A Park on Tap

Thanks to the ingenuity of one brewer and her historic reuse lease, one of the country's smallest national parks has made a big splash in the world of craft beer.

For the 2.1 million visitors to Hot Springs National Park each year, a trip to Superior Bathhouse Brewery offers more than a pint and a plate. The only brewery in the world to use thermal spring water and the first brewery located in a U.S. national park, Superior—which also has a full-service, family-friendly restaurant—has become a reimagined next chapter for a beloved historic spot.

Rose Schweikhart was performing as a professional tuba player in Europe when she noticed that beer took on a place's identity. "I fell in love with different beers from all over the world and enjoyed seeing how unique styles emerged from different regions." After moving to Arkansas in 2011, the home brewer and Hot Springs newcomer became curious about the area's famous thermal water. So curious, in fact, that Schweikhart boldly asked the National Park Service if she could use its waters to brew commercially.

The answer came quickly: Yes, but you have to do so in the park—and we may have just the place.

WATER WORKS

First treasured by Indigenous groups and pioneer settlers, then later protected by Congress, the springs became acclaimed in the late 1800s when tents and mudholes were replaced by bathhouses and hotels. Built in 1916, Superior operated as a bathhouse until 1983, when it was the last spot to close as the resort destination became less popular. Although the bathhouse sat vacant for 28 years, the Park Service had the foresight to not only stabilize the historic buildings, but also entice entrepreneurs via adaptive reuse leases for vacant structures.

Armed with a 90-page lease proposal and historic architecture plan, Schweikhart knew she'd have to balance preserving the original building with the necessary updates to make the brewery and restaurant functional. "I feel honored to have been chosen as the caretaker of the Superior, so this was ultimately an exercise in, 'How would this idea be beneficial not only to the taxpayers and the visitors but also to the park, to the water, and to the town?'"

It's very unusual for thermal hot springs to be drinkable. Bathable beyond Hot Springs? Sometimes. But drinkable beyond Hot Springs? Rarely.

"The water that comes up in the springs today fell as drops of rain 4,000 years ago," says Schweikhart of the geological phenomenon at work in the park. "And this place has been used by humans for as long as humans have known about it."

In Hot Springs, water bubbles out of the ground at a steamy 143 degrees Fahrenheit. Each capped spring is pumped into a centralized tank, tested for safety, and then redistributed to the sites on Bathhouse Row. "While many U.S. breweries use geothermal as an energy saver, Superior's is the only beer in the world that is made with thermal water," says Schweikhart, who, along with her team, has to do very little—if anything—to treat it.

Some styles of beer, such as a hoppier hazy IPA or double IPA, require a few minerals to balance the flavor profile, much like adding salt while you're cooking. "But I try to do as little modification to the water as possible because that's what makes it unique. The flavor of the water and—as weird as it sounds—even the texture of the water, that comes through in our beer and that's what makes it special."

Sample the only beer in the world brewed using thermal spring water at Superior Bathhouse Brewery in Hot Springs National Park.

GATHERING SPOT

Nearly 14 years later, the business has garnered quite a following among park visitors, craft beer enthusiasts, and locals. What's on tap? Eighteen craft beers, including the Beez Kneez, a honey basil blonde; Spicy Ride, a jalapeño ale; Foul Play oatmeal stout; and a bock called Goat Rock. Guests can choose four tasters in a flight or opt for a "Beer Bath" to sample all the beers on tap.

In 2022, Superior launched a wildly popular patio space; in 2023, Schweikhart renovated the cooking and storage spaces; and 2024 will see an expanded upstairs dining area. "I can't make the Superior bigger, so as both Hot Springs and my customer numbers grow, it's my job to think about how we can change to keep everything running smoothly," she explains. "I don't think I'll ever be done with this place—and I don't want to be."

For its nearly seven decades in business, Superior Bathhouse was the people's bathhouse: The smallest on the block, it was a budget option that prided itself on superior service. Honoring the historic site, Schweikhart is continuing that legacy. "The bathhouse gives me access to the water, which I use to make beer, before serving it back to the people. It's a new use for a very old place."

10 BEST PARKS FOR
BUCKET LIST PARK ROOMS

What are some of the most coveted hotel rooms in America's national parks? They range from historic cabins and carriage barns to poolside bungalows where Hollywood stars once frolicked. Hard to get—and impossible to forget—each of these accommodations has a stunning setting, too. Book now to nab a chance at these unforgettable escapes.

1 Grand Teton
NATIONAL PARK
Wyoming

Jenny Lake Lodge, born as a dude ranch in the early 1920s, is now enjoying a second life as a wilderness hotel that features a main building and 37 log cabins. Modern amenities do not detract from the rustic ambience of a suite that features handmade quilts, peeled-wood furniture, and a wood-burning stove for those cool Wyoming nights. **Water Lily** is Jenny Lake Lodge's largest and most secluded cabin. Out back is a porch equipped with rocking chairs and a view through thick forest to a nearby meadow. A hearty breakfast and five-course dinner, included in the room rate, are taken in the lodge dining room. Like the Water Lily Suite, they are magnificent affairs, with entrees such as pan-seared duck breast in a pomegranate glaze with toasted almond couscous. *gtlc.com/lodges/jenny-lake-lodge*

2 Grand Canyon
NATIONAL PARK
Arizona

Not the kind of place for sleepwalking, the wonderfully rustic **Buckey O'Neill Cabin**—the oldest continuously standing structure on the South Rim—is just steps from the edge of the Grand Canyon. Arizona politician turned prospector William "Buckey" O'Neill built the cabin in the 1890s while searching for copper in the region. He later volunteered for Theodore Roosevelt's Rough Riders and died at the Battle of San Juan Hill. Part of **Bright Angel Lodge & Cabins,** the cabin exemplifies its pioneer origins with solid log walls, wood-shingle roof, and old-fashioned green door and window frames. The bygone vibe continues inside with log paneling and a stone fireplace. The suite is very popular with Grand Canyon regulars and must be reserved by phone *(888-29-PARKS). grand canyonlodges.com*

3 Death Valley
NATIONAL PARK
California

"All the advantages of hell without the inconveniences," is how an early 20th-century newspaper report described Death Valley. The same could be said today for

The Oasis at Death Valley resort is a stylish retreat near the Panamint Range.

the fabulous Pool Bungalow at **The Oasis at Death Valley.** Definitely designed with romance in mind, the stand-alone bungalow was made for no more than two, with doors that open onto private stairs that lead to the resort's deliciously sinful pool area. For a little more wiggle room, reserve one of the property's new **Casitas,** which have living quarters and a wet bar, as well as a complimentary golf cart. Opened in 1927 as Furnace Creek Resort, the inn (and its lush Garden of Eden setting) has long drawn the Hollywood crowd to the desert, with a guest list that includes Bette Davis, Jimmy Stewart, and Claudette Colbert. Clark Gable and Carole Lombard honeymooned here. The spring-fed pool is a popular attraction, but there's plenty of other ways to keep busy, including biking, spa treatments, golf, and decadent dining. *oasisatdeathvalley.com*

4 Cuyahoga Valley
NATIONAL PARK
Ohio

One of the few bed-and-breakfasts found inside a national park, the **Inn at Brandywine Falls** recalls the days when Ohio's Cuyahoga River was an integral part of America's industrial revolution. James Wallace, owner of the local mill, constructed the clapboard complex in 1848, and it passed down to modern times seemingly untouched from those halcyon days. Set in the old carriage barn behind the main house, **The Granary** is the inn's largest accommodation, albeit rustic with tasteful touches. The two-story suite is framed in wood-plank floors and hand-hewn ceiling beams, heated with a wood-burning stove, and decorated with family heirlooms and antiques. A king-size bed is tucked up on the sleeping loft. Downstairs on the main floor is a parlor (with a double bed for extra guests) and a small kitchenette with fridge and microwave. The Granary's bathroom boasts a jumbo tub. The staff can bring breakfast in bed, or guests can eat with everyone else in the big house dining room, enjoying a hearty Ohio farm breakfast that varies slightly from day to day. *brandywinefallsinn.com*

5 Cumberland Island
NATIONAL SEASHORE
Georgia

Guests needn't bring their own reading material for a stay at the **Greyfield Inn,** especially if they make plans to sleep in the **Library Suite,** named after the adjacent sitting room and its copious collection of

▶ *The oldest operating hotel in Yellowstone National Park, the "Grand Dame" Lake Yellowstone Hotel opened in 1891 and proudly boasted such cutting-edge amenities as indoor electricity, plumbing, and steam heat.*

bygone novels, historical accounts, and biographies. The Carnegies (of steel fame) once owned much of Georgia's Cumberland Island and built fabulous mansions as both vacation homes and year-round residences. Still owned and operated by Carnegie descendants, Greyfield (erected in 1900 for Margaret Carnegie Ricketson) is as lovingly decorated as it was a century ago with antiques and family mementos, making it an ideal romantic getaway. Located on the house's main floor, the suite has views from the back window over Cumberland Sound and the marsh along Old House Creek, and is also steps away from the cozy bar and the spacious living room where guests gather at night to swap stories about their day on the Georgia barrier island. Each stay includes breakfast, a picnic lunch, and a gourmet three-course dinner. *greyfieldinn.com*

6 Golden Gate
NATIONAL RECREATION AREA
California

San Francisco swank meets vintage Army architecture at **Cavallo Point— the Lodge at the Golden Gate,** set along the Marin County waterfront of Golden Gate National Recreation Area. Once upon a time this was Fort Baker, a post–Civil War outpost. The 24 whitewashed structures around the parade ground were built as housing for the officers and their families. More than a century later, **Frank House** is the resort's most lavish unit, a two-story clapboard manse with more than 1,100 square feet of living space, including two bedrooms, a living room, and a full kitchen. There aren't many places even in hotel-rich San Francisco where guests can wake up with the Golden Gate Bridge hovering above their toes, or where they can lounge in a glass-enclosed sun porch with Alcatraz looming in the distance. Stylish walnut furnishings, leather armchairs, period details, and a funky modern fireplace complete the picture. *cavallopoint.com*

7 Yellowstone
NATIONAL PARK
Wyoming

Living up to its lofty moniker, the **Presidential Suite** at the **Lake Yellowstone Hotel** hosted both Warren Harding (1923) and Calvin Coolidge (1927) while they were on official visits to the Wyoming national park. The commanders in chief no doubt found the spacious second-floor room an oasis of luxury in the middle of the wilderness. The suite wraps around the hotel's southwestern corner, overlooking timberland on one side and the broad sweep of Yellowstone Lake on the other. It was created along with the rest of the hotel in 1891 and refurbished by master park architect Robert Reamer a decade later. The double front door opens onto a large living room flanked by two nice-size bedrooms, each furnished with antiques and reproductions. The artwork reflects Yellowstone wildlife themes, in particular the etched glass panel of grizzly bears fishing for salmon. Although the two bathrooms and kitchenette are modern, the rest of the suite is resolutely unplugged to retain its rustic ambience—yes, even presidents must do without television and air-conditioning. *yellowstone nationalparklodges.com*

8 | Yosemite
NATIONAL PARK
California

Yosemite National Park has spawned plenty of famous men (John Muir, Stephen Mather, Ansel Adams), but author, naturalist, and hotelier Mary Curry Tresidder was undoubtedly the "first lady" of the valley. Her **namesake suite** occupies a private apartment built in the **Ahwahnee Hotel** in 1928 for the Tresidder family. Mary lived here

A Starring Role in Films

Hundreds of movies have been shot on location in America's national parks, with the landscapes playing a starring role in the films. Some of the scenes are the stuff of Hollywood legend, like the edge-of-your-seat finale from Alfred Hitchcock's thriller *North by Northwest*, in which Cary Grant battles enemy agents on the face of Mount Rushmore. The alien rendezvous at the end of *Close Encounters of the Third Kind* famously transpires at the base of Devils Tower. And Charlton Heston's spacecraft crash-landed into Lake Powell at the start of *Planet of the Apes*.

With their rugged outdoor scenery, national parks are especially suited for Westerns. *How the West Was Won* used Bent's Old Fort and the Badlands, which also starred in the 1990 Oscar winner *Dances With Wolves*. Although much of *Butch Cassidy and the Sundance Kid* was set in what is now Canyonlands, key scenes were filmed at Zion. Clint Eastwood, meanwhile, scowled and squinted his way across the desolate dunes of White Sands in *Hang 'Em High*. Just because a national park is old, large, or famous doesn't make it a Hollywood darling. Despite their size and acclaim, Yellowstone and the Grand Canyon have appeared in very few films. On the other hand, the Illinois & Michigan Canal National Heritage Corridor has provided backdrops for at least 20 movies, Glacier more than 35 films, and Death Valley more than 45.

Curl up for a relaxing afternoon—complete with a view of Halibut Cove—in the guest cabins at Kenai Fjords Wilderness Lodge.

the better part of 40 years, running the hotel with a very personal touch and penning books like *Trees of Yosemite*. It was converted into a guest room after Mary's death in 1970. Also called the "Queen's Room" because Queen Elizabeth II and Prince Philip stayed there in 1983, the suite sprawls across the lodge's secluded sixth floor and features fabulous views of Yosemite Valley and its celebrated granite walls. Among the suite's features are Craftsman cabinets and a four-poster canopy bed. A black-and-white portrait of Mary (in a rocking chair) hangs above the suite's sitting area. Her ghost is said to haunt the entire floor, fussing over the guests

in much the same manner as Mary did in real life, making sure everything is pleasing. *travelyosemite.com*

Kenai Fjords
NATIONAL PARK
Alaska

If you're lucky enough to stay at **Kenai Fjords Wilderness Lodge,** then you're lucky indeed. The only accommodation on remote Fox Island, this resort is intentional in its small size and uncompromising in its service. Arrive by boat from Seward and check into one of the resort's eight **water-facing guest cabins,** which come with all-inclusive amenities and a focus on eco-tourism. The

highlight here is unique Fox Island, where marine life abounds and must-dos include guided kayaking tours of blue-green waters. When you return, slip into a wood-fired oceanfront sauna to unwind for the next meal, sure to be another seasonal feast featuring Pacific Northwest fare. *alaskacollection.com*

10 | Bryce Canyon
NATIONAL PARK
Utah

Built in the late 1920s, the **Western Cabins** at the **Lodge at Bryce Canyon** still look much as they did 80 years ago, tucked into the pines about 100 yards from the canyon rim. And there's something to be said for their lasting draw. Gilbert Stanley Underwood's design is classic national park: log slab siding anchored by stone corner piers, rough rubble masonry chimneys, and steep gable roofs with cedar shingles. The interiors retain their original wooden siding and stone fireplaces (although gas has replaced wood). Latter-day improvements include queen-size beds and full bathrooms. Each cabin boasts a private porch with views of the surrounding forest and a wonderful whiff of pine throughout the day. Their location close to the canyon's Rim Trail makes the cabins ideal for those who cherish sunrise and sunset views or photography. *visitbrycecanyon.com*

10 BEST PARKS FOR
CULINARY DELIGHTS

Adding flavor—literally—to any vacation is the local cuisine, be it a spicy Southwestern meal, a pie made of northern Rockies berries, or the freshest Alaskan catch pulled straight from the river. Many of the dining options inside parks emphasize fresh ingredients grown or raised in the region, giving you a memorable taste of the area and its history.

1 Acadia
NATIONAL PARK
Maine

Partake in a tradition that stretches back more than 130 years by munching on **popovers with strawberry jam, butter, and tea** on the back lawn of **Jordan Pond House** in Acadia National Park. Both the house and nearby lake are named after the pioneering Jordan family, who established a farm here in the 1840s. Thirty years later the property was transformed into a pleasant country café to serve the growing number of tourists flocking to Mount Desert Island. Popovers have been a specialty since the very start, and they seem to taste even better when admiring Pemetic Mountain and the Bubbles across the lake. The popovers can be served à la mode with homemade ice cream and are ideal with hot tea, coffee, or a glass of Prosecco. *jordanpondhouse.com*

2 Glacier
NATIONAL PARK
Montana

One of the great debates in Montana (and throughout the northern Rockies) is who makes the best huckleberry pie. **Eddie's Café & Mercantile** in Apgar Village certainly makes a bid for the title and just about everything else one can make with **huckleberries.** Their menu includes buttermilk pancakes with huckleberry syrup, huckleberry milkshakes, huckleberry lemonade and iced tea, huckleberry-flavored coffee, deep-dish huckleberry cobbler, and huckleberry-peach pie à la mode. The berries are so popular, visitors could easily "huckleberry" their way across the whole region: The **Mountain Bar at St. Mary's Lodge** makes a pretty mean huckleberry martini. The **Whistle Stop Restaurant** in East Glacier and **Park Café** in St. Mary are known, respectively, for huckleberry-stuffed French toast

and huckleberry pies. There's even a huckleberry cannery called **The Huckleberry Patch** in Hungry Horse with a factory shop that sells pies, fudge, licorice, taffy, and jams flavored with Montana's favorite berry.

3 Wrangell–St. Elias
NATIONAL PARK & PRESERVE
Alaska

It doesn't get any fresher than this: **Alaska salmon** straight from the Copper River and served with nouvelle cuisine flourish in the dining room at the **Salmon & Bear Restaurant.** This far out, you may think folks would be happy with basic grub. But the restaurant bucks that trend with culinary maestro and co-owner Joshua Slaughter, who combines superfresh local ingredients with sophisticated cooking techniques and presentation. The signature dish is salmon, caught by local anglers and served medium rare; dry aged

No matter the season, farms in the Cuyahoga Valley have fresh produce and locally made goods.

Black Angus is a favorite, too. The rustic lodge has been around since 1916, when the entire building was moved to McCarthy from the coastal town of Katalla, where it was part of a fish cannery. The drinks are acclaimed as well, earning *Wine Spectator* awards since 2020. Plan ahead: The restaurant is intimate and reservations go quickly. *salmon -bear.com*

4 Cuyahoga Valley
NATIONAL PARK
Ohio

Year-round, visitors to fertile Cuyahoga Valley National Park can taste the bounty of this region thanks to the **Cuyahoga Valley Farmers Market.** Started in 2004 and powered by volunteers, the market supports dozens of family farmers, while getting nutritious foods into the community. Find the season's best **fruits, vegetables,** and **herbs,** plus **grass-fed meats, fresh eggs,** and prepared goods like **maple syrup, locally roasted coffee,** and handmade soap. Begun in 1999, the innovative **Countryside Initiative** helped turn the valley's old farmsteads into once again thriving, sustainable farms. Ask for a tour, shop for fresh goods, and see national park agriculture in action at such sites as Greenfield Berry Farm, Oxbow Orchard, and The Spicy Lamb Farm. *cuyahoga valleyfarmersmarket.org* or *nps.gov/cuva/learn/historyculture/countryside.htm*

5 Blue Ridge Parkway
NATIONAL PARKWAY
Virginia

Nothing is more evocative of good old-fashioned Southern cooking than **grits,** the cornmeal porridge that goes with any meal. Founded in 1910 along the Blue Ridge, **Mabry Mill** produces its own stone-ground cornmeal for the restaurant's hush puppies, spoon bread, griddle cakes, muffins, and illustrious grits. Guests can eat the grits piping hot with melted butter or sliced and fried in pork drippings. The dining experience is enhanced on summer and autumn weekends when live bluegrass often envelops the grounds. Guests can also take a self-guided tour of the old historic buildings, including a blacksmith shop, whiskey still, and the much photographed water mill, its gray-plank profile reflected in the adjacent pond. Traditional Appalachian crafts are a Mabry staple, including ranger-led demonstrations on how to grind maize into cornmeal. The gift shop sells cornmeal that can be used to make grits, bread, or cakes back home. *mabrymillrestaurant.com*

▶ *At the start of the 1900s, Ed and Lizzie Mabry built a gristmill, waterwheel, and water supply system. A few years later they added a sawmill, woodworking shop, and blacksmith, rounding out many of the services at Meadows of Dan along the Blue Ridge Parkway.*

6 Hawai'i Volcanoes
NATIONAL PARK
Hawaii

Perched on the 4,000-foot summit of the Kīlauea caldera, the Volcano House dates back to 1846 and is the oldest hotel in the state. Originally a single-room shelter made of ohia wood and grass, it was replaced with a four-bedroom structure in 1866 and again replaced in 1877. In 2013, it underwent an extensive restoration. One of the hotel's highlights is its restaurant, **The Rim.** Even with views of the Halema'uma'u Crater and live music accompaniment vying to steal the show, the menu still wows with such items as seared blackened **ahi sashimi, kona kampachi with seaweed salad,** and **local Hawaii**

Ranchers steaks with wild Hamakua mushrooms and potatoes. *Hawaii volcanohouse.com*

7 Capitol Reef
NATIONAL PARK
Utah

Capitol Reef's canyons, cliffs, and other geological wonders are always a sight to behold, but the park's **orchards** are also well worth your time. Plan a visit in the warmer months, when Fruita Historic District, a burst of green valley within the Waterpocket Fold's stone, is abundant with cherries, peaches, and apples in season. You-pick fruits make a delightful souvenir, and the 1908 **Gifford House Store and Museum** has baked goods, tea, quilts, jams, and handmade crafts. It's best known, though, for **superbly fresh pies**—strawberry-rhubarb, cherry, peach, and mixed-berry. Picnic tables right outside make it easy to gobble down your fare before setting out to explore the park. *nps.gov/care*

8 Yellowstone
NATIONAL PARK
Idaho, Montana & Wyoming

In a rare happenstance for a national park, in Yellowstone you can watch buffalo herds wander around inside the park and sample a bison bratwurst in the same day. (The buffalo served at half a dozen

Regional cuisine shines at the Metate Room restaurant in Mesa Verde National Park.

of the park's restaurants doesn't derive from local herds but rather ranch-raised bison from neighboring states.) At the Old Faithful Snow Lodge, the **Obsidian Dining Room's** finger-licking **short ribs** are braised in a succulent jus and served with buttermilk mashed potatoes or vegetable hash. The kitchen also serves an awesome **bison tartare** appetizer and a hearty **bison chili.** *yellowstone nationalparklodges.com*

9 Mount Rushmore
NATIONAL MEMORIAL
South Dakota

There's a cherished dessert lore that credits Thomas Jefferson with bringing the first written recipe for **ice cream** to the United States. Mount Rushmore's **Memorial Team Ice Cream** serves up scoops of the sweet stuff based on that original formula, which initially called for six egg yolks, one cup of sugar, a quart of heavy whipping cream, two teaspoons of vanilla, and a pinch of salt. Pride Dairy in North Dakota helped the national memorial re-create the dish to serve crowds, and today a scoop of vanilla feels downright patriotic. While you wait for your treat, check out the life-size photos of the baseball players that made up the Mount Rushmore amateur team. When not working to chisel out presidential faces, these men were practicing their

pitches, and the Memorial Team name is in their honor. *mtrushmore nationalmemorial.com*

10 Mesa Verde
NATIONAL PARK
Colorado

Nobody does Southwest cuisine like the **Metate Room.** Perched at around 8,000 feet on the edge of Mesa Verde in Colorado, this upscale eatery at the Fair View Lodge serves endless views and savory dishes in equal measures. The chicken mole is to die for, a quartered chicken with poblano mole, smashed Yukon potatoes, and pepita gremolata. Other modern interpretations of **heritage foods** include smoked salmon green chili mousse, stuffed squash with roasted acorn squash and native rice blend, and a Black Angus rib eye with a house-made butter and special succotash. The modern decor carries Southwest touches with its Native American carpets and pottery. Metate's panorama overlooks the desert canyons to the east of Mesa Verde, a view that nearly exceeds the food. *visitmesaverde.com*

Libation Stations

With ambience and chef's-kiss-level service, these three park bars are *the* spots for kicking up your feet and kicking back a drink.

The Ahwahnee Bar (Yosemite National Park): Cozy and refined, this watering hole specializes in craft cocktails like the Glacier Point, El Capitini, and Yosemite Hot Toddy.

The Whiskey Bar at Cascades Restaurant (Rocky Mountain National Park): After a day of hiking, you'll want to spring for the good stuff, and this lounge's selection of whiskeys and classic cocktails, including the cheeky Redrum Punch, will go down just right.

Arizona Steakhouse in Bright Angel Lodge (Grand Canyon National Park): Find local Arizona craft beer, wine, and spirits, such as the Tower Station IPA from Mother Road Brewing Co. and Javelina cocktail with Western Sage gin.

10 BEST PARKS FOR
TENT CAMPING

One of the best ways to appreciate a park is to camp in the heart of it. Distinguished by their beautiful settings and proximity to fun, the following sites are exceptional spots to pitch a tent. Most require advance reservations, so always check online before setting out.

1 Great Sand Dunes
NATIONAL PARK & PRESERVE
Colorado

Naturally, Great Sand Dunes days are filled with sandboarding, hiking, and splashing in Medano Creek. But overnighters at the Colorado park know that the wonders continue at night, when the International Dark Sky Park quiets and the heavens start their show. Nighttime ranger programs help guide the eye. For ease, pitch your tent just one mile from the visitor center at **Piñon Flats Campground,** with views of dunes and the Sangre de Cristo Mountains. The campground is dotted with—you guessed it—pinyon trees. Enjoy their shade, but skip the hammocks, which damage the trees and make them vulnerable to disease. Sites feature a picnic table and fire grate; restrooms with flush toilets, a dishwashing sink, and potable water spigots are available in all three loops of Piñon Flats. While there are no showers

in the campground, in the summer months guests can rid themselves of sand at the outdoor rinse stations in the dunes parking lot. Meanwhile, drinks, snacks, firewood, and camping supplies can be purchased in the camp store. Deer are frequent visitors and bears have the potential to stop by, so campers are reminded to use the food lockers and never feed the wildlife.

2 Capitol Reef
NATIONAL PARK
Utah

Cradled in the midst of red-rock country, **Fruita Campground** in Capitol Reef National Park has been called an oasis in the desert. Situated alongside the Fremont River just a mile from the visitor center, it's the only developed campground in the park. The 71 sites are serviced with restrooms and water, and picnic tables and fire rings are available at each site. Campers

won't want for things to do: The Fruita Historic District is the heart of Capitol Reef, with orchards and historic sites highlighting the verdant valley. Take a 90-minute scenic drive or tackle the all-levels hikes available throughout the district. Given its popularity, Fruita Campground must be reserved ahead of time from March through October; winter months are filled first come, first served.

3 Assateague Island
NATIONAL SEASHORE
Maryland

Assateague Island National Seashore is a stunning place to camp, and a good site here provides sweeping views of the bay or of the Atlantic fringed by long beaches. The white noise at night is the rush of surf. The bird-watching is spectacular, and lucky campers will glimpse the wild ponies that roam the island, too. There are several

Camp under the stars at one of 71 campsites in the red-rock desert of Capitol Reef National Park.

types of **waterfront camping** available, including oceanside or bayside drive-ins and walk-ins; some accommodate RVs and trailers. All the campsites have a picnic table and fire ring or grill, and there are chemical toilets, cold-water showers, and drinking water located in each camping area. The lovely, tranquil setting attracts people time and time again, but beware the mosquitoes, especially in summer. Come armed with repellent, and since camping on the beach doesn't afford a lot of shade, bring a shade tent or umbrella to deflect the rays during the day.

Hot Springs
NATIONAL PARK
Arkansas

The only camping option within Hot Springs National Park, **Gulpha Gorge Campground** capitalizes on just enough amenities combined with unbeatable access to the city. Sites are situated along Gulpha Gorge Creek and under a canopy of trees, giving the desired faraway feel despite the park's downtown proximity. Historic Bathhouse Row and its museum, shops, and brewery are just a short stroll away. Twenty-six miles of hiking trails explore the reaches of Hot Springs, and a trailhead within the campground increases the convenience. Picnic tables, grills, water, and flush toilets round out the facilities.

▶ *The tip-top open-air observation deck of the Hot Springs Mountain Tower reaches 1,256 feet above sea level, overlooking the national park and surrounding Ouachita Mountains.*

5 | Sleeping Bear Dunes
NATIONAL LAKESHORE
Michigan

The setting of Sleeping Bear Dunes, along 31 miles of beautiful Lake Michigan's southern shoreline, makes it a particularly special place for camping. The popular **D. H. Day Campground** offers a happy medium between roughing it in the backcountry and camping with the ease of modern conveniences: It's rustic camping with vault toilets and spigots in a peaceful lakeshore setting. The 88 campsites are nicely shaded and screened from one another by trees and brush, and from almost all areas of the campground, campers have very easy access to the gorgeous, unspoiled white-sand beaches of Lake Michigan or any of the 21 inland lakes. On a clear day, campers can get up early and take the short walk to

the beach to see the sun rise over the lake. In the evening, they can participate in a ranger-led program on local culture, lake ecology, or the history of logging and shipping in the area. Each campsite has a fire ring that people can gather around to stay cozy in the cool Michigan evenings.

6 | Acadia
NATIONAL PARK
Maine

Acadia National Park's beautiful, classic campgrounds are popular for good reason. The park's two primary campgrounds—**Blackwoods** and **Seawall**—have lovely wooded sites that are within a 10-minute walk of the Maine Atlantic coast. Blackwoods Campground is located on the east side of the island, closer to some of Acadia's major features like the network of historic, broken-stone carriage roads and the Park Loop Road. Seawall, on the west side of Mount Desert Island, is a bit less crowded and provides a more private camping experience closer to the coast. Both campgrounds reserve the majority of their sites for tents, though there are some sites that accommodate RVs and pop-ups. Each campground has running water, flush toilets, picnic tables, and fire rings. The free Island Explorer shuttle bus makes regular stops at both sites, too.

7 Sequoia
NATIONAL PARK
California

Sequoia National Park is known for its sequoia trees so giant and ancient that legend has it a logger in 1888 needed five days and a team of five men to cut down a tree. But the park also has dramatic canyons, rock formations, rivers, and wild elevations with stunning overlooks. At a height of 7,500 feet, the small **Cold Springs Campground** in the Mineral King area assures access to spectacular views and is also near the famed sequoia groves. Open only in the summertime, the area is shaded by aspens and evergreens and the campsites are tent-only— the winding, rocky road to the campground isn't navigable by trailers and RVs. Each campsite has a picnic table and firepit with a grill; there are vault toilets and drinking water in the campground.

8 Lassen Volcanic
NATIONAL PARK
California

Lassen Volcanic National Park is so named because of the 30 volcanic domes within its boundaries, but there are quite a few other natural features to be discovered here. These include lovely Manzanita Lake, which formed when a volcanic dome collapsed and the resulting avalanche dammed up a creek. This site is the setting for

A hiker relaxes after a day on Sequoia National Park's 211-mile-long John Muir Trail.

Manzanita Lake Campground, a striking place from which to base an exploration of the park. To start, there's trail access from the campground to the lake, which affords outstanding views of Lassen Peak opposite. Manzanita Lake is also a very popular spot for fishing (catch and release only), swimming, and boating; the Loomis Museum (which has exhibits and videos about the area) and an amphitheater for evening programs are less than a mile away from the campground. Each campsite has a picnic table and campfire ring, and the campground has flush toilets and drinking water. For other amenities, the nearby camp store has a laundromat, hot showers, and some groceries.

9 | Virgin Islands
NATIONAL PARK
Virgin Islands

Camping, yes—but make it in paradise. Such is the vibe of St. John's **Cinnamon Bay Beach & Campground,** located on the North Shore's longest beach. Two types of campsites can be reserved: bare and eco-tent. Bare sites come with a wooden platform, and tents, linens, and cookware can be rented. Eco-tent sites are especially family-friendly, with a fan, electricity, and the option of bunks. Four shared but modern bathhouses dot the property. The on-site Rain Tree Café offers breakfast and dinner, and a food truck serves up lunch with ice cream and cold drinks. Water-sports rentals are also available, and plenty of trails, snorkeling, and exploring will fill your day. Don't forget bug spray (the no-see-ums are especially naughty at night) and reef-friendly sun protection.

10 | Rocky Mountain
NATIONAL PARK
Colorado

The high-mountain landscape of Rocky Mountain National Park makes for some magnificent overnighting, and **Moraine Campground,** in a ponderosa pine forest at an elevation of 8,100 feet, is a great place to set up a tent. The newly rehabilitated campground can accommodate just about any kind of camping, from trailers and RVs to tent-only walk-ins. There are stunning views of the snowcapped Rockies from many of the sites, so come early, if possible, to claim one—the campsites can be reserved from Memorial Day weekend through September, while the rest of the year is first come, first served only. Each campsite has a picnic table and fire grate; restrooms, water spigots, and vault toilets are located centrally, but there are no showers in the park. Moraine also has a beautifully maintained amphitheater for evening campfire programs in the summer, and hiking trails to Cub, Fern, Bear, and Odessa Lakes originate at or near Moraine.

Camping With Ease

Exploding in popularity, "glamping" is an option for even the most camping averse among us. The following sites so specialize in comfort that they could just as easily be called luxurious.

Clear Sky Resorts (Arizona): Air-conditioned and heated, lavish sky domes feature upscale linens, chic furniture, a modern bathroom, and (no doubt the best part) a panoramic window, all just 20 minutes from the Grand Canyon's South Rim entrance. *clearskyresorts.com*

Under Canvas (nationwide): Found from Acadia to Moab to Glacier and elsewhere, these upscale, safari-inspired canvas tents have optional en suite bathrooms, West Elm furniture, and nightly activities, plus café-style dining and evening s'mores, too. *undercanvas.com*

Capitol Reef Resort (Utah): The living is easy when you can do so in a lavish Conestoga wagon, based on 19th-century wagons, complete with accommodations for up to six and access to a private bathroom. *capitolreefresort.com*

RV CAMPGROUNDS

National parks strike a good balance when it comes to RV camping: Full hookups aren't super common, but what they lack in number they more than compensate for with spectacular scenery. Enjoy ranger programs, walks on the beach, brilliant night skies, and unparalleled beauty, all from the comfort of a rolling home.

1 Grand Teton
NATIONAL PARK
Wyoming

Why bring an RV here instead of opting for a hotel? Because campers at **Flagg Ranch** or **Colter Village** can go to sleep under a blanket of stars that are not dimmed by city lights and then awake to see the first rays of light kissing the Teton Range. In the morning, the east-facing range receives a constantly changing flood of light that first affects the color of the sky and then fills in the crevices and cracks, peaks and pinnacles. Another RV advantage is having easier access to the trails and activities of the park long before outsiders arrive, such as paddling upon Jackson Lake when the surface is still and glassy and having the luxury to experience this place in an unparalleled sense of peace and solitude. In addition to the morning light show, which makes for dramatic photographs, travelers may find themselves sharing their campsite with a passing deer or moose or, to be fair, bear, but the free-roaming menagerie is a vivid reminder that you're living on the land and experiencing it in full.

2 Assateague Island
NATIONAL SEASHORE
Maryland

Although Baltimore, Philadelphia, Norfolk, Richmond, and Washington, D.C., arc around this thin strip of a park in Maryland, it's not necessarily an easy place for urban dwellers to reach. After drivers run the gauntlet and clear the dense commercial traffic heading to Ocean City, a town sprawled across two states (Maryland and Virginia), a lonely country road sweeps into the protected seashore on Assateague Island, where the names of Assateague's two campgrounds—**Bayside** and **Oceanside**—enticingly say it all. Park the RV at Bayside and to the west is the Sinepuxent Bay, a small portion of larger Chincoteague Bay. Nearby along the waterfront are areas for clamming and crabbing and launching a canoe, while on the east side of the entry road is Oceanside Campground where, just beyond the dunes, Atlantic Ocean waves crash onto the shore. Campers frequenting this area can enjoy lazy days stretched out on the sands, walking along forest and dunes trails, warming themselves around a fire ring, or even better, enjoying the night sky by the light of a beachfront blaze.

3 Badlands
NATIONAL PARK
South Dakota

The surreal juxtaposition of Badlands' mixed-grass prairie against sharp buttes, pinnacles, and spires never ceases to impress—except,

Take in the incredible mix of pastel-hued ridges and canyons of Badlands National Park as you make your way to RV-friendly campgrounds.

that is, if you stay overnight and indulge in the trifecta of sunset, night sky, and sunrise. At **Cedar Pass Campground,** RVers can do just that. Managed by Cedar Pass Lodge, the campground has both tent and RV sites, the latter featuring electric and nonelectric services. Campfires aren't permitted here—the fire danger is simply too high—but camp stoves and charcoal grills are allowed. Ideally located, sites are near the Ben Reifel Visitor Center and the amenities of Cedar Pass Lodge, meaning you're never far from ranger programs or a hot meal, groceries, souvenirs, and other supplies. Of course, that's if you can manage to take your eyes off the geological formations just outside your window.

4 | Mount Rainier
NATIONAL PARK
Washington

Located in one of the most picturesque areas of the country, this national park is easily one of the most popular areas for RV camping. That's the good news. The downside is that since the park was created in a time when mobile campers were small, many large RVs won't be able to navigate the tight turns on some of the park roads. At the park's three campgrounds open to RVs, limits on size range from 27 to 35 feet, with trailers limited between 18 and 27 feet. Should the rig fit, prepare for a supremely pleasing experience. In the southeast corner of the park, **Ohanapecosh Campground** is bisected by the crystal clear Ohanapecosh River, a snowfield-fed swift ribbon of water that exemplifies the beauty of the Pacific Northwest. At **Cougar Rock Campground** in the southwest corner of the park, the real attraction is easy access to great recreational activities, including photography, hiking, climbing, fishing, mountain biking, and nightly ranger programs during the summer. **White River Campground** is near the popular Sunrise area. Although weather narrows the window of access to RV camping in Mount Rainier, the range of services and activities—let alone views of the iconic snowcapped mountain surrounded by forests of western

▶ *No matter how you measure Kings Canyon (the depth ranges from 7,700 to 8,200 feet), it still earns its name and beats out Grand Canyon which, at its deepest point, is 6,093 feet.*

hemlock, Douglas fir, and western red cedar—makes Mount Rainier well worth the drive.

5 | Grand Canyon
NATIONAL PARK
Arizona

Anyone who has ever visited Grand Canyon National Park has seen tourists leave their cars, walk to the rim, and depart 10 minutes later, certain that they had fully experienced the Grand Canyon. Not even close. As RVers know, it takes days to even begin to appreciate this natural phenomenon, which is why they prefer to roll into the park's campgrounds, put out the lawn chairs, and sit a spell. On the South Rim, the most popular site is **Mather Campground,** which is within a short shuttle bus ride from the rim and close to services that include an amphitheater, market, and laundry. Attached to the campground, **Trailer Village** offers specialty paved

sites for RVs—one of the rare national park locations where full hookups are offered. A more remote option, **Desert View** is about 25 miles east, and while smaller RVs are welcome, the calendar isn't as accommodating—it's open only between May and October. So, too, is the **North Rim's namesake campground,** which sits among ponderosa pines. Make reservations well in advance, because even though the canyon's immense, the campgrounds are not.

6 | Acadia
NATIONAL PARK
Maine

Acadia National Park shares Maine's Mount Desert Island with pockets of privately owned land and the village of Bar Harbor. It's an interesting situation, and one that benefits RV campers, who can roll into the campgrounds of **Blackwoods** or **Seawall** knowing that they're well within the park, close to the coast, sheltered by red spruce, white pine, and fir trees, and within walking distance of rocky and sandy beaches. RVs up to 35 feet have the advantage of pull-through camping at sites that add an extra layer of privacy to an already secluded area courtesy of lush vegetation. And thanks to the park's layout, when a camper can't handle another dinner of grilled hot dogs and s'mores, it is only a few miles to the village of Bar Harbor, from where, following a meal at a local restaurant, it's only a short drive back to the best room in town: the RV.

7 | Great Smoky Mountains
NATIONAL PARK
North Carolina & Tennessee

Sandwiched between the village of Cherokee on the east and the energetic tourist draws of Gatlinburg and Pigeon Forge on the west, this national park straddling Tennessee and North Carolina allows campers to slip into nature and slip out of range of cell phone signals and fudge shops, giving travelers the chance to escape and do something really different: relax. Approximately **1,000 sites spread across 10 campgrounds** are suitable for RV camping—all but Big Creek (where RVs are not allowed)—and offer better site separation and thicker vegetation than at open-air camps. Get a jump on the day with almost instant access to the activities—fishing, paddling, and swimming in streams, exploring quiet hiking trails, and reaching the Cades Cove Loop which, due to the presence of historic structures left behind by mountain families, will be packed with cars by mid-morning. Savvy RVers will be on the road early, perhaps unloading bicycles and taking advantage of special two-wheels-only access to Cades Cove on Wednesdays in summer. In the high season, many of the campgrounds host evening ranger programs.

8 | Canyonlands
NATIONAL PARK
Utah

Before setting out for Canyonlands, note that only two paved roads lead into this sprawling southeastern Utah park—a sign that RV camping here may be a bit more challenging. The **Willow Flat Campground** in Islands in the Sky features 12 sites (only one of which is pull-through) so count on there being a high demand for these slots in peak season. At the **Needles Campground,** a total of 26 sites make camping in the Needles more popular—that and the fact that there are bathrooms, water, and picnic tables here. But whether the site is seriously primitive or just very primitive, the reason campers come here for an

Camping Sense

The hiker's adage, "take only pictures, leave only footprints," applies to campers as well. For campers, the ideal is "leave no trace," a collection of simple rules that demonstrate respect for the natural environment by limiting our impact on it, all for the betterment of Mother Nature. But camouflaging the impact of multiple overnights can be tricky. Rule one: Don't book a site that's not appropriate for the size of your RV—check online to make sure there's clearance for your unit. Two: Keep your RV atop the gravel pad, leave plastics and aluminum in recycling bins, and dump gray water in the appropriate drains. Three: Do not dig pits, chop down trees, or blaze new trails—for that matter, don't build huge blazes; keep campfires confined on the concrete grill pads. Last, remember where you are: You are a guest in the wilderness. Out of respect to your neighbors—both human and otherwise—hush during "quiet hours" and do not feed curious animals looking for a handout. It's not only illegal but also habituates them to humans, an unwanted situation.

overnight (or more) is for easy access to the park's numerous overlooks, where walking a few dozen yards will reveal a canyon that's 2,000 feet deep and a landscape that stretches out for a hundred miles. Later that evening comes front row seating before one of the darkest night skies in the country, a fact that literally expands the universe. Park rangers host night-sky programs in the summer, revealing to participants with the aid of far-seeing telescopes—or perhaps even the naked eye—new stars and constellations and, in a way, a whole new world.

An RV parks for the night at Moraine Campground in Kings Canyon.

9 Kings Canyon
NATIONAL PARK
California

Part of the draw at Kings Canyon is what the park lacks: hordes of visitors. Their loss, as Kings Canyon is a sight to behold—even called "a rival to Yosemite" by John Muir. The park's two roads both take you to RV campgrounds. Year-round **Grant Grove Village** leads to Azalea, Crystal Springs, and Sunset Campgrounds. Enjoy the convenience of the park visitor center, a market, programming, and plenty of soaring tree cover, and keep watch for wild ones—bears are active in this area. In a more remote region of Kings Canyon, **Cedar Grove Campgrounds** (Sentinel, Canyon View, Moraine, and Sheep Creek) are open early spring to late fall, have the same open

stands of evergreens, and some boast space for much larger RVs. A bike trail connects the campgrounds to Cedar Grove Village services.

10 Glacier
NATIONAL PARK
Montana

There are more than a dozen campgrounds and more than 1,000 sites sprinkled across northwestern Montana's Glacier National Park, a spectacular northern Rockies wilderness. Although only a handful of the campgrounds are off-limits to RVs, an RVer will have trouble choosing a site from the remaining options available, among them the many beautiful lakeside campgrounds along **Lake McDonald** and **St. Mary Lake;** the beautiful forest-set

Avalanche Campground beside the Going-to-the-Sun Road; and **Many Glacier Campground,** near Swiftcurrent Lake. Beware: RV excursions may be affected by size restrictions on the Going-to-the-Sun Road (21 feet long, 8 feet wide, 10 feet high), since snappy turns can limit navigation. Passing that test, overnight guests soon learn that campgrounds are usually given the best placement in the parks and that being here after the daytime crowds are gone is a magical experience—one heightened by the sight of spectacular star formations, followed by sunrises creeping over the Continental Divide, illuminating peaks on the Livingston and Lewis Ranges of the Rocky Mountains. Dare we say that—even better than coffee—*this* is the best part of waking up.

OTHER WONDERS

A diver explores the under-water world of Dry Tortugas National Park.

SAYING "I DO" OR "I WILL"

Many love stories have been written in national parks, where quiet sanctuaries amid the widespread grandeur and lovely settings designed by nature serve as the perfect backdrop for a marriage proposal or wedding. Ceremonies are usually low-key and discreet (permits and restrictions apply), but what grander setting could romance need than these spectacular sights?

1 | Cape Cod
NATIONAL SEASHORE
Massachusetts

Sweet sea breezes, curling waves that have been kissing the sand for millennia, and 44,600 acres of pristine shoreline create a setting that rivals the spiritual peace of the grandest cathedrals. Among the ecosystems of this Massachusetts peninsular preserve are grasslands, heathlands, woodlands, and forests, as well as hidden lakes and ponds. And if Cape Cod's endless **Atlantic coast beaches**—some 40 miles' worth—are the church, the soft sands are nature's pews. Arrive before dawn and, as day breaks, a well-timed proposal can lead to a new chapter that begins in the glow of sunrise. Thanks to nature and John F. Kennedy—the local resident turned president who signed the papers to preserve it—this national seashore and its untamed wilderness will always be here to offer suitors, and their grandchildren, a magical place to create memories.

2 | Virgin Islands
NATIONAL PARK
Virgin Islands

High on the list of exotic wedding locations, tropical **Trunk Bay Beach** on the U.S. Virgin Island of St. John offers the option for a shoes-free ceremony. That's actually a good idea considering that rolling waves can soak the feet of the whole wedding party. Consistently ranked among the world's 10 best beaches, Trunk Bay Beach is sandwiched between crystal clear, blue-green Caribbean waters and a tropical potpourri of coconut palms and seaside grapes. By nature, Trunk Bay weddings are laid-back and natural, and the ceremony could just as easily conclude with a surfing contest as with a kiss. With Mother Nature and human nature setting the pace, often the only guideline couples follow is timing their first kiss to a Technicolor Caribbean sunset.

3 | Grand Teton
NATIONAL PARK
Wyoming

With a ticket in hand and proposal at the ready, arrive at the landing at the Jenny Lake Visitor Center and board an open boat to Mount St. John and a trail that leads to **Inspiration Point.** The walk is easy

The stunning Virgin Islands make for a picture-perfect wedding destination.

and quite picturesque as it passes Hidden Falls and its river rapids before arriving at Inspiration Point (elev. 7,200 feet), an overlook dominated by a view of Jenny Lake with Mount St. John and Grand Teton to the rear. The dramatic vista will be unforgettable, especially because this is where they said yes. Return to seal the deal at Grand Teton, where weddings are often held at one of the **Signal Mountain summit** turnouts, adding unique vistas to photographs. Couples at this Wyoming park can bring the ceremony indoors in the Menors Ferry Historic District, where the **Chapel of the Transfiguration** boasts a broad window that frames the tallest of the Teton peaks.

4 | Joshua Tree
NATIONAL PARK
California

While of course the happy couple can choose their own accents, the boho vibe and earth tone palette of the **Indian Cove Amphitheater** at Joshua Tree National Park is built right into this desert dreamscape. Ideal for groups up to 100 (only one car is allowed, so guests are transported in and out via shuttle), this park highlight is cradled by towering rock formations under a wide-open sky. The Indian Cove area has 101 campsites, plus more than a dozen designed for groups, so it's also ideal for turning a wedding into a weekend. For smaller occasions, various picnic areas throughout Joshua Tree would make lovely venues, while scenic spots like Split Rock and Porcupine Wash use the park's captivating geography as a bridal backdrop.

5 | Shenandoah
NATIONAL PARK
Virginia

Virginia is for lovers, says the state tourism slogan, and what's not to love about Shenandoah National

Joshua Tree National Park offers couples a rustic oasis for their nuptials.

Park? With a variety of venues able to host many guests, **Skyland** and **Big Meadows Lodge** combine natural beauty with the ease of an event facility. Any event with 30 or more people requires an amphitheater or rented facility. But if you're picturing something a little more wild—perhaps at sunrise, in a wildflower meadow, at the base or a waterfall or the top of the mountain—the park has plenty of those, too. **Skyline Drive** has dozens of overlooks that maximize views with no hiking required. Virginia is also known for its wineries, and many bucolic vineyards are a short drive from Shenandoah. Toast to forever at Veritas Vineyard and Winery in Afton (near the Rockfish Gap entrance) or Chester Gap Cellar (near the northernmost Front Royal entrance). Cheers!

Yosemite (Marryin') Sam

For couples who dream of marrying in a chapel in a valley, one national park rises to the top of their list of where to wed: Yosemite. That's because Yosemite offers a chance to be married in a chapel in *the* valley.

Moved to its present site beside Southside Drive in 1901, the circa-1879 Yosemite Valley Chapel offers couples the privilege of exchanging vows in a house of worship in a national park. The simple wooden chapel is an excellent example of early chapel architecture in the Sierra Nevada, one that is as plain as its surroundings are majestic. It looks familiar and comforting. The interior consists solely of several rows of wooden pews, hanging lights, a simple altar, and a single cross. But the sanctuary can be enhanced with candles, flowers, ribbons, and accoutrements. Without frills or adornment, the sincerity of the chapel—the first structure in Yosemite National Park placed on the National Register of Historic Places (1973)—shines through.

6 | Hot Springs
NATIONAL PARK
Arkansas

There aren't many towns in America where one side of the street is a city and the other a national park that shares the city's name. In fact, Hot Springs in Arkansas may be the *only* one. For centuries, Quapaw people tapped into naturally healing, heated, and therapeutic mineral waters that were later utilized by 20th-century entrepreneurs as the centerpiece of a gilded age resort—and those are reasons

enough to visit Hot Springs. Add a lovely grand hotel, the Oaklawn racetrack, a gorgeous circa-1880s promenade, and a commanding view of the Ouachita Mountains from a circuitous drive to the top of 1,060-foot Hot Springs Mountain, and there's every reason to plan a romantic getaway. The National Park Service has preserved several **historic bathhouses:** Buckstaff and Quapaw are in operation, and Fordyce has been restored as a museum. We can't make promises, but following a day of

muscle-melting massages and mineral-rich baths, you just might find yourself taking advantage of the moment to ask, "Will you?"

7 | Yosemite
NATIONAL PARK
California

About a century ago, wealthy tourists invested nearly a week traveling from San Francisco to Yosemite. Then, it would take them *another* week to reach the park's lodges via horseback. Their sheer determination proves that Yosemite—with its 1,200 square miles of forests, fields, mountains, valleys, rivers, and, of course, waterfalls—is the perfect setting for a proposal or, perhaps, even a wedding. **Bridalveil Fall,** which, by name alone, deserves to be a member of the wedding party, is a favorite for photos and a marvelous backdrop for newlyweds. The fall plunges 620 feet down a sheer cliff, exploding in great froths of water at its base and sending spray far and wide. Other popular spots are **Glacier Point** (also good for those with mobility considerations) or the quaint and historic **Yosemite Valley Chapel** (page 413).

8 | Sequoia & Kings Canyon
NATIONAL PARKS
California

Can anything make you feel more wholly secure of the future than the sight of extraordinarily old trees? Humbling and stunning, the soaring **sequoias** at Sequoia and Kings Canyon National Parks are magnificent backdrops to "I dos." Timing is everything here, as the park elevations make snow possible from November through April. Consider tying the knot at the **Giant Forest Museum'**s patio, where built-in seats group guests in a cozy cluster. For elopements and small-scale affairs, **Panoramic Point** is breathtaking. Indoor venues take some of

▶ *In September 1996, Cumberland Island National Seashore pulled off the top-secret wedding of John F. Kennedy, Jr., and Carolyn Bessette. The couple wed in the island's First African Baptist Church and held their reception at the historic Greyfield Inn.*

the guesswork out of weather, and heart-of-the-park lodges like Wuksachi, Grant Grove, and Cedar Grove have dedicated event facilities.

9 | Golden Gate
NATIONAL RECREATION AREA
California

Stretching around the San Francisco Bay Area, the Golden Gate National Recreation Area touches three counties and features an assortment of sites, including beaches, forests, trails, decommissioned military bases, and a lighthouse. This means the odds of finding a romantic place to propose are very high—and the odds approach 100 percent in the **Marin Headlands.** Here, overlooks, a lighthouse, and vista points reveal a montage of the Pacific Ocean,

the Golden Gate Strait, the Golden Gate Bridge, and, in the distance, San Francisco and its iconic skyline. One of the most picturesque settings in America, and just a quick jaunt to fine dining and lodging, the Headlands can help anyone begin their own classic love story.

10 | Great Smoky Mountains
NATIONAL PARK
Tennessee

Prior to the establishment of Great Smoky Mountains National Park in 1934, Cades Cove, Tennessee, was the center of the Smoky Mountains community. Even at the dawn of the modern age, residents still lived a somewhat pioneer lifestyle, but one enjoyed in a pastoral valley surrounded by low and lovely mountains. Three historic churches—**Methodist, Missionary Baptist,** and **Primitive Baptist**—supported the community, and today each is a wonderfully unadorned, no-frills (we're talking no heat or electricity) house of worship ideal for small rustic weddings. Make reservations well in advance (only two weddings are allowed each day at any location), and advise guests to allot extra time: The Cades Cove Loop is a one-way road that becomes quite congested during the park's peak season, so it may take a while for your guests to arrive—and for newlyweds to depart.

10 BEST PARKS FOR
ARTS & GARDENS

Park units may be best known as galleries of natural beauty, but they also preserve and reflect America's human artistry. The inventory of creatively focused parks runs the gamut from manicured formal gardens and musical icons to the homes of celebrated American artists and historic theaters that still stage events.

1 Frederick Law Olmsted
NATIONAL HISTORIC SITE
Massachusetts

As the father of landscape architecture and the nation's paramount parkmaker, **Frederick Law Olmsted** spent the latter half of the 19th century designing and redefining the very essence of urban green space. He is most renowned as the creator of New York's Central Park. But he also designed the nation's first state park (Niagara Falls), created the master plans of both Stanford and the University of California at Berkeley, landscaped the Chicago World's Fair of 1893, helped preserve Yosemite Valley, and designed thousands of green spaces— parks, gardens, cemeteries, residential neighborhoods, college campuses, arboretums, landscaped roadways –in 24 different states. Olmsted's beloved Fairsted house in Brookline, Massachusetts, forms

the nucleus of this historic site. Olmsted founded the world's first professional landscape design firm here in 1883. In addition to living quarters and studio space, the house also contains archives with more than a million original blueprints, photographs, and design-related documents.

2 Longfellow House–Washington's Headquarters
NATIONAL HISTORIC SITE
Massachusetts

By the time **Henry Wadsworth Longfellow,** one of the world's best known poets, moved into this yellow Georgian-style mansion, the home already had a storied history as the site of **George Washington's** headquarters for nine months of the Revolutionary War. Years later the Cambridge, Massachusetts, home became a wedding present to Longfellow and his wife, Fanny Appleton. The couple raised five children here, prioritizing art, music, and literature. In the mid-1800s the Longfellows hired an English landscaper to design an Italian Renaissance formal garden, complete with a Gothic rose window in the center. Following her father's death, in 1903 Alice Longfellow commissioned two prominent female landscape architects to do a full Colonial Revival restoration of the family's garden, as well as add new elements like a sundial and pergola. Today the formal garden and historic grounds are open year-round, free of charge, from dawn to dusk, and guests are invited to borrow games, art supplies, and books from the visitor center.

3 | Marsh-Billings-Rockefeller
NATIONAL HISTORICAL PARK
Vermont

This park in central Vermont is a beacon of early American conservation, environmentalism, and sustainable living. The property originally belonged to mid-19th-century diplomat and linguist **George Perkins Marsh,** author of *Man and Nature,* a landmark work on ecology and mankind's negative impact on the planet. Philanthropist and railroad magnate **Frederick H. Billings** purchased the estate and much of the surrounding land in 1869 and set about rehabilitating its distressed forests and farms along the lines suggested by Marsh. Conservationists **Laurance Spelman Rockefeller** and his wife, **Mary French** (Billings' granddaughter), became stewards of the property in the 1950s and later donated it to the National Park Service. Today's park pays tribute to these landmark environmentalists, as well as the reforestation and progressive

▶ *Mammoth Cave National Park hosts an annual Cave Sing, a holiday tradition and nod to the first Christmas celebration hosted underground there in the winter of 1883.*

farming methods pioneered here. The hub of the park is the old Marsh estate on the north bank of the Ottauquechee River in Woodstock. Rangers lead guided tours of the Marsh-Billings-Rockefeller Mansion (1805) and formal garden, while the history of conservation is explored at the visitor center in the Carriage Barn (1895). You can also hike deep into the woods, climb Mount Tom, or explore the Billings Farm and Museum.

4 | New Orleans Jazz
NATIONAL HISTORICAL PARK
Louisiana

Like the music it seeks to honor and preserve, this park unit in the French Quarter of old New Orleans is energetic, eclectic, and often free-form. The park pays tribute to a unique American invention and revolves around the people and places that shaped the city's **jazz**

heritage, including legends like Louis Armstrong, Charles "Buddy" Bolden, and Sidney Bechet. In Louis Armstrong Park, visitors can join interpretive talks and walks, watch video documentaries, and relish live jazz. The park stages weekly concerts throughout the year, as well as special programs. All events are free. Self-guided audio tours are also available for smartphones.

5 | Saint-Gaudens
NATIONAL HISTORICAL PARK
New Hampshire

As one of the driving forces behind the 19th-century American Renaissance in art and architecture, **Augustus Saint-Gaudens** (1848–1907) created much of what we now consider classic American iconography. From double eagle coins and commemorative medals to presidential busts and dramatic equestrian statues, the Irish-born artist immortalized the American spirit and history in bronze, copper, marble, and gold. His long-time summerhouse, gardens, and studios—and more than a hundred of his works—are the focus of this park in the Connecticut River Valley. The grounds and gardens are open year-round; the sculptor's home and studios are open Memorial Day through October. Ranger-guided programs offered in season include an art walk and house tour. Classical music concerts are staged in

Immerse yourself in the history of jazz in New Orleans.

the Little Studio on many summer Sundays, carrying on a tradition established by Saint-Gaudens himself. The park also hosts the Park Service's oldest artist-in-residence program. Figurative sculptors chosen for the program spend the summer and fall working at Ravine Studio and interacting with the curious.

6 | Ford's Theatre
NATIONAL HISTORIC SITE
District of Columbia

Ford's Theatre is seared into America's collective soul as the place where a great man fell to an assassin's bullet. *Our American Cousin* was playing on April 14, 1865, when celebrated actor John Wilkes Booth, an ardent Confederate sympathizer, slipped into the presidential box and shot **Abraham Lincoln.** The mortally wounded president was carried across Tenth Street to the Petersen boardinghouse, where he was pronounced dead early the following morning. The federal government expropriated the theater shortly after, declaring that Ford's would never again be used for public amusement. In the early 1930s both the theater and Petersen House were turned over to the National Park Service. Despite that earlier decree, Ford's was restored into an active theater in the 1960s and now presents a wide array of stage productions. The basement

After decades spent shuttered, Ford's Theatre, the site of Lincoln's assassination, again stages productions.

Ford's Theatre Museum displays numerous Lincoln presidency and assassination artifacts, such as Booth's derringer pistol, as well as the bloodstained pillow from Lincoln's deathbed.

7 Wolf Trap
NATIONAL PARK FOR THE PERFORMING ARTS
Virginia

Wolf Trap has long held a special place in the hearts of theater and music aficionados in the Washington, D.C., area. According to local legend, wolves really did run wild, once upon a time, in these Northern Virginia woods. **Catherine**

Filene Shouse, an ardent supporter of the arts, donated Wolf Trap to the National Park Service in 1966 along with the funds to develop the property into a cultural node. The hundred-plus shows that take the stage each year range from opera, symphony, and Shakespeare to classic rock, Broadway musicals, and stand-up comedy. In the summer, the park's main stage, the 7,000-capacity Filene Center, offers both covered seats and outdoor seating on a sprawling lawn perfect for picnicking during a performance. Children's events are held in the nearby Theatre-in-the-Woods. The Wolf Trap Foundation, which operates the facility in

partnership with the National Park Service, also organizes wintertime performances at an indoor theater called The Barns at Wolf Trap. Rangers conduct backstage tours of the Filene Center October through April.

8 Thomas Cole
NATIONAL HISTORIC SITE
New York

When 17-year-old Englishman **Thomas Cole** arrived in the United States in 1818, he had trained as an engraver's assistant and calico designer's apprentice. Within a decade, he was the country's leading landscape painter, a founding

Get Art Schooled in the Parks

Want some hands-on artistic experience while surrounded by inspiring natural beauty? These Park Service sites offer both children and adults the chance to indulge their creative sides amid history and nature.

Saint-Gaudens National Historical Park (New Hampshire): Saint-Gaudens offers beginning sculpture workshops for adults, teens, and children, as well as more advanced training in skills like casting, moldmaking, and patination. All classes are conducted by the artist-in-residence.

Weir Farm National Historic Site (Connecticut): Visitors are encouraged to try their hand at plein air painting and sketching as part of Take Part in Art, a program that includes the free distribution of watercolors, pastels, graphite, and colored pencils, as well as special events where professional artists share their techniques.

Wolf Trap National Park for the Performing Arts (Virginia): Wolf Trap tenders a full slate of educational art programs for children, including interactive workshops with visiting performers. Adult instruction is available in a wide range of disciplines, from guitar and African drumming to opera and musical theater.

member of the National Academy of Design, and had launched a new art movement called the **Hudson River School** that came to define the nation's obsession with the great American wilderness. The name derives from Cole's love affair with upstate New York, in particular the Catskills. From 1827 he was a regular at Cedar Grove Farm and later married the owner's niece, Maria Bartow. You can take guided tours of the federal-style Main House (1815) and the Old Studio (1839) where he painted, as well as watch a film on Cole at the visitor center, hike along the Hudson River School Art Trail, or indulge in Cedar Grove's famous Sunday Salons with noted artists, writers, and scholars.

9 | Vanderbilt Mansion
NATIONAL HISTORIC SITE
New York

Although named after one of the nation's most renowned families, this park in the Hudson Valley is actually dedicated to an era in American history: **the gilded age,** the period between the Civil War and the turn of the 20th century when the United States grew into an industrial powerhouse. A spin-off of that success was superwealthy families like the Vanderbilts and fabulous country homes like the Vanderbilt Mansion. Constructed in 1896–99 as a

The Vanderbilt Mansion is located in Hyde Park, home to three Park Service sites.

country home for **Frederick William Vanderbilt,** the sprawling estate includes the ornate, 54-room mansion, copious woodland, and three terraced Italian-style gardens that were Frederick's pride and joy. An avid gardener and flora aficionado, the tycoon often labored in the gardens himself and was a frequent winner of plant and flower awards at horticulture shows. The gardens withered after Vanderbilt's 1938 death but were revived starting in the 1980s by a group of local gardeners called the F. W. Vanderbilt Garden Association. Working in conjunction with the Park Service,

the club organizes guided garden tours, plant sales, alfresco wine tastings, and other events.

10 | Weir Farm
NATIONAL HISTORICAL PARK
Connecticut

American impressionist landscape artist **Julian Alden Weir** purchased this Connecticut property on a whim in 1882. Collector Erwin Davis offered to trade the 153-acre farm for $10 and a single painting from the artist's private collection. Craving a country escape, Weir accepted the deal and moved in a year later with his first wife, Anna Dwight Baker. Over the next 36 years, the farm served as a creative retreat for many of Weir's artist friends, among them **John Singer Sargent, Childe Hassam,** and **John Henry Twachtman.** After Weir's death in 1919, sculptor **Mahonri Mackintosh Young** and painter **Sperry Andrews** also lived and worked at the farm, with Andrews staying until his death in 2005, long after the Park Service had acquired the property. Seasonal ranger-led walks explore the farm's artistic, agrarian, and geological legacies. The visitor center hosts rotating art and history exhibitions. You can meander through the farm buildings on their own or hike bucolic trails to Weir Pond and the nearby Weir Preserve.

WORTH-THE-HIKE VIEWS

There are thousands upon thousands of miles of trails in America's national parks, and hiking them could keep a person happily busy for years. But in the interest of time, and to get a sweeping scope of a park all in one go, here are a few hikes that lead to some of the most distinctive views available.

1 Rocky Mountain
NATIONAL PARK
Colorado

For a classic summit hike, look no further than the **Twin Sisters Peak Trail** in Rocky Mountain National Park, famous for its mountain scenery. The Twin Sisters Peaks are two adjacent summits near the park's eastern boundary. Starting at the Lily Lake Visitor Center, this 3.5-mile trail (one way) ascends to a high saddle between the peaks, climbing 2,338 feet using a series of well-cut switchbacks. The trail ends at an elevation of about 11,400 feet, getting steeper and more rugged the higher it goes. Take breaks on the way up to enjoy views of Longs Peak, Mount Meeker, and Estes Cones. From the saddle, short, rough, and scrambling trails lead to either summit—the west peak is easier to reach. From here, the 360-degree views of the Continental Divide, Great Plains, and Estes Park area are phenomenal. For an easier hike with equally awesome reward, consider the four-mile round-trip hike from the **Bear Lake Trailhead** to Emerald Lake, which journeys past four alpine lakes along the route.

2 Statue of Liberty
NATIONAL MONUMENT
New York

The rigorous, almost vertical—and some might say claustrophobic—**climb to the crown** of Statue of Liberty National Monument rewards the hearty with unparalleled views from the top. The narrow, 354-step spiral staircase that ascends to the crown from the statue's pedestal was closed after the September 11, 2001, attacks and again during the COVID-19 pandemic, but it's once again open to the public. A limited number of people are allowed up to the crown each day, and those privileged few enjoy stunning panoramic views of the Manhattan skyline and New York Harbor. Almost as intriguing is the close-up view afforded of the interior iron framework that supports the massive copper shell of the statue. A gift from France in the late 19th century, the statue was designed by sculptor Frédéric-Auguste Bartholdi and the iron framework was engineered by Gustave Eiffel. So put on light clothing (temperatures in the crown well exceed those on the ground), mentally prepare, and start climbing—burning quads will be forgotten upon catching sight of the views.

3 | Badlands
NATIONAL PARK
South Dakota

The crumpled, distinctive landforms of Badlands National Park create one of the most desolately beautiful landscapes in the American West. Most people view the park through the windows of their car, but a Badlands experience will be much richer on foot, taking advantage of one of the stunning hikes through the bluffs. The trails are generally pretty short, but that doesn't mean a person won't sweat—the ascents are steep. The 1.5-mile **Notch Trail** meanders through a canyon and then ascends so sharply that the Park Service has installed a log ladder to help hikers clamber up. The trail passes numerous jagged gullies, canyons, and spires that water carved out of sandstone to create the contorted Badlands formations. Eventually the trail comes out onto a ledge—the "Notch"—and rewards hikers with sweeping views of the White River Valley basin below, an expanse periodically interrupted by weird upthrust rock formations. This hike should be reserved for the sure of foot without fears of heights. It's Badlands scenery at its most awesome, and everybody who sees it will be glad they ventured out of the car.

▶ *Use technology for good: Consider downloading park maps for offline use. Apps like Google Maps, All Trails, and iNaturalist have highlighted routes available.*

4 | Hawai'i Volcanoes
NATIONAL PARK
Hawaii

Mother Nature makes few promises. And when you set your sights on something as unreliable as watching lava flow into an ocean, she can seem a downright tease. Hawai'i Volcanoes National Park would be the place, though, and a hike to **the end of Chain of Craters Road** would be the way. In 2018 boat tours made a big business out of sightings of Kīlauea's streaming lava, but on foot it's a cheaper, albeit volatile, hike along the flow field to take in the ravaged black cliffs and 90-foot-tall Hōlei Sea Arch. The powerful forces of the Pacific will eventually claim the arch for the sea, so view it while you can. Although Kīlauea was active in 2023, the lava remained within the crater. Visitors can check online for updates and a recent lava flow map, keeping in mind that while this park boasts some of the world's most active

volcanoes, other areas, such as its rainforest and still steaming craters, are A+ backups.

5 | Glacier
NATIONAL PARK
Montana

The **Hidden Lake Nature Trail** in the Logan Pass area of Glacier National Park offers a relaxing three-mile round-trip excursion to the kind of scenery for which the park is renowned. The boardwalk trail begins at the Logan Pass Visitor Center; the slightly raised walkway helps protect the alpine meadow beneath. Keep an eye out for mountain goats, bighorn sheep, and pikas on nearby Clements Mountain and Mount Reynolds along the way. At trail's end is a captivating overlook of Hidden Lake 750 feet below, glistening in its glacier-carved basin, with Bearhat Mountain rising behind it. On a clear day, Sperry Glacier might be visible, a mere remnant of the massive ice sheets that sculpted Glacier's topography. The trail is very popular, so go early to avoid crowds.

6 | Crater Lake
NATIONAL PARK
Oregon

Crater Lake, one of Oregon's crown jewels, was formed when a volcano exploded with such force that its summit collapsed, leaving a caldera

*Lava bursts some 60 feet into
superheated ocean water at
Hawai'i Volcanoes National Park.*

six miles wide that filled with water. There are more than 100 miles of trails within the park, but a couple short ones will take hikers to top overlooks. For panoramic views of the east side of the park, the lake, and even Mount Shasta, take the 2.5-mile trail (one way) up **Mount Scott** to an elevation of 8,929 feet, the highest point in the park. The trailhead for this roughly three-hour round-trip hike is about 14 miles east of park headquarters along Rim Drive. For an easier, shorter hike, try the **Garfield Peak Trail,** which originates east of Crater Lake Lodge along the caldera rim and ascends 1,000 feet to stunning views of the lake and a wonderful vantage point for Phantom Ship, an eerie-looking island rock formation that comprises 400,000-year-old lava flows. The hike is 1.7 miles one way and will take between two and three hours to accomplish round-trip.

7 Lewis & Clark
NATIONAL HISTORIC TRAIL
Montana & Idaho

One of the most challenging mountainous sections Lewis and Clark encountered during their epic expedition across the West was the Lolo Trail, their path across the Bitterroot Mountains for about 120 miles through Montana and Idaho. Today, the **Lolo Motorway/Forest Service Road 500**—a primitive dirt

Zion's Angels Landing is a popular—and strenuous—hike with an unforgettable view.

road built in the 1930s—follows much of the original Lolo Trail and provides access to points that the explorers wrote about in their journals. Enter the Motorway in Idaho off U.S. 12 via Forest Service Road 107, and then drive east about five miles past Devil's Chair and Howard's Camp to the marker for the Indian Post Office. Stop and make the short hike to the mysterious rock cairn—its origin and purpose uncertain—at the site that Lewis and Clark passed in 1805. At 7,033 feet, this is the highest point on the **Lolo Trail;** take a few minutes to admire the stunning views of the Bitterroot Range, an achingly beautiful landscape little changed from when Lewis and Clark passed through.

8 Golden Gate
NATIONAL RECREATION AREA
California

Lands End in Golden Gate National Recreation Area has the San Francisco Bay's most turbulent and rocky coast, with headlands sculpted by driving wind and waves. Treacherous outcroppings, lying in choppy Pacific waters off the coast with only their tips jutting above the surf, have been the cause of many shipwrecks, remnants of which can still be seen from the cliffs above the shore. To see it up close, take the Lands End segment of the **Coastal Trail.** Go at low tide for a chance to see the wrecks of the *Ohioan,* a freighter lost in 1937, and two tankers, the *Lyman A. Stewart* (sank 1922) and the *Frank H. Buck* (sank 1937). To the northeast stands Mile Rock, one of the deadly outcroppings responsible for ships' demises, and Mile Rock Lighthouse. About half a mile along the trail, a set of stairs leads down to Mile Rock Beach—it's worth the climb down (and later back up) for a close-up look at the strength and ferocity of the Pacific waters along this section of the coast. A short trail leads to Mile Rock Lookout, also worth a quick side trip. Head back up to the Coastal Trail and continue north to Eagles Point Overlook, where a wooden viewing platform and staircase on the cliff offer thrilling panoramic views of the entrance to San Francisco Bay.

9 Zion
NATIONAL PARK
Utah

Famed for its jagged rock ledges, yawning gorges, and narrow canyons flanked by striated cliffs, Zion National Park promises plenty of timeless hiking. From the trailhead at the Grotto picnic area, hike the **five-mile trail to Angels Landing.** The trail was cut into solid rock in 1926 and has a series of 21 short switchbacks on the 1,500-foot climb. The scene from the top is spectacular and gives dizzying views into the depths of Zion Canyon, as well as an impressive perspective on the other canyons, cliffs, and gullies that score the landscape passed en route. Angels Landing requires a reservation permitting process, and the hike is not for those with a fear of heights—the route to the top is very steep, and at points, the ground next to the trail falls away to steep drop-offs. Chains anchored along part of the route serve as handholds to help hikers climb. Those who reach the top will understand its name.

10 Sleeping Bear Dunes
NATIONAL LAKESHORE
Michigan

On the banks of Lake Michigan on the state's Lower Peninsula, Sleeping Bear Dunes National Lakeshore is perfectly situated for challenging hikes to rewarding views. For beautiful overlooks of the lake and the surrounding landscape, take the **Sleeping Bear Point Trail**—a 2.8-mile loop over sandy terrain to sweeping sights of the water, surrounding dunes, and South Manitou Island. As far as elevation goes, it isn't a terribly strenuous hike, but scrambling up dunes as sand falls away underfoot can make that 2.8 miles feel like much, much more. The vistas from the top are well worth the effort, though, and if Lake Michigan proves captivating enough, there's a spur off the loop that leads down to the shoreline itself.

Lewis and Clark on the Lolo Pass

Hikers today thrill at the view from the Indian Post Office on the Lolo Trail in the Bitterroot Mountains. But the members of the Corps of Discovery at that same spot back in 1805 were not nearly so euphoric. Although the weather was worsening when the expedition reached the Bitterroot Valley in mid-September, Lewis and Clark knew they had no choice but to start over the mountains. They began their ascent on September 11, and for 10 punishing days battled frostbite, malnutrition, and exhaustion as they struggled through the pass before finally breaking out of the mountains onto Weippe Prairie on September 20. They passed the spot marked by the Indian Post Office on September 16 in knee-deep snow and frigid temps, and Clark wrote in his journal that day, "I have been wet and as cold in every part as I ever was in my life, indeed I was at one time fearfull my feet would freeze in the thin Mockirsons which I wore ... men all wet cold and hungary." Their courage during this ordeal is something to bear in mind when standing in the same spot today and taking in the glorious view of the Bitterroots.

Where Inspiration Strikes

Making art in a park is having a muse like no other. And for dozens of professional makers, it's their job, as part of artist residency programs in the National Park Service. Learn about the oldest one here.

Down a short trail, tucked into the trees and overlooking a lush forest, the glass-walled Ravine Studio was built in 1900 as a stone-carving space and private refuge for Augustus and Augusta Saint-Gaudens. The couple had started summering in Cornish, New Hampshire, just a few years earlier and enjoyed infusing their home and property with their eclectic taste and global vision. While some visiting Saint-Gaudens National Historical Park purposely seek out Ravine Studio, others are surprised to stumble on the intimate site where, from June through October, they'll likely find a sculptor hard at work at their craft.

Honoring the life and art of sculptor Augustus Saint-Gaudens and the Cornish Colony, this national park for the arts includes the home, studios, grounds, and more than 100 works by the acclaimed artist. Its artist residency, the Park Service's oldest, is supported by the nonprofit partner Saint-Gaudens Memorial. Following a competitive selection process, the park's sculptor-in-residence runs an open studio, hosts demonstration workshops, gives talks about the art of sculpture, and participates in other events throughout the park's season.

"It's really compelling for visitors to be able to watch our sculptor-in-residence work and to hear how you go from an idea to a small piece of clay to a monumental sculpture," says Rainey McKenna, the visitor experience and resource stewardship program manager at both Saint-Gaudens and Marsh-Billings-Rockefeller National Historical Park. (The sites, just a half-hour apart, are co-managed.)

It's common for travelers to stop in without prior knowledge of Augustus Saint-Gaudens and the mark he made on the art world. What they find, McKenna says, can be called a type of Zen. "Visitors tend to be enamored with the park. It's such a lovely space—intimate, soothing, and clearly designed by artists."

RESIDENT SCULPTOR & SCULPTOR-IN-RESIDENCE
Internationally renowned as one of America's most influential sculptors, Saint-Gaudens was 13 when he began an apprenticeship as a cameo cutter. His dedication to the skill soon led him to study sculpture in France and Italy, and by 1876 the artist had his first major commission. Throughout his career Saint-Gaudens completed 150 artworks in bronze, plaster, stone, and wood—including designs for U.S. gold coins.

His family began summering in New Hampshire in 1885 and soon formed the Cornish Colony, a community of artists, writers, composers, actors, and other visionaries who congregated in this region. The Saint-Gaudenses permanently settled in Cornish in 1900, and Augustus continued to live and work on the property until he died in 1907.

Vermont-based Sean Hunter Williams first learned about the Saint-Gaudens artist residency from a fellow sculptor who'd held the role earlier. Williams, a professional stone carver whose portfolio includes monuments, historic restoration, and public and private commissions, was seeking a regional opportunity to explore different sides of his craft.

"A residency gives you a chance to create without being under a strict deadline or restricted by commission. It's a chance to explore your own sensibilities and interests, plus, here, to teach and interact with the general public," says Williams.

See inside the artist's private studio space at Saint-Gaudens National Historical Park.

"Saint-Gaudens was not only a member of the community of artists in Cornish, but he was also charismatic and a pillar of that community, and other artists moved up to the area to be around him," explains Williams. "Being next to his work every day ... that's why people go to Rome or Venice or Florence—to be immersed."

While at Saint-Gaudens in 2022 and 2023, Williams focused on portraiture, which was fitting for the setting. "The idea of immersion goes with both visiting the park and being an artist working here—you're in the very environment where another artist was truly inspired every day. I've definitely felt that at Saint-Gaudens," he continues. "You just have to take one walk around and you're inspired to make something—a flower or an animal or a part of one of Saint-Gaudens' sculptures can inspire you to improve your work."

THE NEXT CHAPTER

In recent years the park has added two student artist-in-residence roles that partner younger artists with the professional artists-in-residence at Saint-Gaudens and Marsh-Billings-Rockefeller to help run programs and create immersive experiences for visitors. Art in the Park, for example, is a free opportunity for visitors to use various materials to create their own works.

A summer concert series fills the park with music on Sundays, and local youth events include programs for young families and STEAM (science, technology, engineering, art, and math) camps for teens that integrate hikes, games, nature, and plenty of time to play.

This, McKenna says, is the park's purpose in action. "One of the most important ways that we can share and, better yet, pass on the legacy of this place is to nurture the next generation of artists."

URBAN ESCAPES

Even dedicated urbanites need a break now and then from the concrete jungle, round-the-clock noise, and exhaust fumes. These cities are among those lucky enough to have units of the National Park System right on their doorsteps, offering recreation and natural beauty within easy reach of millions.

1 Gateway
NATIONAL RECREATION AREA
New York & New Jersey

With **three units located around New York Harbor** in both New York and New Jersey, Gateway is just a subway or bus ride away for residents of America's largest metropolis. Some come to swim or sun on beaches, others to bike or jog miles of trails, to fish or kayak, to visit historic sites, or to go on a wildflower walk. And these activities are just the beginning of the year-round offerings at Gateway, a getaway perfect for everything from chilling out to learning a new skill at one of dozens of ranger-led programs. Camping areas allow city folk to sample life in the wild, while buildings at Fort Tilden in Queens have been converted into an arts center. Fort Tilden is just one of six decommissioned forts or military airfields open to the public within Gateway; the Sandy Hook Proving Ground was the first U.S. Army weapons testing grounds, established in 1874. Brooklyn's Jamaica Bay Wildlife Refuge ranks among the most popular bird-watching sites in the region, with marsh, fields, woods, and ponds. Other sites within the park offer golf and even places for flying model airplanes. If you can't find something to do at Gateway, you're just not looking closely enough.

2 Golden Gate
NATIONAL RECREATION AREA
California

Want to hike through one of the world's most beautiful forests? Want to visit the legendary prison where Al Capone and "Machine Gun" Kelly were locked up? Want to go hang gliding, enjoy a quiet picnic, admire an art collection, or visit a Cold War–era missile site? All of this and a lot more is possible at Golden Gate, a collection of some **three dozen separate units stretching for 70 miles** on or near the Pacific Coast, both north and south of San Francisco. A short list of attractions here would include Marin Headlands, a natural area at the north end of the Golden Gate Bridge; Alcatraz Island, former home of an infamous federal prison; Crissy Field, a restored tidal marsh and renowned windsurfing center; Ocean Beach, the longest beach in the Bay Area and a mecca for serious surfers; Point Bonita Lighthouse, a still active beacon; the reconstructed Cliff House and the Sutro Baths, the latter now in ruins; and Muir Woods National Monument, a grove of magnificent coast redwood trees. Bay Area residents treasure these and other park sites, all contributing to making one of the nation's most interesting cities even more appealing.

Come spring, the Gateway Arch is surrounded by beautiful blooming cherry trees.

3 | Gateway Arch
NATIONAL PARK
Missouri

The nation's tallest monument was established in 1935 to mark the Louisiana Purchase and the westward flow of American pioneers. Originally called the Jefferson National Expansion Memorial, its 2018 change in status to Gateway Arch National Park included an increase in walkable riverside green space—with grass, trees, ponds, and walkways encouraging people to explore and enjoy the city-center park. St. Louis visitors love the popular arch tram, as well as the underground museum devoted to the area's Indigenous heritage, European settlement, and eventual construction process of the arch. Currently under renovation, but also significant to St. Louis' story, the adjacent Old Courthouse was made famous by the Dred Scott case, where the refusal of the court to allow citizenship to the enslaved Black man contributed to the fervor that started the Civil War. For locals, the park's appeal is less about its structures and more about its space: Yoga classes, family-friendly events, and regular concerts take place on

Theodore Roosevelt Island features a memorial dedicated to the 26th president.

the lawn. **Acres along the Mississippi River** shifted from roadway to walkways, with benches, gardens, and meadows intentionally landscaped to keep the country's smallest national park an urban oasis.

4 Santa Monica Mountains
NATIONAL RECREATION AREA
California

The world's largest urban national park, Santa Monica Mountains National Recreation Area spans almost **40 miles from Los Angeles westward into Ventura County,** its many separate units covering nearly 240 square miles. Within

the designated area are state and local parks, ranging from historic sites to wild areas offering solitude in a region of freeways, shopping malls, and near-endless suburbs. The **65-mile-long Backbone Trail System** traverses the rugged Santa Monica Mountains, its entire length open to hikers, while some segments are available for mountain biking and horseback riding. Located in the land of Hollywood and Beverly Hills, the park has inevitable movie and television connections: Many movies and TV shows were filmed at Paramount Ranch in Agoura Hills, now home to fine hiking trails; Peter Strauss Ranch, along Mulholland Highway,

was donated to the park by actor Strauss, who wanted to preserve its natural beauty; and Will Rogers State Historic Park, just off Sunset Boulevard, preserves the estate of the famed humorist and actor who died in 1935. One of the park's most fascinating historical attractions is Rancho Sierra Vista/Satwiwa in Newbury Park, where the Satwiwa Native American Indian Culture Center interprets the heritage of the Chumash and Gabrielino/Tongva people who once lived here. In the northern part of the national recreation area, Cheeseboro Canyon boasts extensive hiking trails and woodlands of the imposing valley oak, a tree species found only in California.

5 Theodore Roosevelt Island
NATIONAL MEMORIAL
District of Columbia

This District of Columbia **park in the Potomac River** is a treasured green getaway, a peaceful respite from city life for residents of the nation's capital. Acquired in 1932 to honor our greatest conservationist president, the 88-acre Theodore Roosevelt Island comprises 2.5 miles of hiking trails through woods that seem removed from civilization, as well as a statue of Roosevelt and stone monuments inscribed with some of his quotations. Had this park been accessible during his presidency, there's no doubt that T. R. would have skipped out on Cabinet meetings now and then to enjoy nature here, just as he enjoyed roaming nearby Rock Creek Park to watch birds and partake in the "strenuous life" he always advocated.

6 | Cuyahoga Valley
NATIONAL PARK
Ohio

Connecting the Ohio cities of Cleveland and Akron, Cuyahoga Valley boasts **a combination of attractions** that can justifiably be called unique in the National Park System. Consider that visitors can ride a restored railroad train, attend a summer concert by the famed Cleveland Orchestra, go cross-country skiing, drive a scenic byway, jog or bike alongside the historic Ohio & Erie Canal, attend a theater performance, play golf, go horseback riding, or admire some of the most beautiful waterfalls in the eastern United States—all within the boundaries of Cuyahoga Valley National Park. Running for 22 miles through the heart of the park, the Cuyahoga River once served as a poster child for pollution (it actually caught fire several times when massive oil slicks ignited). The river has recovered enough that fish, birds, and other wildlife now abound along its tree-shaded length. Several villages within the park provide

▶ *Before receiving its modern name in 1933, Theodore Roosevelt Island was called the following: Analostan, by the Indigenous Neocostins; My Lord's Island (1632); Barbados (1680); and Mason's Island (1717).*

amenities such as dining, shopping, and bike rental. In the southwestern part of the park, Hale Farm & Village is a living history attraction that re-creates 19th-century farm life in the Cuyahoga Valley.

7 | Boston Harbor Islands
NATIONAL RECREATION AREA
Massachusetts

In a world where open space is hard to come by near cities, the National Park Service finds innovative ways to bring nature and recreational activities to urban residents. One good example is Boston Harbor Islands National Recreation Area, **a collection of 34 islands and peninsulas close to the historic Massachusetts capital,** administered through a partnership among federal, state, and local government agencies and private businesses.

"Minutes away, worlds apart" was once the park's slogan, and a visitor sea kayaking around Grape Island or hiking a trail at Worlds End, a 244-acre peninsula overlooking Hingham Harbor, would agree that the congestion of downtown Boston seems far removed. For great views of the Boston skyline, head to Spectacle Island, which has the highest point in the park, 157 feet above sea level, as well as five miles of hiking trails. History is represented at places such as Little Brewster Island, home to the country's oldest light station, and 19th-century Army post Fort Warren on Georges Island.

8 | Indiana Dunes
NATIONAL PARK
Indiana

A very high percentage of the nearly two million annual visitors to Indiana Dunes come simply to sunbathe or swim along **15 miles of sandy Lake Michigan shore.** In itself, that resource makes the park a treasured getaway for residents of Chicago and nearby cities such as Gary and Michigan City, Indiana. Beyond the beach, though, trails wind through natural habitats of surprising biodiversity, with rare plants and butterflies living among the dunes, savannas, marshes, prairies, and woodlands. Miller Woods, Cowles Bog, Heron Rookery, and Ly-co-ki-we are among the best trails for nature lovers. For a glimpse into the area's past, visit restored Chellberg Farm, where three generations of a family of Swedish farmers lived.

9 Chattahoochee River
NATIONAL RECREATION AREA
Georgia

Residents of Atlanta benefit today from planning decisions made in the 1970s, when public officials began working to protect **48 miles** of the Chattahoochee River on the edge of the city. As Atlanta has grown, the various units of the national recreation area endure as green spaces for picnicking, hiking, mountain biking, horseback riding, and enjoying nature. The river itself is extremely popular for canoeing, kayaking, tubing, and rafting, as well as fishing for trout, bass, and catfish. (Because river water comes from the dam on Lake Sidney Lanier, it's mostly too cold for comfortable swimming even in summer.) Of the many separate units within the park, the Cochran Shoals area may be the most popular, with a three-mile trail and a wetlands boardwalk. Concessionaires rent kayaks, canoes, and inner tubes at several locations along the Chattahoochee.

10 Mississippi
NATIONAL RIVER & RECREATION AREA
Minnesota

With **various units spanning 72 miles of the Mississippi River in the vicinity of the Twin Cities of Minneapolis and St. Paul,** this park is actually a consortium of state, regional, county, and municipal areas, ranging from city parks to museums and from wildlife refuges to historic sites. The Park Service owns only 67 of the recreation area's 54,000 acres, but coordinates activities and assists travelers from its visitor center in downtown St. Paul (in the lobby of the Science Museum of Minnesota). Citizens of the Twin Cities happily avail themselves of the area's opportunities for hiking, cross-country skiing (Fort Snelling State Park is a favorite), and canoeing. The 12.7-mile stretch of the Mississippi between the Crow River boat ramp and the Coon Rapids Dam offers excellent scenery and wildlife far different from the commercial waterway downstream.

Urban National Parks

National park? Isn't that a huge tract of wilderness far removed from civilization, with perhaps high mountains and vast forests? That was a common perception until the 1960s and '70s, when the first nature- and recreation-oriented "urban" national park units were established (as distinguished from, for example, historic sites such as Statue of Liberty National Monument). Fire Island National Seashore (near New York City), Indiana Dunes National Park, Gateway National Recreation Area, and Golden Gate National Recreation Area were among the first sites dedicated to the idea that urban residents, too, deserve national parks. These and similar National Park System sites across the country now provide a taste of the natural world to countless people who might otherwise experience it only on television or the Internet.

ISLAND OASES

Solitude, romance, adventure, excitement—this is what comes to mind when travelers think of island getaways. The National Park System encompasses hundreds of islands, some large and some tiny, and those featured on this short list of memorable spots beckon visitors to enjoy their charms.

1 | Isle Royale
NATIONAL PARK
Michigan

Wild and remote, yet not forbiddingly so in either category, Isle Royale offers a true North Woods experience to those who venture across Lake Superior's frigid waters to reach it. **The largest island in the world's largest freshwater lake,** Isle Royale provides commercial lodging, developed campgrounds, and unspoiled wilderness during its season of mid-April through October. (The park closes completely in winter.) Most people access the island via one of the ferries that leave from Minnesota or Michigan, though some opt to take a half-hour seaplane flight. Travelers can hike through mixed coniferous and hardwood forest, kayak along rocky shores, or camp beside gorgeous lakes, falling asleep to the howling of wolves and waking to the "laughter" of loons. Various boat services and ranger-guided trips give you the chance to explore Isle Royale in countless ways, from roughing it to (comparatively) leisurely tours. A day trip to Isle Royale is possible, but such a visit leaves only a few hours to enjoy the park; plan for a sojourn of at least a few days.

2 | Cumberland Island
NATIONAL SEASHORE
Georgia

Cumberland's beautiful beaches and a wilderness area of wetlands and woodlands make this **18-mile-long Georgia barrier island** a great choice for a relaxing getaway or an adventure trip. There's plenty for history buffs, too, beginning with Native American shell mounds and continuing through European settlements, slavery and post-emancipation communities, and mansions from a period when the island was a favored retreat of the wealthy Carnegie family. Regular ferry service provides access to Cumberland Island from St. Marys, Georgia; boats can also anchor near the island. There's still private property on Cumberland, and an elegant inn providing accommodations, but extensive wilderness and sandy Atlantic beaches make this locale special for those who bike its roads or hike its trails.

3 | Channel Islands
NATIONAL PARK
California

Despite its location off the coast of heavily populated Southern California, Channel Islands National Park receives relatively light visitation. Those people who do make the boat (or seaplane) trip across the Santa Barbara Channel can explore **five major island groups** where biodiversity is so globally significant that the park has earned

Trunk Bay, in the Virgin Islands, is named after the leatherback turtle, known locally as a "trunk."

the "Galápagos of North America" nickname. From ultrarare birds to blue whales (the world's largest animal), Channel Islands is a wildlife film come to life. All the islands are reachable for day-trippers (the closest, **Anacapa,** is a 90-minute boat trip from the mainland), and even a brief visit will bring sightings of seals, sea lions, sea otters, whales (more than two dozen species have been seen in the waters off the islands), and the largest colonies of seabirds in Southern California. Primitive camping can be found on all five islands, and backcountry camping is allowed on Santa Cruz and Santa Rosa. For a real adventure, sea kayakers can explore striking sea cliffs and Painted Cave, one of the world's largest and deepest sea caves. Found on Santa Cruz, the cave is nearly a quarter mile long.

4 | Virgin Islands
NATIONAL PARK
Virgin Islands

If you look beyond the white-sand beaches of the Caribbean's St. John, you'll find **hiking trails through tropical forests, ancient petroglyphs, and historic plantations.** Of course, if you can't look beyond the white-sand beaches, we don't blame you for that either: Virgin Islands National Park has some of the most stunning locations for swimming, snorkeling, and soaking in the tropics. However you spend the visit,

▶ *Found nowhere else on Earth, the island fox lives on six of the islands in Channel Islands National Park. The four-pound carnivore is approximately the size of a house cat.*

don't miss the chance to browse the park's Cruz Bay Visitor Center, where exhibits share the island's history of enslavement as well as the future of its renewing wildlife.

5 | Dry Tortugas
NATIONAL PARK
Florida

West of Key West by about 70 miles, remote Dry Tortugas National Park is a series of **seven small islands** and blue open water accessible by seaplane or ferry. Land makes up just one percent of the park, with the rest being the stunning Gulf of Mexico and its coral reefs, sea turtles, and other rich sea life. Of the seven isolated islands, **Garden Key** is the only one with camping, and it's also the most popular, as its massive 19th-century Fort Jefferson is a big draw. Camping is first come, first served here, and visitors are reminded that they must bring all necessary supplies, including water. Aboard a commercial tour or your

own boat, be sure to visit 30-acre **Loggerhead Key,** which is popular for its many shipwrecks. Take note of the 150-foot lighthouse that stands in the center. One of the most awe-inspiring times to visit is in spring, when millions of birds swarm these islands, taking a brief rest before continuing their northward migration.

6 | Apostle Islands
NATIONAL LAKESHORE
Wisconsin

After getting information at the visitor center in the small northern town of Bayfield, Wisconsin, take a boat ride out to explore some of the **21 islands of this Lake Superior park.** Eighty percent of Apostle Islands National Lakeshore is designated wilderness, which means it's an unspoiled island environment. Eighteen of the islands allow camping, and six boast historic lighthouses. Sea kayaking is a popular way to travel around the islands (rentals are available from park concessionaires), but Lake Superior's notoriously changeable weather and rough water mean paddling experience is strongly recommended. If interested in a true escape, consider this: Almost half the camping on the Apostle Islands takes place on just one of them, **Stockton Island.** Wilderness camping is available, and the park has established a camping zone system to assure solitude.

Part of the Great Florida Birding and Wildlife Trail, Gulf Islands National Seashore is home to more than 300 species of birds, including the great blue heron.

7 | Cape Lookout
NATIONAL SEASHORE
North Carolina

The Outer Banks of North Carolina are deservedly popular: a series of Atlantic Ocean barrier islands with expansive beaches, great fishing, historic beacons, and excellent wildlife-watching opportunities. Cape Lookout National Seashore, comprising **three main islands—North Core Banks, South Core Banks, and Shackleford Banks**—can be reached only by private boat or commercial ferries from a few communities on the North Carolina mainland, and has only minimal facilities. (Vehicles can be transported on some ferries.) As a result, the islands of Cape Lookout draw fewer visitors and offer more solitude than other Outer Banks islands. Need other reasons to visit? How about seeing the 163-foot-high lighthouse, touring historic Portsmouth Village, admiring the wild horses, collecting seashells, camping in a primitive site far from other people, or simply enjoying the sun, sand, and salt breeze of a nearly deserted beach.

8 | Gulf Islands
NATIONAL SEASHORE
Mississippi

This Gulf Coast park includes 12 separate units in both Florida and Mississippi, on the mainland and on islands. But true island lovers will be most interested in the Mississippi islands where boat-in primitive camping is allowed: **Horn, Petit Bois,** and part of **Cat.** With stunning white-sand beaches—originating as eroded quartz in the Appalachian Mountains and washed down to the Gulf of Mexico by rivers—these undeveloped Gulf Islands offer beauty, nature, and solitude for visitors who have their own boats or who take charter boats for the 12-mile trip from the mainland. (Watch for bottlenose dolphins during the crossing.) Swimming, fishing, hiking, bird-watching, and beachcombing are all popular activities. Campers must bring their own food and supplies (plus extras in case weather delays a return to the mainland), as well as the usual insect repellent, mosquito netting, sunblock, and first-aid gear. It's strictly a policy of "pack it in, pack it out" on these remote islands, but the wild and lonely place makes the planning and preparation worthwhile.

9 | Sleeping Bear Dunes
NATIONAL LAKESHORE
Michigan

Most people visit this park on Lake Michigan's eastern shore to swim, play on the tall dunes, or see historic sites on the mainland. Some, however, have discovered Sleeping Bear's **two wild islands, North and South Manitou.** The former, about 15,000 acres, is managed as wilderness open for backpacking, except

Wizard Island

One of the most striking and fascinating islands in the National Park System is 320-acre Wizard Island, which rises 765 feet above the brilliant blue water of Oregon's Crater Lake. The lake itself was formed when water collected in a caldera, or collapsed volcanic summit. Wizard Island (named for a fancied resemblance to a sorcerer's hat) is a classic cinder cone, built of material ejected from the caldera floor after the main volcanic explosion. Seasonal boat tours take visitors to the island, where a 0.9-mile trail ascends to a small crater at the top.

CRATER LAKE
NATIONAL PARK

for a 20-acre area around a small village. South Manitou, about three-by-three miles, is more developed, with camping allowed only in three official park campgrounds. Along with several historic buildings, South Manitou has a 104-foot-tall lighthouse, dating from 1871, that offers a panoramic view of Lake Michigan. Commercial ferries leave from the town of Leland to reach both islands.

10 | Biscayne
NATIONAL PARK
Florida

The watery wonderland of Biscayne National Park—95 percent of this Miami-area park is composed of Biscayne Bay and the Atlantic—has more to offer than just scuba diving, snorkeling, and boating. You can

take a commercial tour boat to see Elliott Key, once home to a thriving maritime community and now a park site with camping, swimming, and hiking. **Boca Chica Key,** also reachable by tour boat, offers camping and a 65-foot-high ornamental lighthouse built by a businessman who once owned the island. When open, the lighthouse observation deck provides a great panorama of the bay and the city skylines beyond. Those with their own boats can see day-use **Adams Key,** once the site of an exclusive fishing club visited by several U.S. presidents. The best way to explore Biscayne's islands is via kayak or canoe, which allows you to traverse shallow channels and lagoons to see wildlife such as sharks, rays, wading birds, and possibly a manatee or sea turtle.

UNDER-THE-RADAR PARKS

Yellowstone and the Grand Canyon deserve the appreciation they receive, but the National Park System includes more than 425 units with scores of lesser known sites that well reward a visit—and that are far less crowded than their big-name park siblings.

1 National Park of American Samoa
NATIONAL PARK
American Samoa

As the only park unit south of the equator, National Park of American Samoa no doubt requires some planning. Hawaiian Airlines is a major U.S. carrier to American Samoa, and smaller planes can assist with interisland flights. Once you've arrived at this tropical paradise, you'll wonder why more visitors don't prioritize the passage. The tropical rainforest and coral reefs are likely the country's finest, and its **five volcanic islands** (three are park units) explode with natural resources. **Ofu's** acclaimed beach, trails, and rainforest are popular for visitors, while the largest island, **Tutuila,** has a trail leading to the tip-top of Mount Alava.

Before your travels, take time to learn about the Samoans and their local culture and customs—these environmental caregivers have protected the islands for 3,000 years and are proud to show the impact of conservation.

2 Chaco Culture
NATIONAL HISTORICAL PARK
New Mexico

Other Native American sites in the Southwest are more famous—including Mesa Verde in Colorado and Canyon de Chelly in Arizona—but none was more important in its day than the communities that developed in Chaco Canyon, in what is now northwestern New Mexico. Only this park's remoteness and distance from major highways have kept it from greater renown

as a historical destination. From around A.D. 850 to 1250, Chaco Canyon was a major ceremonial, trade, and administrative center for the ancestral Puebloans, with elaborate and spectacular public architecture that matched its highly developed social organization. For reasons unknown, the center's inhabitants abandoned the site in the mid-13th century. In recognition of its importance, Chaco Culture National Historical Park and associated locations (including **Aztec Ruins National Monument,** 60 miles north) have been designated World Heritage sites. The park's nine-mile **Canyon Loop Drive** accesses six major archaeological sites, including "great houses": very large multistory public buildings with adjacent kivas, or ceremonial rooms. Truly awe-inspiring in both

Majestic El Capitan rises above beds of pepperweed in Guadalupe Mountains National Park (p. 441).

its appearance and historical significance, Chaco Culture is well worth the journey to see it.

3 | Great Basin
NATIONAL PARK
Nevada

Feel the urge to climb a 13,000-foot mountain? Want to tour a beautiful cave? Eager to backpack to alpine lakes and through high desert where few others venture? Want to marvel at some of the oldest living things on Earth? All that and more awaits at this park in eastern Nevada, where glacier-sculpted, 13,063-foot **Wheeler Peak** rises like an outpost of the Rocky Mountains in the Great Basin. A beautifully scenic drive climbs to more than 10,000 feet on Wheeler's flank, providing fabulous views with minimal effort. At the road's end, an easy hike leads to groves of picturesquely contorted bristlecone pines, some more than 3,000 years old. Wheeler is also home to **one of the southernmost glaciers in the Northern Hemisphere** (small, but geologically unique). After exploring the mountains, take a ranger-guided tour of **Lehman Caves,** where paths snake through marble and limestone passages featuring strange helictites, delicate aragonite crystals, and rare formations called cave shields. Travelers to Great Basin National Park can find these attractions in a place that sees only about 140,000 visitors a year.

A soft blanket of snow covers the landscape at Great Basin National Park.

4 Guadalupe Mountains
NATIONAL PARK
Texas

On the list of Southwest wonders, **El Capitan** in Guadalupe Mountains National Park ranks high up there—and it's only the eighth highest peak in Texas. The four highest points in the state are also in this wilderness park, where 25 million years of tectonic activity has lifted an ancient fossil reef from the ocean floor thousands of feet above sea level. But it isn't just height that matters here—**arid lowlands** offer streams, cacti, wildflowers, and a whole host of interesting animal life. Largely roadless, the park doesn't have a scenic drive to carry visitors to its many highlights, but most would argue that that makes setting off on foot or hoof all the more glorious. While fall and spring do have seasonal surges, the overall isolation of Guadalupe keeps the traffic down. Pack well and enjoy the wild.

5 Jean Lafitte
NATIONAL HISTORICAL PARK & PRESERVE
Louisiana

This multifaceted park is named for Jean Lafitte, a notorious 19th-century French pirate and smuggler (of both enslaved people and goods) who lived outside the law in the swamps and bayous surrounding

south Louisiana. Like the region itself, Lafitte is a complicated figure. While he may have played fast and loose with America's laws and participated in the slave trade, he and his men did help defend New Orleans from the British in 1815, which resulted in pardons from President James Madison. As it was even then, the region's melting pot still reflects the influence of pre-Columbus America, France, Spain, England, Africa, the Canary Islands, Germany, the Caribbean, the Confederate States, and many other places. You can learn a lot about local history at the **six separate sites** in this park, which spread from the swamps of the Mississippi River Delta to the prairies farther west. Take a history walk in the **French Quarter of New**

Orleans, see abundant wildlife in a **nature preserve,** attend a live radio broadcast of Cajun music in an **old-time theater,** visit the site of the famed **1815 Battle of New Orleans,** watch a **Cajun-cooking demonstration,** and enjoy many other activities. In few places in America do unique traditions endure as they do along the bayous and marshes of southern Louisiana.

6 Colorado
NATIONAL MONUMENT
Colorado

The red-rock country of the Southwest is justifiably famous for its spectacular canyons, buttes, spires, and other sandstone formations. Colorado National Monument offers a splendid array of eroded cliffs and pinnacles in an accessible and easily toured location, without the crowds and long-distance drives associated with more famous sites in the region. Much of the park can be seen on the 23-mile **Rim Rock Drive,** which can be reached just a few minutes off I-70 near Grand Junction, Colorado. Take a copy of the geology tour pamphlet (downloaded online or picked up from the visitor center) to maximize the drive. Trails here range from the short and easy **Window Rock Trail** to more strenuous hikes such as the route into **Monument Canyon,** which passes major rock sculptures including **Independence**

Wrangell–St. Elias National Park boasts the largest glacial system in the United States, including well-known examples like Hubbard, Nabesna, and Malaspina Glaciers.

Monument (a 450-foot-high sandstone monolith), **Kissing Couple,** and the **Coke Ovens.**

7 | Wrangell–St. Elias
NATIONAL PARK & PRESERVE
Alaska

Although Denali National Park and Preserve in Alaska draws more visitors, the state's Wrangell–St. Elias deserves recognition. It is **the largest national park in the United States** (by far!) and the site of many of the highest mountains in the country. Many outfitters and concessionaires offer adventure tours into the park, from river rafting and sea kayaking to scenic flights and guided glacier hikes. There's another option, too: Travelers can (with care and depending on weather) explore the park on two mostly unpaved roads, which provide access to trailheads, campsites, and historic sites. **McCarthy Road** runs 61 miles through boreal forest to the tiny community of McCarthy, following an old railroad right-of-way and crossing rivers on high trestles. Nearby are the remains of a once thriving copper mine, now a national historic landmark. The 42-mile **Nabesna Road** passes through spectacular mountain landscape and offers the chance to see Dall sheep, among other wildlife. Lucky visitors might come across caribou, moose, grizzly and black bears, mountain goats, gray wolves, coyotes, red foxes, wolverines, and porcupines. Both these drives are challenging and require planning and preparation, but offer thrilling experiences as a reward. And with less than 70,000 visitors per year, you're likely to find that thrill in solitude.

8 | Sunset Crater Volcano
NATIONAL MONUMENT
Arizona

Too few of the millions of people who visit Arizona's Grand Canyon each year detour to this fascinating park just northeast of Flagstaff. Set in a landscape full of dozens of symmetrical cones and other evidence of **six million years of volcanic activity,** Sunset Crater is a classic example of a cinder cone, named for the reddish oxidized material at its top. Many volcanic features can be seen along trails, including lava "squeeze-ups," spatter cones, and lava bubbles. No climbing is allowed on Sunset Crater Volcano, which is so symmetrical that it looks like every child's drawing of a volcano.

9 | Buffalo
NATIONAL RIVER
Arkansas

This beautiful northwestern Arkansas stream was designated America's first official national river in 1972 after years of controversy pitting those who wanted to protect its crystal clear water, lush forests, and spectacular bluffs against those who wanted the river dammed to form a sprawling reservoir. People who canoe or raft the 135-mile Ozark river today—the spring white water on the upper parts or the gentle flat water of its lower reaches—give thanks that the conservationists won. **Three official wilderness areas** along its length add to the "wild" quality of the river. At places like Steel Creek, sheer sandstone cliffs rise 400 feet from the water's edge, and all along the river are gravel bars for primitive camping, swimming holes for cooling off on a summer day, and rewarding hiking trails. There's even a ghost town on the lower river: Rush, where zinc was once mined. All in all, Buffalo National River preserves some of the finest scenery in the central United States, as well as one of the country's best float streams.

10 Wind Cave
NATIONAL PARK
South Dakota

The film shown in the visitor center at this South Dakota site is entitled "One Park, Two Worlds"—an apt slogan. Not only does the park feature Wind Cave, with passages full of unusual and beautiful formations, the 44-square-mile landscape aboveground is home to a diverse array of wildlife. In 1903, this site became **the first park in the world created to protect a cave,** noted primarily for its outstanding display of boxwork, an unusual formation of thin calcite fins resembling honeycombs. Eventually Wind Cave came to be recognized as one of the world's largest. Various ranger-guided tours are offered, including one suitable for people with physical limitations. Before or after a cave tour, driving or hiking in the park can bring sightings of bison, elk, pronghorns, mule deer, coyotes, and prairie dogs. The park's approximately 350 bison move around as they graze; it's easier to see the appealing little prairie dogs, whose colonies are visible near roads.

Park Attendance by the Numbers

In 2022 the least visited U.S. national park was National Park of American Samoa. There's a simple explanation for that: With its location in a remote part of the South Pacific, it's simply a challenge to get to. In the same year, 12.9 million people took in the fresh air at Great Smoky Mountains National Park, and the Blue Ridge Parkway won out as the most visited Park Service unit, with a whopping 15.7 million visitors. All told, 2022 was a banner year for the National Park System, with just shy of 312 million recreation visits in total.

July is the most popular month for visitors to hit the parks, in line with the peak of summer, school breaks, and warm days. More than 45 million people set foot in a Park Service property in July 2019.

Visitation is also divided up fairly evenly between the park units. Thirty-eight percent of visitors explore recreation parks, 32 percent opt for historical and cultural parks, and 30 percent go for a nature park.

But the coolest fact of all? Since 1904, more than 15 *billion* people have played at, slept in, or just enjoyed the great outdoors somewhere in the National Park System.

The Washington Monument as seen from the Lincoln Memorial in Washington, D.C.

NATIONAL PARK SERVICE
CONTACTS

National Park Service
1849 C St, NW / Washington, DC 20240-0001
202-208-6843 / *www.nps.gov*

A

Abraham Lincoln Birthplace National Historical Park
2995 Lincoln Farm Road
Hodgenville, KY 42748-9707
270-358-3137
www.nps.gov/abli

Acadia National Park
PO Box 177
Bar Harbor, ME 04609-0177
207-288-3338
www.nps.gov/acad

Adams National Historical Park
135 Adams Street
Quincy, MA 02169-1749
617-773-1177
www.nps.gov/adam

African Burial Ground National Monument
290 Broadway, First Floor
New York, NY 10007-1823
212-238-4367
www.nps.gov/afbg

Agate Fossil Beds National Monument
301 River Road
Harrison, NE 69346-2734
308-665-4113
www.nps.gov/agfo

Alagnak Wild River
PO Box 245
King Salmon, AK 99613-0245
907-246-3305
www.nps.gov/alag

Alibates Flint Quarries National Monument
PO Box 1460
Fritch, TX 79036-1460
806-857-6680
www.nps.gov/alfl

Allegheny Portage Railroad National Historic Site
110 Federal Park Road
Gallitzin, PA 16641-2000
814-886-6150
www.nps.gov/alpo

National Park of American Samoa
MHJ Building, 2nd Floor
Pago Pago, AS 96799
684-633-7082
www.nps.gov/npsa

Amistad National Recreation Area
10477 Highway 90 West
Del Rio, TX 78840-9350
830-775-7491
www.nps.gov/amis

Andersonville National Historic Site
496 Cemetery Road
Andersonville, GA 31711-9707
229-924-0343
www.nps.gov/ande

Andrew Johnson National Historic Site
121 Monument Avenue
Greeneville, TN 37743-5552
423-638-3551
www.nps.gov/anjo

Aniakchak National Monument & Preserve
PO Box 245
King Salmon, AK 99613-0245
907-246-3305
www.nps.gov/ania

Antietam National Battlefield
PO Box 158
Sharpsburg, MD 21782-0158
301-432-5124
www.nps.gov/anti

Apostle Islands National Lakeshore
415 Washington Avenue
Bayfield, WI 54814-0770
715-779-3397
www.nps.gov/apis

Appalachian National Scenic Trail
Appalachian Trail Park Office
PO Box 50
Harpers Ferry, WV 25425-0807
304-535-6278
www.nps.gov/appa

Appomattox Court House National Historical Park
PO Box 218
Appomattox, VA 24522-0218
434-352-8987
www.nps.gov/apco

Arches National Park
PO Box 907
Moab, UT 84532-0907
435-719-2299
www.nps.gov/arch

Arkansas Post National Memorial
1741 Old Post Road
Gillett, AR 72055-9707
870-548-2207
www.nps.gov/arpo

Arlington House, The Robert E. Lee Memorial
c/o Turkey Run Park
700 George Washington Memorial Parkway
McLean, VA 22101-1716
703-235-1530
www.nps.gov/arho

Assateague Island National Seashore
7206 National Seashore Lane
Berlin, MD 21811-2540
410-641-1441
www.nps.gov/asis

Aztec Ruins National Monument
725 Ruins Road
Aztec, NM 87410-9715
505-334-6174
www.nps.gov/azru

B

Badlands National Park
25216 Ben Reifel Road
Interior, SD 57750-0006
605-433-5361
www.nps.gov/badl

Bandelier National Monument
15 Entrance Road
Los Alamos, NM 87544-9508
505-672-3861
www.nps.gov/band

Belmont-Paul Women's Equality National Monument
144 Constitution Avenue, NE
Washington, DC 20002
bepa_info@nps.gov
www.nps.gov/bepa

Bent's Old Fort National Historic Site
35110 State Highway 194 East
La Junta, CO 81050-9523
719-383-5010
www.nps.gov/beol

Bering Land Bridge National Preserve
PO Box 220
Nome, AK 99762-0220
907-443-2522
www.nps.gov/bela

Big Bend National Park
PO Box 129
Big Bend National Park
TX 79834-0129
432-477-2251
www.nps.gov/bibe

Big Cypress National Preserve
33100 Tamiami Trail East
Ochopee, FL 34141-1000
239-695-2000
www.nps.gov/bicy

Big Hole National Battlefield
PO Box 237
Wisdom, MT 59761-0237
406-689-3155
www.nps.gov/biho

Big South Fork National River & Recreation Area
4564 Leatherwood Road
Oneida, TN 37841-9544
423-569-9778
www.nps.gov/biso

Big Thicket National Preserve
6102 FM 420
Kountze, TX 77625-7842
409-951-6700
www.nps.gov/bith

Bighorn Canyon National Recreation Area
Bighorn Canyon NRA Visitor Center
South District
20 US Highway 14A East
Lovell, WY 82431-9626
307-548-5406
www.nps.gov/bica

Birmingham Civil Rights National Monument
1914 4th Avenue North, Suite 440
Birmingham, AL 35203-3517
205-679-0065
www.nps.gov/bicr

Biscayne National Park
9700 SW 328th Street
Sir Lancelot Jones Way
Homestead, FL 33033-5634
305-230-1144
www.nps.gov/bisc

Black Canyon of the Gunnison National Park
102 Elk Creek
Gunnison, CO 81230-9304
970-641-2337
www.nps.gov/blca

Blackstone River Valley National Historical Park
67 Roosevelt Avenue
Pawtucket, RI 02860-2127
401-725-8638
www.nps.gov/blrv

Blue Ridge Parkway
199 Hemphill Knob Road
Asheville, NC 28803-8686
828-348-3400
www.nps.gov/blri

Bluestone National Scenic River
PO Box 246
Glen Jean, WV 25846-0246
304-465-0508
www.nps.gov/blue

Booker T. Washington National Monument
12130 Booker T. Washington Highway
Hardy, VA 24101-9688
540-721-2094
www.nps.gov/bowa

Boston African American National Historic Site
15 State Street
Boston, MA 02109
617-429-6760
www.nps.gov/boaf

Boston Harbor Islands National Recreation Area
15 State Street, 4th Floor
Boston, MA 02109-3502
617-223-8666
www.nps.gov/boha

Boston National Historical Park
21 Second Avenue
Charlestown Navy Yard
Boston, MA 02129-4543
617-242-5601
www.nps.gov/bost

Brices Cross Roads National Battlefield Site
2680 Natchez Trace Parkway
Tupelo, MS 38804-9718
1-800-305-7417
www.nps.gov/brcr

Brown v. Board of Education National Historical Park
1515 SE Monroe Street
Topeka, KS 66612-1143
785-354-4273
www.nps.gov/brvb

Bryce Canyon National Park
PO Box 640201
Bryce Canyon, UT 84764-0201
435-834-5322
www.nps.gov/brca

Buck Island Reef National Monument
2100 Church Street, #100
Christiansted, St. Croix, VI 00820-4611
340-773-1460
www.nps.gov/buis

Buffalo National River
402 North Walnut Street, Suite 136
Harrison, AR 72601-1173
870-439-2502
www.nps.gov/buff

C

Cabrillo National Monument
1800 Cabrillo Memorial Drive
San Diego, CA 92106-3601
619-523-4285
www.nps.gov/cabr

Camp Nelson National Monument
6614 Danville Road Loop 2
Nicholasville, KY 40356-9593
859-881-5716
www.nps.gov/cane

Canaveral National Seashore
212 South Washington Avenue
Titusville, FL 32796-3521
386-428-3384
www.nps.gov/cana

Cane River Creole National Historical Park
PO Box 925
Natchitoches, LA 71458-0925
318-352-0383
www.nps.gov/cari

Canyon de Chelly National Monument
PO Box 588
Chinle, AZ 86503-0588
928-674-5500
www.nps.gov/cach

Canyonlands National Park
2282 Resource Boulevard
Moab, UT 84532-3406
435-719-2313
www.nps.gov/cany

Cape Cod National Seashore
99 Marconi Site Road
Wellfleet, MA 02667-8142
508-255-3421
www.nps.gov/caco

Cape Hatteras National Seashore
1401 National Park Drive
Manteo, NC 27954-2708
252-473-2111
www.nps.gov/caha

Cape Krusenstern National Monument
PO Box 1029
Kotzebue, AK 99752-1029
907-442-3890
www.nps.gov/cakr

Cape Lookout National Seashore
131 Charles Street
Harkers Island, NC 28531-9702
252-728-2250
www.nps.gov/calo

Capitol Reef National Park
HC 70, Box 15
Torrey, UT 84775-9602
435-425-3791
www.nps.gov/care

Capulin Volcano National Monument
46 Volcano Highway
Capulin, NM 88414-4412
575-278-2201
www.nps.gov/cavo

Carl Sandburg Home National Historic Site
81 Carl Sandburg Lane
Flat Rock, NC 28731-8635
828-693-4178
www.nps.gov/carl

Carlsbad Caverns National Park
3225 National Parks Highway
Carlsbad, NM 88220-5354
575-785-2232
www.nps.gov/cave

Carter G. Woodson Home National Historic Site
1900 Anacostia Drive, SE
Washington, DC 20020-6722
202-690-5185
www.nps.gov/cawo

Casa Grande Ruins National Monument
1100 West Ruins Drive
Coolidge, AZ 85128-3200
520-723-3172
www.nps.gov/cagr

Castillo de San Marcos National Monument
1 South Castillo Drive
St. Augustine, FL 32084-3252
904-829-6506
www.nps.gov/casa

Castle Clinton National Monument
26 Wall Street
New York, NY 10005-1996
212-344-7220
www.nps.gov/cacl

Castle Mountains National Monument
2701 Barstow Road
Barstow, CA 92311-6609
760-252-6100
www.nps.gov/camo

Catoctin Mountain Park
6602 Foxville Road
Thurmont, MD 21788-1598
301-663-9388
www.nps.gov/cato

Cedar Breaks National Monument
2390 West Highway 56, Suite #11
Cedar City, UT 84720-4151
435-986-7120
www.nps.gov/cebr

Cedar Creek & Belle Grove National Historical Park
PO Box 700
Middletown, VA 22645-0700
540-869-3051
www.nps.gov/cebe

César E. Chávez National Monument
PO Box 201
Keene, CA 93531-0201
661-823-6134
www.nps.gov/cech

Chaco Culture National Historical Park
PO Box 220
Nageezi, NM 87037-0220
505-786-7014
www.nps.gov/chcu

Chamizal National Memorial
800 South San Marcial Street
El Paso, TX 79905-4123
915-532-7273
www.nps.gov/cham

Channel Islands National Park
1901 Spinnaker Drive
Ventura, CA 93001-4354
805-658-5730
www.nps.gov/chis

Charles Pinckney National Historic Site
1214 Middle Street
Sullivan's Island, SC 29482-9748
843-577-0242
www.nps.gov/chpi

Charles Young Buffalo Soldiers National Monument
PO Box 428
Wilberforce, OH 45384-0428
937-352-6757
www.nps.gov/chyo

Chattahoochee River National Recreation Area
1978 Island Ford Parkway
Sandy Springs, GA 30350-3400
678-538-1200
www.nps.gov/chat

Chesapeake & Ohio Canal National Historical Park
142 West Potomac Street
Williamsport, MD 21795-1039
301-739-4200
www.nps.gov/choh

Chickamauga & Chattanooga National Military Park
3370 LaFayette Road
Fort Oglethorpe, GA 30742-4265
706-866-9241
www.nps.gov/chch

Chickasaw National Recreation Area
901 West 1st Street
Sulphur, OK 73086-4822
580-622-7234
www.nps.gov/chic

Chimney Rock National Historic Site
9822 County Road 75
Bayard, NE 69334-9393
308-586-2581
www.nps.gov/places/000/chimney-rock-national-historic-site.htm

Chiricahua National Monument
12856 East Rhyolite Creek Road
Willcox, AZ 85643-4722
520-824-3560
www.nps.gov/chir

Christiansted National Historic Site
2100 Church Street, #100
Christiansted, VI 00820-4611
340-773-1460
www.nps.gov/chri

City of Rocks National Reserve
PO Box 169
Almo, ID 83312-0169
208 824-5901
www.nps.gov/ciro

Clara Barton National Historic Site
5801 Oxford Road
Glen Echo, MD 20812-1201
301-320-1410
www.nps.gov/clba

Colonial National Historical Park
PO Box 210
Yorktown, VA 23690-0210
757-898-2410
www.nps.gov/colo

Colorado National Monument
1750 Rim Rock Drive
Fruita, CO 81521-0001
970-858-2800
www.nps.gov/colm

Congaree National Park
100 National Park Road
Hopkins, SC 29061-9118
803-776-4396
www.nps.gov/cong

Constitution Gardens
c/o National Mall & Memorial Parks
1100 Ohio Drive, SW
Washington, DC 20242-0001
202-426-6841
www.nps.gov/coga

Coronado National Memorial
4101 East Montezuma Canyon Road
Hereford, AZ 85615-9376
520-366-5515
www.nps.gov/coro

Cowpens National Battlefield
338 New Pleasant Road
Gaffney, SC 29341-4522
864-461-2828
www.nps.gov/cowp

Crater Lake National Park
PO Box 7
Crater Lake, OR 97604-0007
541-594-3000
www.nps.gov/crla

Craters of the Moon National Monument & Preserve
PO Box 29
Arco, ID 83213-0029
208-527-1300
www.nps.gov/crmo

Cumberland Gap National Historical Park
91 Bartlett Park Road
Middlesboro, KY 40965-5011
606-248-2817
www.nps.gov/cuga

Cumberland Island National Seashore
101 Wheeler Street
St. Marys, GA 31558-8421
912-882-4336
www.nps.gov/cuis

Curecanti National Recreation Area
102 Elk Creek
Gunnison, CO 81230-9304
970-641-2337
www.nps.gov/cure

Cuyahoga Valley National Park
15610 Vaughn Road
Brecksville, OH 44141-3018
440-717-3890
www.nps.gov/cuva

D

Dayton Aviation Heritage National Historical Park
16 South Williams Street
Dayton, OH 45402-8235
937-225-7705
www.nps.gov/daav

De Soto National Memorial
8300 De Soto Memorial Highway
Bradenton, FL 34209-9748
941-792-0458
www.nps.gov/deso

Death Valley National Park
PO Box 579
Death Valley, CA 92328-0579
760-786-3200
www.nps.gov/deva

Delaware Water Gap National Recreation Area
1978 River Road
Bushkill, PA 18324-0002
570-426-2452
www.nps.gov/dewa

Denali National Park & Preserve
PO Box 9
Denali Park, AK 99755-0009
907-683-9532
www.nps.gov/dena

Devils Postpile National Monument
PO Box 3999
Mammoth Lakes, CA 93546-3999
760-934-2289
www.nps.gov/depo

Devils Tower National Monument
PO Box 10
Devils Tower, WY 82714-0010
307-467-5283
www.nps.gov/deto

Dinosaur National Monument
4545 East Highway 40
Dinosaur, CO 81610-9724
435-781-7700
www.nps.gov/dino

Dry Tortugas National Park
40001 State Road 9336
Homestead, FL 33034-6733
305-242-7700
www.nps.gov/drto

Dwight D. Eisenhower Memorial
c/o National Mall & Memorial Parks
1100 Ohio Drive, SW
Washington, DC 20242-0001
202-426-6841
www.nps.gov/ddem

E

Ebey's Landing National Historical Reserve
PO Box 774
Coupeville, WA 98239-0774
360-678-6084
www.nps.gov/ebla

Edgar Allan Poe National Historic Site
c/o Independence National Historical Park
143 South 3rd Street
Philadelphia, PA 19106-2818
215-965-2305
www.nps.gov/edal

Effigy Mounds National Monument
151 Highway 76
Harpers Ferry, IA 52146-7519
563-873-3491
www.nps.gov/efmo

Eisenhower National Historic Site
243 Eisenhower Farm Road
Gettysburg, PA 17325-7034
717-338-9114
www.nps.gov/eise

El Malpais National Monument
1900 East Santa Fe Avenue
Grants, NM 87020-4014
505-876-2783
www.nps.gov/elma

El Morro National Monument
HC 61 Box 43
Ramah, NM 87321-9603
505-783-4226
www.nps.gov/elmo

Eleanor Roosevelt National Historic Site
4097 Albany Post Road
Hyde Park, NY 12538-1997
845-229-6225
www.nps.gov/elro

Emmett Till and Mamie Till-Mobley National Monument
PO Box 361
Sumner, MS 38957-0361
662-483-1231
www.nps.gov/till

Eugene O'Neill National Historic Site
PO Box 280
Danville, CA 94526-0280
925-228-8860
www.nps.gov/euon

Everglades National Park
40001 State Road 9336
Homestead, FL 33034-6733
305-242-7700
www.nps.gov/ever

F

Federal Hall National Memorial
26 Wall Street
New York, NY 10005-1996
212-825-6990
www.nps.gov/feha

Fire Island National Seashore
120 Laurel Street
Patchogue, NY 11772-3596
631-569-2100
www.nps.gov/fiis

First Ladies National Historic Site
205 Market Avenue South
Canton, OH 44702-2107
330-452-0876
www.nps.gov/fila

First State National Historical Park
c/o New Castle Court House
Museum
211 Delaware Street
New Castle, DE 19720-4815
302-478-2769
www.nps.gov/frst

Flight 93 National Memorial
PO Box 911
Shanksville, PA 15560-0911
814-893-6322
www.nps.gov/flni

Florissant Fossil Beds National Monument
PO Box 185
Florissant, CO 80816-0185
719-748-3253
www.nps.gov/flfo

Ford's Theatre National Historic Site
c/o National Mall & Memorial Parks
1100 Ohio Drive, SW
Washington, DC 20242-0001
202-426-6924
www.nps.gov/foth

Fort Bowie National Historic Site
PO Box 158
Bowie, AZ 85605-0158
520-847-2500
www.nps.gov/fobo

Fort Caroline National Memorial
12713 Fort Caroline Road
Jacksonville, FL 32225-1240
904-641-7155
www.nps.gov/timu/learn/
historyculture/foca.htm

Fort Davis National Historic Site
PO Box 1379
Fort Davis, TX 79734-1379
432-426-3224
www.nps.gov/foda

Fort Donelson National Battlefield
174 National Cemetery Drive
PO Box 434
Dover, TN 37058-0434
931-232-5706
www.nps.gov/fodo

Fort Frederica National Monument
6515 Frederica Road
St. Simons Island, GA 31522-9727
912-638-3639
www.nps.gov/fofr

Fort Laramie National Historic Site
965 Gray Rocks Road
Fort Laramie, WY 82212-7625
307-837-2221
www.nps.gov/fola

Fort Larned National Historic Site
1767 KS Highway 156
Larned, KS 67550-9321
620-285-6911
www.nps.gov/fols

Fort Matanzas National Monument
8635 A1A South
St. Augustine, FL 32080-8411
904-471-0116
www.nps.gov/foma

Fort Monroe National Monument
41 Bernard Road, Building 17
Fort Monroe, VA 23651-1001
757-722-3678
www.nps.gov/fomr

Fort McHenry National Monument & Historic Shrine
2400 East Fort Avenue
Baltimore, MD 21230-5393
410-962-4290
www.nps.gov/fomc

Fort Necessity National Battlefield
1 Washington Parkway
Farmington, PA 15437-9514
724-329-5512
www.nps.gov/fone

Fort Point National Historic Site
Building 201, Fort Mason
San Francisco, CA 94123-0022
415-561-4959
www.nps.gov/fopo

Fort Pulaski National Monument
41 Cockspur Island Road
Savannah, GA 31410-1199
912-219-4233
www.nps.gov/fopu

Fort Raleigh National Historic Site
1401 National Park Drive
Manteo, NC 27954-9451
252-473-2111
www.nps.gov/fora

Fort Scott National Historic Site
PO Box 918
Fort Scott, KS 66701-0918
620-223-0310
www.nps.gov/fosc

Fort Smith National Historic Site
301 Parker Avenue
Fort Smith, AR 72901-1938
479-783-3961
www.nps.gov/fosm

Fort Stanwix National Monument
112 East Park Street
Rome, NY 13440-5816
315-338-7730
www.nps.gov/fost

Fort Sumter and Fort Moultrie National Historical Park
1214 Middle Street
Sullivan's Island, SC 29482-9748
843-577-0242
www.nps.gov/fosu

Fort Union National Monument
PO Box 127
Watrous, NM 87753-0127
505-425-8025
www.nps.gov/foun

Fort Union Trading Post National Historic Site
15550 Highway 1804
Williston, ND 58801-8680
701-572-9083
www.nps.gov/fous

Fort Vancouver National Historic Site
800 Hathaway Road, Building 722
Vancouver, WA 98661-3899
360-816-6230
www.nps.gov/fova

Fort Washington Park
13551 Fort Washington Road
Fort Washington, MD 20744-7044
771-208-1555
www.nps.gov/fowa

Fossil Butte National Monument
PO Box 592
Kemmerer, WY 83101-0592
307-877-4455
www.nps.gov/fobu

Franklin Delano Roosevelt Memorial
c/o National Mall & Memorial Parks
1100 Ohio Drive, SW
Washington, DC 20242-0001
202-426-6841
www.nps.gov/fdrm

Frederick Douglass National Historic Site
1411 W Street, SE
Washington, DC 20020-4813
771-208-1499
www.nps.gov/frdo

Frederick Law Olmsted National Historic Site
99 Warren Street
Brookline, MA 02445-5930
617-566-1689
www.nps.gov/frla

Fredericksburg & Spotsylvania National Military Park
120 Chatham Lane
Fredericksburg, VA 22405-2508
540-693-3200
www.nps.gov/frsp

Freedom Riders National Monument
1302 Noble Street, Suite 3G
Anniston, AL 36201-4678
256-715-9189
www.nps.gov/frri

Friendship Hill National Historic Site
c/o Fort Necessity National Battlefield
1 Washington Parkway
Farmington, PA 15437-9514
724-329-2501
www.nps.gov/frhi

G

Gates of the Arctic National Park & Preserve
101 Dunkel Street, Suite 110
Fairbanks, AK 99701-4806
907-459-3730
www.nps.gov/gaar

Gateway Arch National Park
11 North 4th Street
St. Louis, MO 63102-1810
314-655-1600
www.nps.gov/jeff

Gateway National Recreation Area
210 New York Avenue
Staten Island, NY 10305-5019
718-354-4606
www.nps.gov/gate

Gauley River National Recreation Area
PO Box 246
Glen Jean, WV 25846-0246
304-465-0508
www.nps.gov/gari

General Grant National Memorial
26 Wall Street
New York, NY 10005-1996
646-670-7251
www.nps.gov/gegr

George Rogers Clark National Historical Park
401 South Second Street
Vincennes, IN 47591-1001
812-882-1776
www.nps.gov/gero

George Washington Carver National Monument
5646 Carver Road
Diamond, MO 64840-8314
417-325-4151
www.nps.gov/gwca

George Washington Memorial Parkway
700 George Washington Memorial Parkway
McLean, VA 22101-1716
703-289-2500
www.nps.gov/gwmp

George Washington Birthplace National Monument
1732 Popes Creek Road
Colonial Beach, VA 22443-5115
804-224-1732
www.nps.gov/gewa

Gettysburg National Military Park
1195 Baltimore Pike, Suite 100
Gettysburg, PA 17325-2804
717-334-1124
www.nps.gov/gett

Gila Cliff Dwellings National Monument
26 Jim Bradford Trail
Mimbres, NM 88049-8071
575-536-9461
www.nps.gov/gicl

Glacier National Park
PO Box 128
West Glacier, MT 59936-0128
406-888-7800
www.nps.gov/glac

Glacier Bay National Park & Preserve
PO Box 140
Gustavus, AK 99826-0140
907-697-2230
www.nps.gov/glba

Glen Canyon National Recreation Area
PO Box 1507
Page, AZ 86040-1507
928-608-6200
www.nps.gov/glca

Golden Gate National Recreation Area
Building 201, Fort Mason
San Francisco, CA 94123-0022
415-561-4700
www.nps.gov/goga

Golden Spike National Historical Park
PO Box 897
Brigham City, UT 84302-0897
435-471-2209
www.nps.gov/gosp

Governors Island National Monument
10 South Street
New York, NY 10004-1921
212-825-3054
www.nps.gov/gois

Grand Canyon National Park
PO Box 129
Grand Canyon, AZ 86023-0129
928-638-7779
www.nps.gov/grca

Grand Portage National Monument
PO Box 426
Portage, MN 55605-0426
218-475-0123
www.nps.gov/grpo

Grand Teton National Park
PO Box 170
Moose, WY 83012-0170
307-739-3399
www.nps.gov/grte

Grant-Kohrs Ranch National Historic Site
266 Warren Lane
Deer Lodge, MT 59722-1002
406-846-2070
www.nps.gov/grko

Great Basin National Park
100 Great Basin National Park
Baker, NV 89311-9700
775-234-7331
www.nps.gov/grba

Great Egg Harbor National Scenic & Recreational River
200 Chestnut Street
Philadelphia, PA 19106-2912
215-597-5823
www.nps.gov/greg

Great Sand Dunes National Park & Preserve
11999 State Highway 150
Mosca, CO 81146-9798
719-378-6395
www.nps.gov/grsa

Great Smoky Mountains National Park
107 Park Headquarters Road
Gatlinburg, TN 37738-4102
865-436-1200
www.nps.gov/grsm

Greenbelt Park
6565 Greenbelt Road
Greenbelt, MD 20770-3207
771-208-1588
www.nps.gov/gree

Guadalupe Mountains National Park
400 Pine Canyon
Salt Flat, TX 79847-9400
915-828-3251
www.nps.gov/gumo

Guilford Courthouse National Military Park
2332 New Garden Road
Greensboro, NC 27410-2355
336-288-1776
www.nps.gov/guco

Gulf Islands National Seashore, Florida
1801 Gulf Breeze Parkway
Gulf Breeze, FL 32563-5000
228-230-4121
www.nps.gov/guis

Gulf Islands National Seashore, Mississippi
3500 Park Road
Ocean Springs, MS 39564-9709
228-230-4121
www.nps.gov/guis

H

Hagerman Fossil Beds National Monument
PO Box 570
Hagerman, ID 83332-0570
208-933-4105
www.nps.gov/hafo

Haleakalā National Park
PO Box 369
Makawao, HI 96768-0369
808-572-4400
www.nps.gov/hale

Hamilton Grange National Memorial
414 West 141st Street
New York, NY 10031-9138
646-548-2310
www.nps.gov/hagr

Hampton National Historic Site
535 Hampton Lane
Towson, MD 21286-1397
410-962-4290
www.nps.gov/hamp

Harpers Ferry National Historical Park
PO Box 65
Harpers Ferry, WV 25425-0065
304-535-6029
www.nps.gov/hafe

Harriet Tubman National Historical Park
PO Box 769
Auburn, NY 13021-0769
315-568-0024
www.nps.gov/hart

Harriet Tubman Underground Railroad National Historical Park
4068 Golden Hill Road
Church Creek, MD 21622-1102
410-221-2290
www.nps.gov/hatu

Harry S. Truman National Historic Site
223 North Main Street
Independence, MO 64050-2804
816-254-9929
www.nps.gov/hstr

Hawai'i Volcanoes National Park
PO Box 52
Hawaii National Park, HI 96718-0052
808-985-6011
www.nps.gov/havo

Herbert Hoover National Historic Site
PO Box 607
West Branch, IA 52358-0607
319-643-2541
www.nps.gov/heho

Hohokam Pima National Monument
c/o Casa Grande Ruins National Monument
1100 West Ruins Drive
Coolidge, AZ 85228-3200
520-723-3172
www.nps.gov/cagr

Home of Franklin D. Roosevelt National Historic Site
4097 Albany Post Road
Hyde Park, NY 12538-1917
845-229-5320
www.nps.gov/hofr

Homestead National Historical Park
8523 West State Highway 4
Beatrice, NE 68310-6743
402-223-3514
www.nps.gov/home

Hopewell Culture National Historical Park
16062 State Route 104
Chillicothe, OH 45601-8694
740-774-1125
www.nps.gov/hocu

Hopewell Furnace National Historic Site
2 Mark Bird Lane
Elverson, PA 19520-9535
610-582-8773
www.nps.gov/hofu

Horseshoe Bend National Military Park
11288 Horseshoe Bend Road
Daviston, AL 36256-6524
256-234-7111
www.nps.gov/hobe

Hot Springs National Park
101 Reserve Street
Hot Springs, AR 71901-4195
501-620-6715
www.nps.gov/hosp

Hovenweep National Monument
McElmo Route
Cortez, CO 81321-8901
970-562-4282
www.nps.gov/hove

Hubbell Trading Post National Historic Site
PO Box 150
Ganado, AZ 86505-0150
928-755-3475
www.nps.gov/hutr

I

Independence National Historical Park
143 South 3rd Street
Philadelphia, PA 19106-2818
215-965-2305
www.nps.gov/inde

Indiana Dunes National Park
1100 North Mineral Springs Road
Porter, IN 46304-1299
219-395-1882
www.nps.gov/indu

Isle Royale National Park
800 East Lakeshore Drive
Houghton, MI 49931-1896
906-482-0984
www.nps.gov/isro

J

James A. Garfield National Historic Site
8095 Mentor Avenue
Mentor, OH 44060-5753
440-255-8722
www.nps.gov/jaga

Jean Lafitte National Historical Park & Preserve
419 Decatur Street
New Orleans, LA 70130-1035
504-589-3882
www.nps.gov/jela

Jewel Cave National Monument
11149 U.S. Highway 16, Building B12
Custer, SD 57730-8167
605-673-8300
www.nps.gov/jeca

Jimmy Carter National Historical Park
300 North Bond Street
Plains, GA 31780-5562
229-824-4104
www.nps.gov/jica

John D. Rockefeller, Jr. Memorial Parkway
c/o Grand Teton National Park
PO Box 170
Moose, WY 83012-0170
307-739-3399
www.nps.gov/grte/planyourvisit/jodr.htm

John Day Fossil Beds National Monument
32651 Highway 19
Kimberly, OR 97848-9701
541-987-2333
www.nps.gov/joda

John Fitzgerald Kennedy National Historic Site
83 Beals Street
Brookline, MA 02446-6010
617-566-7937
www.nps.gov/jofi

John Muir National Historic Site
4202 Alhambra Avenue
Martinez, CA 94553-3826
925-228-8860
www.nps.gov/jomu

Johnstown Flood National Memorial
733 Lake Road
South Fork, PA 15956-3602
814-886-6171
www.nps.gov/jofl

Joshua Tree National Park
74485 National Park Drive
Twentynine Palms, CA 92277-3597
760-367-5500
www.nps.gov/jotr

K

Kalaupapa National Historical Park
PO Box 2222
Kalaupapa, HI 96742-0040
808-567-6802
www.nps.gov/kala

Kaloko-Honokōhau National Historical Park
73-4786 Kanalani Street, #14
Kailua-Kona, HI 96740-2600
808-329-6881
www.nps.gov/kaho

Katahdin Woods and Waters National Monument
PO Box 446
Patten, ME 04765-0446
207-456-6001
www.nps.gov/kaww

Katmai National Park & Preserve
PO Box 7
1000 Silver Street, Building 603
King Salmon, AK 99613-0007
907-246-3305
www.nps.gov/katm

Kenai Fjords National Park
PO Box 1727
Seward, AK 99664-1727
907-422-0500
www.nps.gov/kefj

Kennesaw Mountain National Battlefield Park
900 Kennesaw Mountain Drive
Kennesaw, GA 30152-4855
770-427-4686
www.nps.gov/kemo

Keweenaw National Historical Park
25970 Red Jacket Road
Calumet, MI 49913-2948
906-337-3168
www.nps.gov/kewe

Kings Canyon National Park
47050 Generals Highway
Three Rivers, CA 93271-9700
559-565-3341
www.nps.gov/seki

Kings Mountain National Military Park
2625 Park Road
Blacksburg, SC 29702-7325
864-936-7921
www.nps.gov/kimo

Klondike Gold Rush National Historical Park – Seattle Unit
319 Second Avenue South
Seattle, WA 98104-2618
206-220-4240
www.nps.gov/klse

Klondike Gold Rush National Historical Park
PO Box 517
Skagway, AK 99840-0517
907-983-9200
www.nps.gov/klgo

Knife River Indian Villages National Historic Site
PO Box 9
Stanton, ND 58571-0009
701-745-3300
www.nps.gov/knri

Kobuk Valley National Park
PO Box 1029
Kotzebue, AK 99752-1029
907-442-3890
www.nps.gov/kova

Korean War Veterans Memorial
c/o National Mall & Memorial Parks
1100 Ohio Drive, SW
Washington, DC 20242-0001
202-426-6841
www.nps.gov/kwvm

L

Lake Chelan National Recreation Area
810 State Route 20
Sedro-Woolley, WA 98284-1263
360-854-7200
www.nps.gov/noca

Lake Clark National Park & Preserve
240 West 5th Avenue, Suite 236
Anchorage, AK 99501-2327
907-644-3626
www.nps.gov/lacl

Lake Mead National Recreation Area
601 Nevada Way
Boulder City, NV 89005-2426
702-293-8990
www.nps.gov/lake

Lake Meredith National Recreation Area
PO Box 1460
Fritch, TX 79036-1460
806-857-3151
www.nps.gov/lamr

Lake Roosevelt National Recreation Area
1008 Crest Drive
Coulee Dam, WA 99116-0037
509-754-7800
www.nps.gov/laro

Lassen Volcanic National Park
PO Box 100
Mineral, CA 96063-0100
530-595-4480
www.nps.gov/lavo

Lava Beds National Monument
PO Box 1240
Tulelake, CA 96134-1240
530-667-8113
www.nps.gov/labe

Lewis and Clark National Historical Park
92343 Fort Clatsop Road
Astoria, OR 97103-9197
503-861-2471
www.nps.gov/lewi

Lincoln Boyhood National Memorial
3027 East South Street
PO Box 1816
Lincoln City, IN 47552-9722
812-937-4541
www.nps.gov/libo

Lincoln Home National Historic Site
413 South 8th Street
Springfield, IL 62701-1905
217-492-4241
www.nps.gov/liho

Lincoln Memorial
c/o National Mall & Memorial Parks
1100 Ohio Drive, SW
Washington, DC 20242-0001
202-426-6841
www.nps.gov/linc

Little Bighorn Battlefield National Monument
PO Box 39
Crow Agency, MT 59022-0039
406-638-3236
www.nps.gov/libi

Little River Canyon National Preserve
4322 Little River Trail NE, Suite 100
Fort Payne, AL 35967-9300
256-845-9605
www.nps.gov/liri

Little Rock Central High School National Historic Site
2120 West Daisy L. Gatson Bates Drive
Little Rock, AR 72202-5212
501-374-1957
www.nps.gov/chsc

Longfellow House–Washington's Headquarters National Historic Site
105 Brattle Street
Cambridge, MA 02138-3407
617-876-4491
www.nps.gov/long

Lowell National Historical Park
67 Kirk Street
Lowell, MA 01852-1029
978-970-5000
www.nps.gov/lowe

Lower Delaware National Wild & Scenic River
1234 Market Street
Philadelphia, PA 19107-3727
617-981-0466
www.nps.gov/lode

Lyndon B. Johnson National Historical Park
PO Box 329
Johnson City, TX 78636-0329
830-868-7128
www.nps.gov/lyjo

Lyndon Baines Johnson Memorial Grove on the Potomac
c/o Turkey Run Park
700 George Washington Memorial Parkway
McLean, VA 22101-1716
703-235-1530
www.nps.gov/lyba

M

Maggie L. Walker National Historic Site
3215 East Broad Street
Richmond, VA 23223-7517
804-226-5041
www.nps.gov/mawa

Mammoth Cave National Park
PO Box 7
Mammoth Cave, KY 42259-0007
270-758-2180
www.nps.gov/maca

Manassas National Battlefield Park
12521 Lee Highway
Manassas, VA 20109-2005
703-361-1339
www.nps.gov/mana

Manhattan Project National Historical Park
c/o NPS Intermountain Regional Office
PO Box 25287
Denver, CO 80225-0287
303-969-2700
www.nps.gov/mapr

Manzanar National Historic Site
PO Box 426
5001 Highway 395
Independence, CA 93526-0426
760-878-2194
www.nps.gov/manz

Marsh-Billings-Rockefeller National Historical Park
54 Elm Street
Woodstock, VT 05091-1023
802-457-3368
www.nps.gov/mabi

Martin Luther King, Jr. Memorial
c/o National Mall & Memorial Parks
1100 Ohio Drive, SW
Washington, DC 20242-0001
202-426-6841
www.nps.gov/mlkm

Martin Luther King, Jr. National Historical Park
450 Auburn Avenue, NE
Atlanta, GA 30312-1525
404-331-5190
www.nps.gov/malu

Martin Van Buren National Historic Site
1013 Old Post Road
Kinderhook, NY 12106-3605
518-758-9689
www.nps.gov/mava

Mary McLeod Bethune Council House National Historic Site
1318 Vermont Avenue, NW
Washington, DC 20005-3607
771-208-1593
www.nps.gov/mamc

Medgar and Myrlie Evers Home National Monument
c/o Mississippi Civil Rights Museum
222 North Street, #2205
Jackson, MS 39201-1808
601-345-7211
www.nps.gov/memy

Mesa Verde National Park
PO Box 8
Mesa Verde National Park, CO 81330-0008
970-529-4465
www.nps.gov/meve

Mill Springs Battlefield National Monument
9020 West Highway 80
Nancy, KY 42544-7747
606-636-4045
www.nps.gov/misp

Minidoka National Historic Site
1428 Hunt Road
Jerome, ID 83338-7010
208-825-4169
www.nps.gov/miin

Minute Man National Historical Park
174 Liberty Street
Concord, MA 01742-1705
978-369-6993
www.nps.gov/mima

Minuteman Missile National Historic Site
24545 Cottonwood Road
Philip, SD 57567-7002
605-433-5552
www.nps.gov/mimi

Mississippi National River & Recreation Area
111 East Kellogg Boulevard, Suite 105
St. Paul, MN 55101-1256
651-293-0200
www.nps.gov/miss

Missouri National Recreational River
508 East 2nd Street
Yankton, SD 57078-4422
605-665-0209
www.nps.gov/mnrr

Mojave National Preserve
2701 Barstow Road
Barstow, CA 92311-6609
760-252-6100
www.nps.gov/moja

Monocacy National Battlefield
4632 Araby Church Road
Frederick, MD 21704-7705
301-662-3515
www.nps.gov/mono

Montezuma Castle National Monument
PO Box 219
Camp Verde, AZ 86322-0219
928-567-3322
www.nps.gov/moca

Moores Creek National Battlefield
40 Patriots Hall Drive
Currie, NC 28435-0069
910-283-9272
www.nps.gov/mocr

Morristown National Historical Park
30 Washington Place
Morristown, NJ 07960-4259
973-539-2016
www.nps.gov/morr

Mount Rainier National Park
55210 238th Avenue East
Ashford, WA 98304-9751
360-569-2211
www.nps.gov/mora

Mount Rushmore National Memorial
13000 Highway 244, Building 31, Suite 1
Keystone, SD 57751-0268
605-574-2523
www.nps.gov/moru

Muir Woods National Monument
1 Muir Woods Road
Mill Valley, CA 94941-2696
415-561-2850
www.nps.gov/muwo

N

Natchez National Historical Park
640 South Canal Street, Suite E
Natchez, MS 39120-3801
601-446-5790
www.nps.gov/natc

Natchez Trace National Scenic Trail
2680 Natchez Trace Parkway
Tupelo, MS 38804-9718
1-800-305-7417
www.nps.gov/natt

Natchez Trace Parkway
2680 Natchez Trace Parkway
Tupelo, MS 38804-9718
800-305-7417
www.nps.gov/natr

National Capital Parks-East
1900 Anacostia Dr., SE
Washington, DC 20020-6722
202-690-5127
www.nps.gov/nace

National Mall and Memorial Parks
1100 Ohio Drive, SW
Washington, DC 20242-0001
202-426-6841
www.nps.gov/nama

Natural Bridges National Monument
HC-60 Box 1
Lake Powell, UT 84533-0001
435-692-1234
www.nps.gov/nabr

Navajo National Monument
PO Box 7717
Shonto, AZ 86054-7717
928-672-2700
www.nps.gov/nava

New Bedford Whaling National Historical Park
33 William Street
New Bedford, MA 02740-6222
508-996-4095
www.nps.gov/nebe

New Orleans Jazz National Historical Park
419 Decatur Street
New Orleans, LA 70130-1035
504-589-3882
www.nps.gov/jazz

New Philadelphia National Historic Site
National Park Service
Regions 3, 4, and 5 Office
601 Riverfront Drive
Omaha, NE 68102-4226
402-661-1520
www.nps.gov/neph

New River Gorge National Park & Preserve
104 Main Street
PO Box 246
Glen Jean, WV 25846-0246
304-465-0508
www.nps.gov/neri

Nez Perce National Historical Park
39063 US Highway 95
Lapwai, ID 83540-9715
208-843-7001
www.nps.gov/nepe

Nicodemus National Historic Site
304 Washington Avenue
Nicodemus, KS 67625-3015
785-839-4233
www.nps.gov/nico

Ninety Six National Historic Site
1103 Highway 248
Ninety Six, SC 29666-8611
864-543-4068
www.nps.gov/nisi

Niobrara National Scenic River
214 West US Highway 20
Valentine, NE 69201-2005
402-376-1901
www.nps.gov/niob

Noatak National Preserve
PO Box 1029
Kotzebue, AK 99752-1029
907-442-3890
www.nps.gov/noat

North Cascades National Park
810 State Route 20
Sedro-Woolley, WA 98284-1263
360-854-7200
www.nps.gov/noca

O

Obed Wild & Scenic River
PO Box 429
Wartburg, TN 37887-0429
423-346-6294
www.nps.gov/obed

Ocmulgee Mounds National Historical Park
1207 Emery Highway
Macon, GA 31217-4320
478-752-8257
www.nps.gov/ocmu

Olympic National Park
600 East Park Avenue
Port Angeles, WA 98362-6798
360-565-3130
www.nps.gov/olym

Oregon Caves National Monument & Preserve
19000 Caves Highway
Cave Junction, OR 97523-9716
541-592-2100
www.nps.gov/orca

Organ Pipe Cactus National Monument
10 Organ Pipe Drive
Ajo, AZ 85321-9626
520-387-6849
www.nps.gov/orpi

Ozark National Scenic Riverways
PO Box 490
Van Buren, MO 63965-0490
573-323-4236
www.nps.gov/ozar

P

Padre Island National Seashore
PO Box 181300
Corpus Christi, TX 78480-1300
361-949-8068
www.nps.gov/pais

Palo Alto Battlefield National Historical Park
600 East Harrison Street, Room 1006
Brownsville, TX 78520-7176
956-541-2785
www.nps.gov/paal

Paterson Great Falls National Historical Park
72 McBride Avenue Extension
Paterson, NJ 07501-2660
973-523-0370
www.nps.gov/pagr

Pea Ridge National Military Park
15930 National Park Drive
Garfield, AR 72732-9532
479-451-8122
www.nps.gov/peri

Pearl Harbor National Memorial
1 Arizona Memorial Place
Honolulu, HI 96818-3145
808-422-3399
www.nps.gov/perl

Pecos National Historical Park
PO Box 418
Pecos, NM 87552-0418
505-757-7241
www.nps.gov/peco

Pennsylvania Avenue National Historic Site
c/o National Mall & Memorial Parks
1100 Ohio Drive, SW
Washington, DC 20242-0001
202-426-6841
www.nps.gov/paav

Perry's Victory & International Peace Memorial
93 Delaware Avenue
PO Box 549
Put-in-Bay, OH 43456-0549
419-285-2184
www.nps.gov/pevi

Petersburg National Battlefield
1539 Hickory Hill Road
Petersburg, VA 23803-4721
804-732-3531
www.nps.gov/pete

Petrified Forest National Park
PO Box 2217
Petrified Forest, AZ 86028-2217
928-524-6228
www.nps.gov/pefo

Petroglyph National Monument
6001 Unser Boulevard, NW
Albuquerque, NM 87120-2069
505-899-0205
www.nps.gov/petr

Pictured Rocks National Lakeshore
PO Box 40
Munising, MI 49862-0040
906-387-3700
www.nps.gov/piro

Pinnacles National Park
5000 East Entrance Road
Paicines, CA 95043-9770
831-389-4486
www.nps.gov/pinn

Pipe Spring National Monument
HC 65, Box 5
406 Pipe Springs Road
Fredonia, AZ 86022-9600
928-643-7105
www.nps.gov/pisp

Pipestone National Monument
36 Reservation Avenue
Pipestone, MN 56164-1269
507-825-5464
www.nps.gov/pipe

Piscataway Park
c/o Fort Washington Park
13551 Fort Washington Road
Fort Washington, MD 20744-7044
771-208-1555
www.nps.gov/pisc

Point Reyes National Seashore
1 Bear Valley Road
Point Reyes Station, CA 94956-9799
415-464-5100
www.nps.gov/pore

Port Chicago Naval Magazine National Memorial
4202 Alhambra Avenue
Martinez, CA 94553-3826
925-228-8860
www.nps.gov/poch

Potomac Heritage National Scenic Trail
c/o Chesapeake & Ohio Canal National Historical Park
142 West Potomac Street
Williamsport, MD 21795-1039
301-739-4200
www.nps.gov/pohe

Poverty Point National Monument
6859 Highway 577
Pioneer, LA 71266-8933
318-926-5492
www.nps.gov/popo

President William Jefferson Clinton Birthplace Home National Historic Site
117 South Hervey Street
Hope, AR 71801-4208
870-777-4455
www.nps.gov/wicl

Prince William Forest Park
18100 Park Headquarters Road
Triangle, VA 22172-1644
703-221-7181
www.nps.gov/prwi

Pullman National Historical Park
610 East 111th Street
Chicago, IL 60628-4651
773-928-7257
www.nps.gov/pull

Puʻuhonua o Hōnaunau National Historical Park
PO Box 129
Hōnaunau, HI 96726-0129
808-328-2326
www.nps.gov/puho

Puʻukoholā Heiau National Historic Site
62-3601 Kawaihae Road
Kawaihae, HI 96743-9720
808-882-7218
www.nps.gov/puhe

R

Rainbow Bridge National Monument
c/o Glen Canyon National Recreation Area
PO Box 1507
691 Scenic View Drive
Page, AZ 86040-1507
928-608-6200
www.nps.gov/rabr

Reconstruction Era National Historical Park
706 Craven Street
Beaufort, SC 29902-5571
843-962-0039
www.nps.gov/reer

Redwood National Park
1111 Second Street
Crescent City, CA 95531-4198
707-464-6101
www.nps.gov/redw

Richmond National Battlefield Park
3215 East Broad Street
Richmond, VA 23223-7517
804-226-1981
www.nps.gov/rich

Rio Grande Wild & Scenic River
c/o Big Bend National Park
PO Box 129
Big Bend National Park, TX 79834-0129
432-477-2251
www.nps.gov/rigr

River Raisin National Battlefield Park
333 North Dixie Highway
Monroe, MI 48162-2578
734-243-7136
www.nps.gov/rira

Rock Creek Park
5200 Glover Road, NW
Washington, DC 20015-1008
202-895-6000
www.nps.gov/rocr

Rocky Mountain National Park
1000 US Highway 36
Estes Park, CO 80517-8397
970-586-1206
www.nps.gov/romo

Roger Williams National Memorial
282 North Main Street
Providence, RI 02903-1240
401-521-7266
www.nps.gov/rowi

Rosie the Riveter World War II Home Front National Historical Park
1414 Harbour Way South, Suite 3000
Richmond, CA 94804-3694
510-232-5050
www.nps.gov/rori

Ross Lake National Recreation Area
810 State Route 20
Sedro-Woolley, WA 98284-1263
360-854-7200
www.nps.gov/noca

Russell Cave National Monument
3729 County Road 98
Bridgeport, AL 35740-6825
205-495-2672
www.nps.gov/ruca

S

Sagamore Hill National Historic Site
20 Sagamore Hill Road
Oyster Bay, NY 11771-1899
516-922-4788
www.nps.gov/sahi

Saguaro National Park
3693 South Old Spanish Trail
Tucson, AZ 85730-5601
520-733-5153
www.nps.gov/sagu

St. Croix Island International Historic Site
PO Box 247
Calais, ME 04619-0247
207-454-3871
www.nps.gov/sacr

St. Croix National Scenic Riverway
401 North Hamilton Street
St. Croix Falls, WI 54024-0708
715-483-2274
www.nps.gov/sacn

St. Paul's Church National Historic Site
897 South Columbus Avenue
Mount Vernon, NY 10550-5018
914-667-4116
www.nps.gov/sapa

Ste. Geneviève National Historical Park
339 St. Marys Road
Ste. Genevieve, MO 63670-1643
573-880-7189
www.nps.gov/stge

Saint Gaudens National Historical Park
139 Saint Gaudens Road
Cornish, NH 03745-9704
603-675-2175
www.nps.gov/saga

Salem Maritime National Historic Site
160 Derby Street
Salem, MA 01970-5643
978-740-1650
www.nps.gov/sama

Salinas Pueblo Missions National Monument
PO Box 517
Mountainair, NM 87036-0517
505-847-2585
www.nps.gov/sapu

Salt River Bay National Historical Park & Ecological Preserve
2100 Church Street, #100
Christiansted, VI 00820-4611
340-773-1460
www.nps.gov/sari

San Antonio Missions National Historical Park
2202 Roosevelt Avenue
San Antonio, TX 78210-4919
210-932-1001
www.nps.gov/saan

San Francisco Maritime National Historical Park
2 Marina Boulevard, Building E, 2nd Floor
San Francisco, CA 94123-1284
415-561-7000
www.nps.gov/safr

San Juan Island National Historical Park
PO Box 429
Friday Harbor, WA 98250-0429
360-378-2240
www.nps.gov/sajh

San Juan National Historic Site
501 Norzagaray Street
San Juan, PR 00901-1213
787-729-6777
www.nps.gov/saju

Sand Creek Massacre National Historic Site
PO Box 249
Eads, CO 81036-0249
719-438-5916
www.nps.gov/sand

Santa Monica Mountains National Recreation Area
26876 Mulholland Highway
Calabasas, CA 91302
805-370-2301
www.nps.gov/samo

Saratoga National Historical Park
648 Route 32
Stillwater, NY 12170-1604
518-670-2985
www.nps.gov/sara

Saugus Iron Works National Historic Site
244 Central Street
Saugus, MA 01906-2107
781-233-0050
www.nps.gov/sair

Scotts Bluff National Monument
PO Box 27
Gering, NE 69341-0027
308-436-9700
www.nps.gov/scbl

Selma to Montgomery National Historic Trail
PO Box 595
Hayneville, AL 36040-0595
334-293-0597
www.nps.gov/semo

Sequoia National Park
47050 Generals Highway
Three Rivers, CA 93271-9700
559-565-3341
www.nps.gov/seki

Shenandoah National Park
3655 US Highway 211 East
Luray, VA 22835-4702
540-999-3500
www.nps.gov/shen

Shiloh National Military Park
1055 Pittsburg Landing Road
Shiloh, TN 38376-4331
731-689-5696
www.nps.gov/shil

Sitka National Historical Park
103 Monastery Street
Sitka, AK 99835-7617
907-747-0110
www.nps.gov/sitk

Sleeping Bear Dunes National Lakeshore
9922 Front Street
Empire, MI 49630-9797
231-326-4700
www.nps.gov/slbe

Springfield Armory National Historic Site
One Armory Square, Suite 2
Springfield, MA 01105-1299
413-734-8551
www.nps.gov/spar

Statue of Liberty National Monument
Liberty Island
New York, NY 10004-1467
212-363-3200
www.nps.gov/stli

Steamtown National Historic Site
150 South Washington Avenue
Scranton, PA 18503-2018
570-445-1898
www.nps.gov/stea

Stones River National Battlefield
3501 Old Nashville Highway
Murfreesboro, TN 37129-3094
615-893-9501
www.nps.gov/stri

Stonewall National Monument
c/o Federal Hall National Memorial
26 Wall Street
New York, NY 10005-1996
212-668-2577
www.nps.gov/ston

Sunset Crater Volcano National Monument
6400 North Highway 89
Flagstaff, AZ 86004-2759
928-526-0502
www.nps.gov/sucr

T

Tallgrass Prairie National Preserve
2480B KS Highway 177
Strong City, KS 66869-9829
620-273-8494
www.nps.gov/tapr

Thaddeus Kosciuszko National Memorial
c/o Independence National
Historical Park
143 South 3rd Street
Philadelphia, PA 19106-2818
215-965-2305
www.nps.gov/thko

Theodore Roosevelt Birthplace National Historic Site
28 East 20th Street
New York, NY 10003-1311
718-551-6978
www.nps.gov/thrb

Theodore Roosevelt Inaugural National Historic Site
641 Delaware Avenue
Buffalo, NY 14202-1001
716-884-0095
www.nps.gov/thri

Theodore Roosevelt Island
c/o Turkey Run Park
700 George Washington Memorial
Parkway
McLean, VA 22101-1716
703-289-2500
www.nps.gov/this

Theodore Roosevelt National Park
PO Box 7
Medora, ND 58645-0007
701-623-4466
www.nps.gov/thro

Thomas Cole National Historic Site
218 Spring Street
Catskill, NY 12414-1027
518-943-7465
www.nps.gov/thco

Thomas Edison National Historical Park
211 Main Street
West Orange, NJ 07052-5612
973-736-0550
www.nps.gov/edis

Thomas Jefferson Memorial
c/o National Mall & Memorial Parks
1100 Ohio Drive, SW
Washington, DC 20242-0001
202-426-6841
www.nps.gov/thje

Thomas Stone National Historic Site
6655 Rose Hill Road
Port Tobacco, MD 20677-3400
804-224-1732
www.nps.gov/thst

Timpanogos Cave National Monument
2038 West Alpine Loop Road
American Fork, UT 84003-9803
801-756-5239
www.nps.gov/tica

Timucuan Ecological & Historic Preserve
12713 Fort Caroline Road
Jacksonville, FL 32225-1240
904-641-7155
www.nps.gov/timu

Tonto National Monument
26260 North AZ Highway 188, Lot 2
Roosevelt, AZ 85545-8148
928-467-2241
www.nps.gov/tont

Touro Synagogue National Historic Site
85 Touro Street
Newport, RI 02840-2969
401-847-4794
www.nps.gov/tosy

Tule Lake National Monument
PO Box 1240
Tulelake, CA 96134-1240
530-664-4015 or 530-667-8113
www.nps.gov/tule

Tule Springs Fossil Beds National Monument
601 Nevada Way
Boulder City, NV 89005-2426
702-293-8853
www.nps.gov/tusk

Tumacácori National Historical Park
PO Box 8067
Tumacacori, AZ 85640-0067
520-377-5060
www.nps.gov/tuma

Tupelo National Battlefield
2680 Natchez Trace Parkway
Tupelo, MS 38804-9718
1-800-305-7417
www.nps.gov/tupe

Tuskegee Airmen National Historic Site
1616 Chappie James Avenue
Tuskegee, AL 36083-2985
334-724-0922
www.nps.gov/tuai

Tuskegee Institute National Historic Site
1212 West Montgomery Road
Tuskegee Institute, AL 36088-1923
334-727-3200
www.nps.gov/tuin

Tuzigoot National Monument
PO Box 219
Camp Verde, AZ 86322-0219
928-634-5564
www.nps.gov/tuzi

U

Ulysses S. Grant National Historic Site
7400 Grant Road
St. Louis, MO 63123-1801
314-842-1867
www.nps.gov/ulsg

Upper Delaware Scenic & Recreational River
274 River Road
Beach Lake, PA 18405-9737
570-685-4871
www.nps.gov/upde

V

Valles Caldera National Preserve
PO Box 359
Jemez Springs, NM 87025-0359
505-670-1612
www.nps.gov/vall

Valley Forge National Historical Park
1400 North Outer Line Drive
King of Prussia, PA 19406-1009
610-783-1000
www.nps.gov/vafo

Vanderbilt Mansion National Historic Site
4097 Albany Post Road
Hyde Park, NY 12538-1997
845-229-7770
www.nps.gov/vama

Vicksburg National Military Park
3201 Clay Street
Vicksburg, MS 39183-3495
601-636-0583
www.nps.gov/vick

Vietnam Veterans Memorial
c/o National Mall & Memorial Parks
1100 Ohio Drive, SW
Washington, DC 20242-0001
202-426-6841
www.nps.gov/vive

Virgin Islands Coral Reef National Monument
1300 Cruz Bay Creek
St. John, VI 00830-6108
340-776-6201
www.nps.gov/vicr

Virgin Islands National Park
1300 Cruz Bay Creek
St. John, VI 00830-6108
340-776-6201
www.nps.gov/viis

Voyageurs National Park
360 Highway 11 East
International Falls, MN 56649-8904
218-283-6600
www.nps.gov/voya

W

Waco Mammoth National Monument
6220 Steinbeck Bend Drive
Waco, TX 76708-5338
254-750-7946
www.nps.gov/waco

Walnut Canyon National Monument
6400 North Highway 89
Flagstaff, AZ 86004-2759
928-526-3367
www.nps.gov/waca

War in the Pacific National Historical Park
135 Murray Boulevard, Suite 100
Hagåtña, GU 96910-5104
671-333-4050
www.nps.gov/wapa

Washington Monument
c/o National Mall & Memorial Parks
1100 Ohio Drive, SW
Washington, DC 20242-0001
202-426-6841
www.nps.gov/wamo

Washita Battlefield National Historic Site
18555 Highway 47A, Suite A
Cheyenne, OK 73628-6405
580-497-2742
www.nps.gov/waba

Weir Farm National Historical Park
735 Nod Hill Road
Wilton, CT 06897-1309
203-834-1896
www.nps.gov/wefa

Whiskeytown National Recreation Area
PO Box 188
Whiskeytown, CA 96095-0188
530-242-3400
www.nps.gov/whis

The White House & President's Park
1849 C Street NW, Room 1426
Washington, DC 20240-0001
202-208-1631
www.nps.gov/whho

White Sands National Park
PO Box 1086
Holloman AFB, NM 88330-1086
575-479-6124
www.nps.gov/whsa

Whitman Mission National Historic Site
328 Whitman Mission Road
Walla Walla, WA 99362-7299
509-522-6360
www.nps.gov/whmi

William Howard Taft National Historic Site
2038 Auburn Avenue
Cincinnati, OH 45219-3025
513-684-3262
www.nps.gov/wiho

Wolf Trap National Park for the Performing Arts
1551 Trap Road
Vienna, VA 22182-1643
703-255-1800
www.nps.gov/wotr

Women's Rights National Historical Park
136 Fall Street
Seneca Falls, NY 13148-1517
315-568-0024
www.nps.gov/wori

World War I Memorial
1493 Pennsylvania Avenue, NW
Washington, DC 20004
202-426-6841
www.nps.gov/wwim

World War II Memorial
c/o National Mall & Memorial Parks
1100 Ohio Drive, SW
Washington, DC 20242-0001
202-426-6841
www.nps.gov/nwwm

Wrangell–St. Elias National Park & Preserve
PO Box 439
Copper Center, AK 99573-0439
907-822-5234
www.nps.gov/wrst

Wright Brothers National Memorial
1401 National Park Drive
Manteo, NC 27954-2708
252-473-2111
www.nps.gov/wrbr

Wupatki National Monument
6400 North Highway 89
Flagstaff, AZ 86004-2759
928-679-2365 or 928-856-1705
www.nps.gov/wupa

Y

Yellowstone National Park
PO Box 168
Yellowstone National Park, WY 82190-0168
307-344-7381
www.nps.gov/yell

Yosemite National Park
PO Box 577
Yosemite, CA 95389-0577
209-372-0200
www.nps.gov/yose

Yucca House National Monument
PO Box 8
Mesa Verde, CO 81330-0008
970-529-4465
www.nps.gov/yuho

Yukon–Charley Rivers National Preserve
101 Dunkel Street, Suite 110
Fairbanks, AK 99701-4806
907-459-3730
www.nps.gov/yuch

Z

Zion National Park
1 Zion Park Boulevard
Springdale, UT 84767-1099
435 772-3256
www.nps.gov/zion

ACKNOWLEDGMENTS

Thank you to the talented writers who lent their research and voices to these pages: Lauren Eberle, Robert E. Howells, Olivia Garnett, Gary McKechnie, Jeremy Schmidt, Mel White, and Joe Yogerst.

The 10 Best of Everything National Parks would not have been possible without the wonderful team at National Geographic Books. In addition to those who contributed to the first edition, thank you to senior editor Allyson Johnson, creative director Elisa Gibson, editorial project manager Ashley Leath, designer Kay Hankins, photo editor Matt Propert, senior cartographer Greg Ugiansky, senior production editor Michael O'Connor, production editor Becca Saltzman, editorial assistant Margo Rosenbaum, and many others.

ILLUSTRATIONS & MAP CREDITS

Map: Park data provided by NPS.

Cover: (UP LE), Darlyne Murawski/National Geographic Image Collection; (UP CTR), skiserge1/Adobe Stock; (UP RT), Barrett Hedges/National Geographic Image Collection; (LO LE), Imtiyas Khan/Shutterstock; (LO CTR), Andrew Peacock/Cavan Images; (LO RT), Zak Zeinert/Adobe Stock. Back cover: spasticlizard/iStock. 1, Library of Congress, #LC-USZC2-833; 2, Matteo Colombo/Getty Images; 6, Neale Haynes/Buzz Pictures/Alamy Stock Photo; 10, Tim Fitzharris/Minden Pictures; 11, Krzysztof Wiktor/Adobe Stock; 12, Skip Brown/robertharding; 13, NPS/S. Muether; 14, Victor Nikitin/Alamy Stock Photo; 15, Sean Pavone/Alamy Stock Photo; 16–7, haveseen/Adobe Stock; 19, Jonathan Larsen/iStock; 20, Keith Ladzinski/National Geographic Image Collection; 21, Keith Crowley/Alamy Stock Photo; 23, Gary Schultz/Design Pics; 25, DnDavis/Shutterstock; 26, Imagic Elements/Alamy Stock Photo; 27, Virrage Images/Shutterstock; 28, Carlton Ward Jr.; 30, Melissa Farlow/National Geographic Image Collection; 31, Bogdan Boev/Shutterstock; 32, Andrew Coleman/National Geographic Image Collection; 34, Kovac/Adobe Stock; 35, Christian Heinrich/imageBROKER/Alamy Stock Photo; 36, Library of Congress, #LC-DIG-ppmsca-13398; 37, Paul Moore/Adobe Stock; 38, Atmosphere/Adobe Stock; 39, Susan E. Degginger/Alamy Stock Photo; 40, NPS/© Janice Wei; 43, Andrew Coleman/National Geographic Image Collection; 44, Abbie Warnock-Matthews/Shutterstock; 45, Manuel Cohen/Art Resource, NY; 47, Ed Callaert/Alamy Stock Photo; 49, aheflln/Adobe Stock; 50, Natalia Bratslavsky/Shutterstock; 51, raclro/iStock; 53, Library of Congress, #LC-USZC4-4243; 54, Stephen Alvarez/National Geographic Image Collection; 56, IrinaK/Shutterstock; 57, Geerati Nilkaew/Alamy Stock Photo; 59, donyanedomam/Adobe Stock; 60, Phil Schermeister/National Geographic Image Collection; 61, Ljupco Smokovski/Adobe Stock; 62, Clint Farlinger/Alamy Stock Photo; 65, Heeb Photos/eStock Photo; 66, Patrick J Endres/AlaskaPhotoGraphics; 67, Jason Edwards/National Geographic Image Collection; 69, Lijuan Guo/Shutterstock; 70–1, diversbelow/Alamy Stock Photo; 73, Sierralara/RooM the Agency/Alamy Stock Photo; 74, alxpin/iStock; 76, Greg Vaughn/Alamy Stock Photo; 79, Timothy Mulholland/Alamy Stock Photo; 80, David/Adobe Stock; 81, Pat & Chuck Blackley/Alamy Stock Photo; 83, Jerry Monkman/Cavan Images; 85, sandsun/iStock; 86, Spring Images/Alamy Stock Photo; 88, Luminis/Shutterstock; 89, lunamarina/Adobe Stock; 91, Chuck Haney/DanitaDelimont/Alamy Stock Photo; 93, David S. Boyer and Arlan R. Wiker/National Geographic Image Collection; 94, FotoRequest/Adobe Stock; 95, ExploringandLiving/Adobe Stock; 96, James Hager/robertharding; 99, Christian Kober/Alamy Stock Photo; 100, Egor Dranichnikov/Adobe Stock; 102, Zach/Adobe Stock; 103, Library of Congress, #LC-DIG-ppmsca-13400; 105, spasticlizard/iStock; 106, William Silver/Alamy Stock Photo; 107, Frau aus UA/Shutterstock; 108, Tony Waltham/robertharding; 111, Tobin Akehurst/Shutterstock; 112, Daniel Sohner/TandemStock; 114, Kerry Hargrove/Adobe Stock; 116, National Park Service; 117, Pat & Chuck Blackley/Alamy Stock Photo; 119, Cavan Images/Sarah Cazares/Alamy Stock Photo; 121, James P. Blair/National Geographic Image Collection; 123, Harrison Shull/Cavan Images, 125, IlexImage/iStock; 126, Cavan Images/Corey Rich/Alamy Stock Photo; 128, TMI/Alamy Stock Photo; 130, Chris Bennett/Cavan Images, 131, National Park Service; 133, Sundry Photography/Adobe Stock; 134, Kyle/Adobe Stock; 135, Free Oscillation/Shutterstock; 136, Oksana Perkins/Shutterstock; 139, Jessica Mentz/Adobe Stock; 140, Chase D'Animulls/Adobe Stock; 142, Ben Herndon/TandemStock;

INDEX

Since 1888, the National Geographic Society has funded more than 14,000 research, conservation, education, and storytelling projects around the world. National Geographic Partners distributes a portion of the funds it receives from your purchase to National Geographic Society to support programs including the conservation of animals and their habitats.

National Geographic Partners, LLC
1145 17th Street NW
Washington, DC 20036-4688 USA

Get closer to National Geographic Explorers and photographers, and connect with our global community. Join us today at nationalgeographic.org/joinus

For rights or permissions inquiries, please contact National Geographic Books Subsidiary Rights: bookrights@natgeo.com

Library of Congress cataloged the first edition as follows:
The 10 best of everything national parks : 800 top picks from parks coast to coast.
p. cm.
Includes index.
ISBN 978-1-4262-0734-1
1. National parks and reserves--United States. I. National Geographic Society (U.S.)
II. Title: Ten best of everything national parks.
E160.A142 2011
917.3--dc22
2010039588

The information in this book has been carefully checked and to the best of our knowledge is accurate. However, details are subject to change, and the publisher cannot be responsible for such changes, or for errors or omissions. Assessments of sites, hotels, and restaurants are based on the author's subjective opinions, which do not necessarily reflect the publisher's opinion.

ISBN: 978-1-4262-2321-1

Printed in China

24/RRDH/1

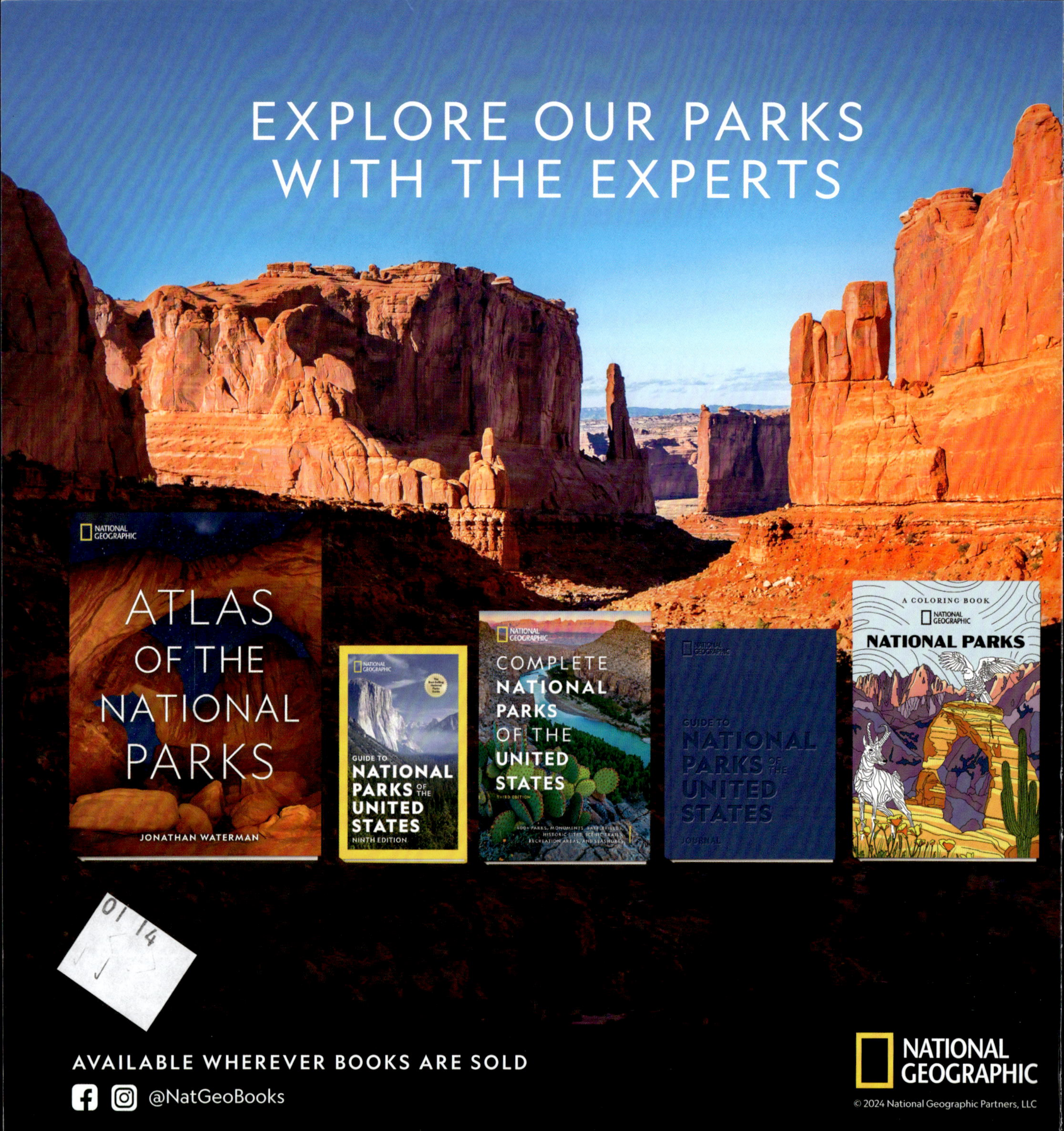
EXPLORE OUR PARKS
WITH THE EXPERTS
NATIONAL GEOGRAPHIC
ATLAS OF THE NATIONAL PARKS
JONATHAN WATERMAN
GUIDE TO NATIONAL PARKS OF THE UNITED STATES
NINTH EDITION
COMPLETE NATIONAL PARKS OF THE UNITED STATES
GUIDE TO NATIONAL PARKS OF THE UNITED STATES
JOURNAL
A COLORING BOOK
NATIONAL PARKS
AVAILABLE WHEREVER BOOKS ARE SOLD
@NatGeoBooks
NATIONAL GEOGRAPHIC
© 2024 National Geographic Partners, LLC

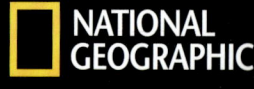